Close Encounters of the Microbial Kind

Michael Wilson • Philippa J. K. Wilson

Close Encounters
of the Microbial Kind

Everything You Need to Know About
Common Infections

 Springer

Michael Wilson
University College London
London, UK

Philippa J. K. Wilson
Brighton, UK

ISBN 978-3-030-56977-8   ISBN 978-3-030-56978-5   (eBook)
https://doi.org/10.1007/978-3-030-56978-5

This Springer imprint is published by the registered company Springer Nature Switzerland AG
The registered company address is: Gewerbestrasse 11, 6330 Cham, Switzerland

MW would like to dedicate this book to his friends and colleagues at University College London; particularly Brian, Colin, Dave, Derren, Elaine, Geoff, Ian, Ivan, Jon, Pete, Ramya, Sandy, Sean, Stephen and Wil.

PW would like to dedicate this book to John for his support and encouragement; to Bernie for all the laughter and fun; to Kirsty, Liam and Sarah who have shown me how to be a good friend and a good doctor; and especially to Dad who has shown me how to be a good person.

Disclaimer

The content of this book is provided by the authors solely for informational and educational purposes. The information provided should not be considered medical advice and is not a substitute for consultation with a qualified medical professional. It is the authors' advice that medical opinion should always be sought if an infectious disease is suspected. References are provided for informational purposes only and do not constitute endorsement of any websites or other sources. Readers should be aware that the content of those websites listed in this book may change.

Preface

Last week I (MW) was cycling along a country road on a lovely sunny day when all of a sudden I felt something hit my face. Then I felt a sharp, stabbing pain just below my left eye and within minutes my eyes were watering, my nose was running and my face was swelling up. I'd been stung by a wasp! For the first time in my life! A few hours later, after the pain had subsided somewhat, I started thinking about a number of things. First of all I felt relief that I hadn't been stung in the eye – I could have been blinded! Then I felt very annoyed – why on earth did it sting me? I hadn't done it any harm. Then I started thinking more about wasps and their lives – what do they get up to when they're not going around stinging humans? And then I wondered about the actual stinging mechanism – how do they do that to people? And finally, why did my body react in that way to the sting?

In this book we'll address similar questions that could be asked about the "wasp sting" of an infectious disease. After all, it's a similar situation. We go about our daily lives and then all of a sudden we're attacked by something without warning and start to feel very ill. Why on earth do microbes attack us? We haven't done them any harm. What are these tiny creatures like? What do they do with themselves when they aren't bothering us? How do they make us ill? How does our body protect us from them and respond to their attacks? What can we do when we are infected? What can we do to avoid being infected?

Importantly, we're going to restrict ourselves to discussing only those infectious diseases that are common in developed countries. We're all too aware of headline-grabbing diseases such as Ebola and cholera, but these diseases, fortunately, aren't likely to affect us unless we travel to exotic places. This book, therefore, will concentrate on those infectious diseases that we're likely to catch while living a normal, unadventurous life in a developed country. These are the infections we need to learn more about so that we can become familiar with the microbes that cause them, find out how they damage our body, learn how they are treated and what we can do to avoid them. Importantly, it will help you to understand why taking antibiotics is not always necessary for the treatment of many infections.

Just as we were putting the finishing touches to this book, the most serious outbreak of an infectious disease in living memory occurred and spread across the globe, COVID-19. While many developing countries are familiar with such threats, this has been a new experience for people in developed nations. The last time anything similar happened was the great influenza pandemic of 1918 (see Box 11.2 in Chap. 11). It's true to say that our lives have been turned upside down. We have had to avoid social contact, minimise travel, change working and learning patterns, and learn to cope with isolation. Surely we can do more than look on helplessly as dedicated health professionals battle against the deadly effects of this virus? This dreadful disease serves as a wakeup call alerting us to the delicate, uneasy relationship we have with the microbial world.

Part 1 of this book consists of an introductory chapter that provides background material on microbiology and infectious diseases. We hope that many of you will be interested in reading this to get a better understanding of the microbes responsible for infections, how we study them, how they cause disease and the ways in which our body fights back against them. However, it's not essential to read Part 1 in order to have an understanding of the subsequent six parts which consist of 35 chapters that deal with nearly 60 diseases that we are all likely to catch. So, if you're interested only in finding out more about a particular disease, you can jump straight to the appropriate chapter.

London, UK   Michael Wilson
Brighton, UK   Philippa J. K. Wilson

Contents

About the Authors

Michael Wilson is an Emeritus Professor of Microbiology at University College London and is a Fellow of the Royal College of Pathologists. He has written 337 scientific papers and 12 books on microbiology and infectious diseases and has supervised the research projects of 35 PhD and 46 MSc students. As well as carrying out research and teaching, he has 33 years' experience in the laboratory diagnosis of infectious diseases.

Philippa J. K. Wilson graduated in medicine from Nottingham University Medical School in 1995 and then worked in hospitals for 5 years before becoming a General Practitioner in 2001. She passed the exams for membership of the Royal College of General Practitioners in 2009. She currently works on the south coast of England as a Primary Care Physician/General Practitioner.

Part I

**Introduction to Microbiology
and Infectious Diseases**

1

Microbes and Infectious Diseases

Abstract Infectious diseases are a major cause of death and disability. They are becoming increasingly difficult to deal with because many microbes have developed resistance to the drugs (antibiotics) used to treat infections. There are six major types of microbes – bacteria, viruses, protozoa, fungi, algae and archaea – but only the first four of these are known to cause disease in humans. These four types of microbes differ greatly in their size, structure and means of reproduction and how they grow. In order to treat a particular infection in a patient, we need to identify which microbe is responsible. Protozoa and algae are identified mainly on the basis of their appearance under the microscope, but the identification of bacteria and viruses needs a different approach. Bacteria (and some fungi) are identified mainly by growing them in the laboratory and studying their appearance and growth. However, analysis of the DNA of bacteria is increasingly being used to identify them. Viruses are identified mainly by analysis of their genetic material which may be DNA or RNA.

All of the surfaces of our bodies that are exposed to the external environment are colonised by microbes – these communities are known collectively as the "human microbiota". Each site on the body has a different microbial community, and we generally live in harmony with them – together we constitute a symbiosis in which both partners benefit. This harmonious co-existence is due to an elaborate set of defence systems that protect us against microbial diseases – these are known as the innate and acquired immune systems. However, some of the microbes that live on us (known as endogenous pathogens or pathobionts) can, under certain circumstances, cause disease. Furthermore, some of the microbes that live in the external environment (known as exogenous pathogens), or on other animals, can overcome our

© Springer Nature Switzerland AG 2021
M. Wilson, P. J. K. Wilson, *Close Encounters of the Microbial Kind*,
https://doi.org/10.1007/978-3-030-56978-5_1

antimicrobial defences and do us harm if we come into contact with them. Mobilisation of our antimicrobial defence systems during an infection often results in inflammation, and this gives rise to four characteristic symptoms – redness, heat, swelling and pain. Every pathogen fights back against, and tries to overcome, our defences using a particular set of weapons which we call "virulence factors". Many of these virulence factors, which include toxins, also damage our tissues. In order to help our natural defence systems in their fight against pathogens, we use a variety of drugs such as antibiotics. Unfortunately, many pathogens have developed resistance to these drugs, and the rate of discovery of new antibiotics is decreasing. We have also developed a number of general ways of preventing infections, and these include the use of antiseptics, disinfectants and various hygiene measures. We can also protect ourselves against certain infections by vaccination, and this has had a dramatic effect on some of the deadliest diseases of humankind, saving or improving the lives of millions of people around the world.

Infectious diseases (also described as "communicable diseases") are a major cause of death and disability for most human beings. Figure 1.1 shows that in 2016 lower respiratory infections, diarrhoeal diseases and tuberculosis were the fourth, ninth and tenth major causes of death, respectively.

However, those of us who live in developed countries often regard infectious diseases as no more than a minor nuisance and nothing to worry about – after all, we've got antibiotics that can deal with them. So what's the problem? One of the main problems is antibiotic resistance. Microbes are increasingly becoming

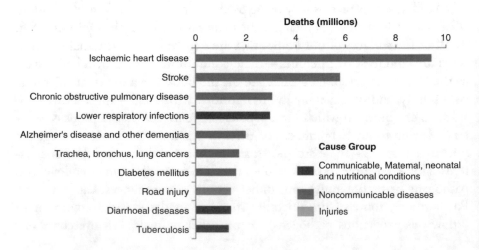

Fig. 1.1 The main causes of global death in 2016. (https://www.who.int/news-room/fact-sheets/detail/the-top-10-causes-of-death)

resistant to antibiotics, and, as a result, we're facing a future in which infectious diseases will once again be a major cause of death in the developed world – we're now in the era of antibiotic-resistant "superbugs" (Box 1.1). Then of course there's the problem of the emergence of new diseases such as the current COVID-19 pandemic. So, all the more reason to learn about the microbes that cause common infections, how these organisms damage us, how these diseases can be treated effectively and how we can avoid becoming infected.

Box 1.1 Superbugs, the Scourge of Modern Medicine: Are the Microbes Winning?

An antibiotic is a chemical that can be used to treat infections in humans or other animals. It does this by killing, or preventing the growth of, the infecting microbe without harming the animal it has infected. In order to do this, it has to be able to target some process or structure in the microbe that isn't present in the animal. The term antibiotic usually refers to a chemical that is effective against bacteria. Chemicals that are used to treat infectious diseases caused by other types of microbe (collectively known as "antimicrobial agents") are usually described as being anti-viral, anti-fungal or anti-protozoal agents.

Following the widescale use of the first antibiotic (penicillin) in the 1940s, it became apparent that bacteria could develop resistance to its action and antibiotic-resistant strains of bacteria began to emerge. What we are seeing here is the process of natural selection. Bacteria can multiply rapidly (some can reproduce every 20 minutes), and every time they do so, they produce a very small number of random mutants. Some of these mutants are completely by chance resistant to the antibiotic in their environment and will, therefore, be unaffected by it and will survive and reproduce in that location. In an environment such as a hospital where the antibiotic is used widely, the antibiotic-resistant mutant will become established (an example of the "survival of the fittest") and will be able to infect other individuals. Medical staff will have to use a different antibiotic if a patient becomes infected with our antibiotic-resistant mutant, and the process may then be repeated to produce a mutant that is now resistant to two different antibiotics. The whole sequence of events may then continue to produce a bacterium that's resistant to many antibiotics. Some bacteria, such as *Staphylococcus aureus*, *Pseudomonas aeruginosa*, *Neisseria gonorrhoeae* and *Enterococcus faecalis*, are particularly good at developing resistance to antibiotics; strains of these organisms resistant to multiple antibiotics are known as "superbugs".

What Are Microbes?

First things first. What do we mean by the term "microbe"? The word comes from the Greek "mikros" and "bios" which mean small and life, respectively. Other terms used for a microbe include "microscopic organism" or "microorganism" which remind us that they are so small that they can only be seen with the help of a microscope.

Fig. 1.2 Portrait of Antonie van Leeuwenhoek by Jan Verkolje. (Jan Verkolje [Public domain], via Wikimedia Commons)

The first person to see a microbe was Antonie van Leeuwenhoek (Fig. 1.2) who, using a microscope he made himself, described in 1676 the microbes that he'd seen in water. He called the tiny creatures he saw "animalcules" and revealed to us an exciting new world that was invisible to the naked eye. He's considered to be the father of the science of microbiology.

Van Leeuwenhoek described these animalcules in letters to the Royal Society of London and, to everyone's surprise, discovered that they also lurked in the mouths of humans. A drawing of the microbes he saw in his mouth was published by the Royal Society and is shown in Fig. 1.3.

OK, so a microbe is a microbe, is a microbe…. Not so. There's so much more to this microbe business. There are six major types of microbe – bacteria, viruses, protozoa, fungi, algae and archaea. The only thing that these very different types of organisms have in common is that they are microscopic (well, most of them are). Other than that, they differ enormously in their appearance, structure and **physiology*** .

Journalists continually irritate microbiologists by getting this wrong. Important pathogenic bacteria such as *Staphylococcus aureus* (see Chap. 3) and *Escherichia coli* (Chap. 25) are regularly reported in the media as being "viruses" which, to a microbiologist, is as ridiculous as saying that an oak tree is a type of bird. A typical example is an article written in a very reputable national daily newspaper on January 2, 2019.

---

* words in bold are defined in the glossary.

PLATE XXIV

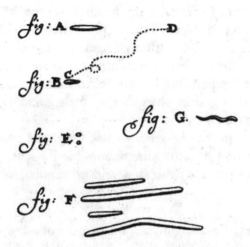

Fig. 1.3 Drawings made by van Leeuwenhoek of the microbes he observed in a sample of dental plaque taken from his mouth

The article was headlined "Wife dies on Caribbean break from rare bacterial infection". However, in the article the reporter says that "Mrs. **XXX** was initially taken to a medical centre in Punta Cana where *Entamoeba histolytica* – a virus that infects 50 million people worldwide each year, killing about 55,000 – was diagnosed". *Entamoeba histolytica* is NOT a virus; it's a protozoan. Protozoa are usually at least 50 times bigger than a virus and are considerably more complex. Later in the article the reporter says that "When she was transferred to a larger hospital in Santo Domingo, the country's capital, doctors changed the diagnosis to meningococcemia, which is caused by a virus associated with meningitis". Meningococcemia is NOT caused by a virus; it's caused by a bacterium called *Neisseria meningitidis*.

Why do reporters and commentators keep making these mistakes? Knowing to which of these six groups a microbe belongs is very important because it tells us an awful lot about it, which makes life easier for microbiologists as it's been estimated that there are at least 10^{12} different microbial species on planet Earth. It's also of great medical importance because if we know that an infection is due to a virus rather than a bacterium, for example, then we'll realise immediately that there's no point in treating it with an antibiotic – it won't work!

What Are the Main Types of Microbe?

Some important, distinguishing features of the six different types of microbes are summarised in Table 1.1. Of these six groups, the algae aren't thought to infect humans, although algal blooms can cause illness by releasing toxins into the environment. Currently there's also little evidence that the archaea are responsible for disease in humans. Although we can find them on our skin and in our gut, very little is known about what they're actually doing there.

So that leaves us with bacteria, viruses, fungi and protozoa to think about. Of these, bacteria and viruses are responsible for most of the infectious diseases that afflict humans living in developed countries.

Table 1.1 Important characteristics of the six types of microbe (a micrometre is one millionth of a metre, i.e. one thousand times smaller than a millimetre: DNA **deoxyribonucleic acid**; RNA **ribonucleic acid**)

Type of microbe	Size	Main features
Bacterium	About 1–2 µm in diameter	• Exist as a single cell (i.e. **unicellular**) • Reproduce by splitting in two • Most have a cell wall • Some are mobile • A few can form spores • Genetic material is DNA
Virus	Usually less than 0.5 µm in diameter	• Don't have a cellular structure • Can't reproduce by themselves • Genetic material can be DNA or RNA • Aren't mobile
Fungus	Most are much larger than bacteria; several micrometres in diameter but can be very long (some are very large, e.g. mushrooms)	• Usually **multicellular** and form filaments • Have a cell wall • Reproduce by forming spores • Genetic material is DNA
Protozoan	Most are much larger than bacteria; often tens of micrometres in diameter	• Unicellular • Reproduce by splitting in two • Don't have a cell wall • Have a flexible shape • Most are mobile • Genetic material is DNA

(continued)

Table 1.1 (continued)

Type of microbe	Size	Main features
Alga	Most are much larger than bacteria; several micrometres in diameter but can be very long (some are very large, e.g. seaweeds)	• Unicellular or multicellular • Have a cell wall • Get their energy from sunlight • Genetic material is DNA
Archaea	About 1–2 μm in diameter	• Unicellular • Have a cell wall • Reproduce by splitting in two • Genetic material is DNA • Most are found in extreme environments

What Do Microbes Look Like?

Examples of each of the six types of microbes are shown in Fig. 1.4. In order to get an idea of the relative sizes of these organisms, it's important to pay careful attention to the magnification used in each image.

Looking at the images in Fig. 1.4, it's easy to understand that even with a high-powered light microscope (i.e. one that can magnify objects 1000 times), we are unlikely to learn much about the structure of these tiny organisms, particularly bacteria and viruses. However, thanks to the electron microscope (commercially available since 1938) with a magnification of up to 10,000,000 times, we now know considerably more about them (Fig. 1.5).

What it looks like through a microscope is still an important way to identify some microbes – particularly the fungi, protozoa and algae. But this is less important for the bacteria, archaea and viruses. Although bacteria are a very diverse group of microbes (there are approximately 10^9 species on Earth), they don't vary much in their shape. They are either spherical (known as cocci) or rod-shaped (known as bacilli) or have a spiral shape (Fig. 1.6). Similarly, archaea and viruses don't really vary much in their appearance.

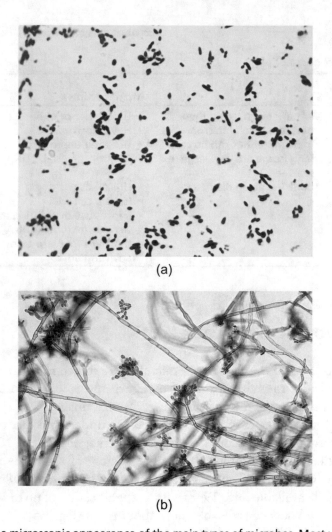

(a)

(b)

Fig. 1.4 The microscopic appearance of the main types of microbes. Most viruses can't be seen through even the most powerful light microscope, but electron microscopy (Fig. 1.5) can be used to study these organisms. (a) **Photomicrograph** of *Streptococcus pneumoniae*, a bacterium found in the human respiratory tract (magnification ×1500). It consists of oval-shaped cells, approximately 1.0 μm in diameter, and they are often in pairs. (Image courtesy of Arnold Kaufman, Centers for Disease Control and Prevention, USA). (b) Photomicrograph of a *Cladosporium* species, a fungus often present in the colon of healthy individuals (×400). The organism grows as **multicellular** filaments with a diameter of 2–6 μm and a length of 40–300 μm. Some form large structures (mushrooms, toadstools) that are visible to the naked eye. (Image courtesy of Dr. Libero Ajello, Centers for Disease Control and Prevention, USA). (c) Photomicrograph of *Entamoeba coli*, a protozoan often found in the colon of healthy individuals (×1150). The usual size of the organism is 15–50 μm. (Image courtesy of Dr. Green, Centers for Disease Control and Prevention, USA). (d) Photomicrograph of *Chlorella vulgaris*, a unicellular alga found in fresh water (magnification ×1300). The diameter of each cell is 3–10 μm. Many algae are multicellular, and some (the seaweeds) are visible to the naked eye. (Andrei Savitsky [CC BY 4.0 (https://creativecommons.org/licenses/by/4.0)]). (e) Photomicrograph of a *Methanobrevibacter* species which is an archaeon. Archaea are similar in size to bacteria and have a diameter of approximately 1.0 μm. (The complete genome sequence of the rumen methanogen Methanobrevibacter millerae SM9. Kelly WJ, Pacheco DM, Li D, Attwood GT, Altermann E, Leahy SC. *Standards in Genomic Sciences* 2016;11:49. This article is distributed under the terms of the Creative Commons Attribution 4.0 International License (http://creativecommons.org/licenses/by/4.0/))

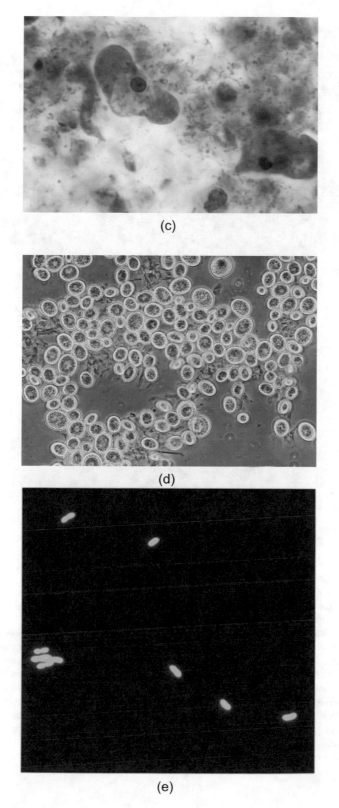

(c)

(d)

(e)

Fig. 1.4 (continued)

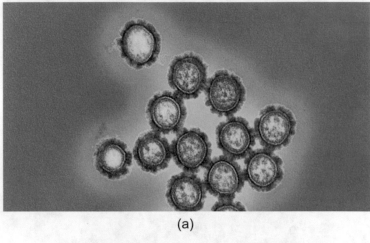

(a)

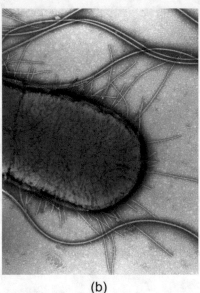

(b)

Fig. 1.5 Appearance of microbes as revealed by electron microscopy. (a) The influenza virus as seen with the aid of an electron microscope. Twelve viruses are shown; each of these has a diameter of approximately 100 nm which is approximately one tenth of the diameter of a typical bacterium. (Image courtesy of the National Institute of Allergy and Infectious Diseases, USA). (b) Part of a cell of *Escherichia coli* as seen through an electron microscope. This shows details of the internal structure of the bacterium as well as the long, whip-like flagella (used to propel the bacterium) and shorter, hair-like fimbriae (enabling the bacterium to attach to surfaces). (Credit: David Gregory & Debbie Marshall. CC BY 4.0). (c) *Giardia lamblia*, a protozoan that can cause diarrhoea as seen through an electron microscope. Typically, the microbe is 10–20 μm long and 7–10 μm wide. (Credit: David Gregory & Debbie Marshall. CC BY 4.0)

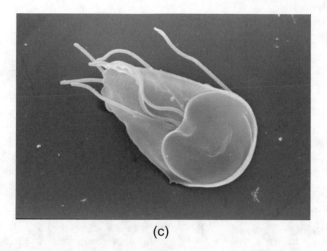

(c)

Fig. 1.5 (continued)

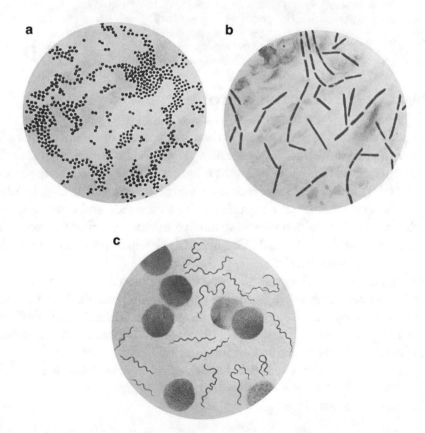

Fig. 1.6 Drawings of photomicrographs showing the three main shapes of bacteria:
(a) cocci, (b) bacilli and (c) spiral-shaped bacteria with red blood cells. (All three images
courtesy of Centers for Disease Control and Prevention, Atlanta, USA)

Box 1.2 How Are Microbes Named?

Microbes, like all organisms, are named using the binomial system (i.e. the name is comprised of two words), and classification is based on the traditional hierarchical system that biologists use to categorise all life on Earth.

The first term of the organism's name refers to the genus to which it belongs, while the second denotes the species – both terms are always italicised, for example, *Staphylococcus aureus*. The genus is often abbreviated, e.g. *S. aureus* or *Staph. aureus*. Species with similar characteristics are considered to belong to the same genus, e.g. *Staphylococcus epidermidis* has many properties in common with *S. aureus*. Genera (the plural of "genus") with similar properties are grouped into a family, similar families into an order, similar orders into a class and similar classes into a phylum.

One of the annoying habits of microbiologists is that they're forever changing the names of microbes. In fairness to them, this is usually because recently acquired information about a particular microbe often means that its original name isn't appropriate. But this can cause confusion when you delve into the past to try and trace the history of a particular microbe. In the early days of the science of microbiology, which was largely focussed on infectious diseases, microbes were often given names based on the disease they caused. However, as knowledge of the structure (i.e. shape and anatomy), physiology (i.e. how it functions) and genetics of the organism accumulated, the name was often altered later.

How Do We Identify Microbes?

We'll start with the bacteria, because these cause many of the infections we'll be covering in this book. Once we know its shape, the next most important characteristic used to identify it is what it looks like after it's been stained using a process developed by a Danish scientist, Christian Gram (Fig. 1.7), in 1884.

This involves smearing the organism onto a glass slide and treating it with purple and red dyes. When viewed through a microscope, those bacteria that are coloured purple are termed "Gram-positive", while those that are red are said to be "Gram-negative" (Fig. 1.8).

Although this procedure was developed nearly one and a half centuries ago, it's still an important step in identifying a bacterium nowadays. This is because the Gram staining behaviour of a bacterium (which determines its ability to retain the purple dye) is related to many of its important characteristics including its cell wall structure and its susceptibility to antibiotics.

Other important characteristics that are useful in identifying it are how it grows, what it feeds on and what waste materials it produces.

Key questions include:

- Does it need oxygen to grow? If it does, then it's called an "aerobe". If it doesn't, it's an "anaerobe". If it's not bothered, then it's a "facultative anaerobe".
- What sugars, carbohydrates, proteins or fats can it use as nutrients?
- Can it grow in acidic conditions?

Fig. 1.7 Portrait of Hans Christian Gram (1900s). (Unknown author [Public domain], via Wikimedia Commons)

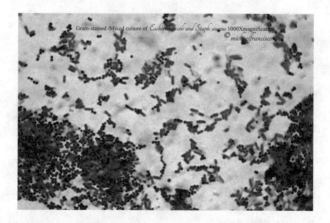

Fig. 1.8 Photomicrograph of a Gram stain of a mixture of Gram-positive cocci (*Staphylococcus aureus* – purple) and Gram-negative bacilli (*Escherichia coli* – red) Magnification ×1000. (Michael R. Francisco from France/CC BY (https://creativecommons.org/licenses/by/2.0))

- Can it grow in high concentrations of salt (sodium chloride)?
- What waste products does it produce?

Such characteristics (together with many others that haven't been listed) of an organism are also important in determining where on planet Earth it can live. Just as an oak tree requires a certain type of soil and particular climatic conditions (temperature, rainfall, etc.) in order to grow, so does each type of microbe. Growing a microbe in the laboratory is an important stage in identifying it and is achieved by methods developed by Robert Koch in 1881 (Box 1.3).

Box 1.3 How Do We Grow Microbes?

A huge problem confronting early microbiologists was how to grow in the laboratory creatures that could only be seen through a microscope. Fortunately, many microbes can reproduce very rapidly (they simply split into two – a process known as binary fission), and this has helped us to develop means of growing them. Imagine placing a single microbe on a surface and supplying its ideal growth conditions. These would include the right nutrients as well as the correct temperature, humidity, atmosphere, pH, etc. If our microbe can reproduce every 20 minutes (which many microbes are able to do), then after only 12 hours it will have reproduced 30 times, and this will have produced a total of 68,719,476,736 cells. The smallest bacteria are approximately 1 μm in diameter and have a volume of 0.524 μm³. The volume of 68,719,476,736 bacteria is 36,009,005,810 μm³, and this will form a cluster with a diameter of 4.1 mm and so is visible to the naked eye – such a cluster consists of identical cells that have all come from a single organism (i.e. they are clones) and is called a colony. In 1881 the German physician and microbiologist Robert Koch developed a simple way to grow colonies of bacteria. He used a solution containing all of the nutrients needed by bacteria (this is called a "medium") and solidified this with gelatin. If bacteria were spread over the surface of this solid, jelly-like medium (using a sterilised loop of wire), then each bacterium would grow into a visible colony after it had been kept (i.e. "incubated") at a suitable temperature in an appropriate environment.

Figure (a). Robert Koch. (Wilhelm Fechner [Public domain] via Wikimedia Commons)

This was a major breakthrough in practical microbiology and forms the basis of all subsequent culture-based methods of isolating, purifying and identifying those microbes that cause disease in humans. Other members of Koch's laboratory then improved his approach by replacing gelatin with agar (in 1882) and by introducing glass containers with removable lids (known as petri dishes – named after the German bacteriologist Julius Petri) instead of glass plates. The nutrient medium was a watery extract of boiled meat. Media have now been designed to grow as many different microbes as possible, and sterile animal blood is often added to enrich them and increase the range of species that can grow. Nowadays, most petri dishes are made of plastic.

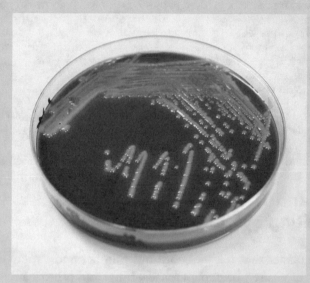

Figure (b). Plastic petri dish containing a nutrient medium solidified with agar. In this image, the lid has been removed, and colonies of a bacterium growing on the surface of the agar can be clearly seen. (Megan Mathias and J. Todd Parker, Centers for Disease Control and Prevention, USA)

Fungi can be grown in a similar way, but viruses require living cells in order to reproduce. Viruses are grown in the laboratory on living cells or tissues using techniques known as "cell culture" and "tissue culture", respectively. These techniques are slow and are expensive to carry out.

Nowadays, thanks to the science of molecular biology, we can identify bacteria and other microbes by analysing their genetic material which is usually DNA (in the case of bacteria, fungi and some viruses) or RNA in some viruses (see Box 1.4). However, fungi and protozoa are still identified mainly on the basis of their microscopic appearance.

Box 1.4 Modern Approaches Used to Identify Microbes

Molecular biological techniques have revolutionised our approach to identifying bacteria and other microbes. No longer do we have to spend days growing the unknown microbe and studying its shape, what it grows on and what it produces. Now all we have to do is extract its genetic material (DNA, or RNA in the case of some viruses) and put this into a machine (known as a "sequencer") that will give us the sequence of bases (i.e. the chemical building blocks of DNA and

RNA) in one of its genes or maybe all of its genes (known as its "genome"). This sequence can be compared and matched with the sequences of the same gene(s) from all known microbes, and so we can identify the unknown microbe. The whole process can be accomplished with the help of a computer in hours rather than days. However, one problem with this approach is that the amount of genetic material present in a sample is usually very small so a technique for increasing (amplifying) the amount present must first be carried out. This is known as nucleic acid amplification (NAA) and is often accomplished by a process known as the polymerase chain reaction (PCR) which uses an enzyme to produce huge numbers of copies of the original genetic material.

In another approach, the proteins present in the microbe can be rapidly analysed in a machine called a mass spectrometer. This can be compared with the protein composition of all known microbes held in databases so enabling identification of the unknown microbe.

Humans and Their Microbiota: A Fascinating Symbiosis

We live on a planet that has an enormous number and variety of microbes. The numbers of viruses, bacteria, archaea, fungi and protozoa have been estimated to be 10^{31}, 10^{30}, 10^{29}, 10^{27} and 10^{27}, respectively. It's difficult to imagine such incredibly large numbers, but the following will give you some idea of their magnitude: if all of the 10^{31} viruses present on our planet were lain end to end, they would stretch for 100 million light years. Microbes comprise an astonishing 18% of the total mass of living creatures (biomass) on our planet (Fig. 1.9). In contrast, animals comprise less than 1% of the total

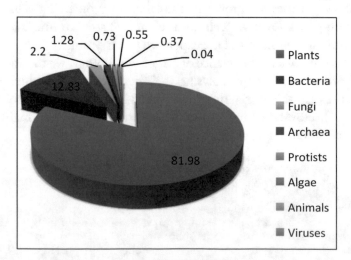

Fig. 1.9 Proportions (%) of the total mass of living organisms (the "biomass") of various types of animal, plant and microbe on planet Earth

biomass. As for their variety, we mentioned previously that there are at least 10^{12} different microbial species.

Given the enormous numbers of microbes in our environment, it needs little imagination to realise that our skin, being exposed to the outside world, must be covered in microbes. Furthermore, the air we breathe also contains microbes (there can be as many as 2000 bacteria per m^3 in indoor air), so they'll be able to get into our nose, throat and lungs. Also, some items of food are covered in microbes (an apple has approximately 20 million bacteria on its surface), which means that they'll be able to reach all parts of our digestive system – mouth, stomach and intestines. The microbes that live with us harmoniously on our various body surfaces are known collectively as our "microbiota" (Fig. 1.10). The microbiota of humans consists of a large number and variety of microbes – approximately 10,000 different species. All of the main groups of microbes, except algae, are found living and growing on our body surfaces.

The next question to ask is "which microbes do we find on our bodies?" Are they the same as the microbes we find in the outside world, i.e. in air, water, food, soil, etc.? The answer to this question is, surprisingly, no. The microbes on our bodies are generally very different from those that are found in the outside world (these are usually referred to as "environmental microbes"). This means that, although we're exposed every day to an enormous variety of environmental microbes, the vast majority of them aren't able to colonise our bodies and join our microbiota. This is because our bodies have different regions that provide a variety of environments and conditions which will encourage particular microbes to thrive and not others. This is reinforced by our defence and immune systems which deter or expel unwanted arrivals – our bodies have a strict set of "immigration rules". The microbes themselves are also active in keeping their special environment favourable to themselves and deterring unwanted environmental microbes.

The next obvious question is "do the various parts of the body all have the same types of microbes?" The answer is no. Each part of the body supports very different microbial communities, and so it's necessary to talk about the "skin microbiota", the "oral microbiota", etc. Furthermore, if we look at these microbial communities in greater detail, we find that they can differ appreciably at different sites within a particular body region. In the case of the skin, for example, the microbiota living on the forehead is very different from that found in the armpits and on the legs. It seems, therefore, that each body site has its own particular set of "immigration rules". These rules arise from the nature of the environment at that particular body site (which is called a "habitat") and are known as "environmental selection factors" or "ecological determinants". These

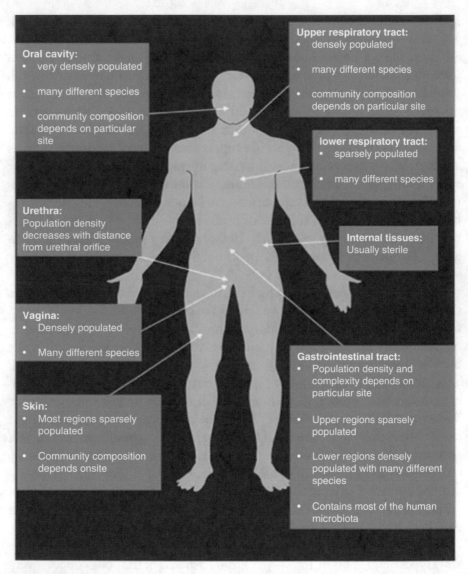

Fig. 1.10 The main sites in a human that are colonised by microbes. The microbial communities found at these sites are collectively known as the "human microbiota"

rules are no different from those that govern which macro-organisms (such as elephants, daffodils, etc.) can become established within a habitat on planet Earth (such as a desert, tropical rainforest, etc.). The only difference is one of scale – in order to understand the microbial communities that live on our bodies, we have to consider what's going on at a microscopic level, i.e. we are now in the realms of micro-ecology rather than macro-ecology.

Microbes and Disease

So, what about microbes and infectious diseases? It wasn't until the 1860s that Louis Pasteur (Fig. 1.11) demonstrated that microbes were able to cause disease.

If we're surrounded by so many microbes and, worse still, have huge numbers of them actually living on us, then why are we all not constantly suffering from some infectious disease? This is a very reasonable question given that, on the whole, the general public has generally been led to believe that all microbes are evil and out to do us harm.

The media bombard us with advertisements for disinfectants (Box 1.5) for use in the kitchen, in the bathroom, on floors, etc. Manufacturers of toys, socks, chopping boards, etc. have incorporated antimicrobial agents into their products in order to make them "safe" for us and our children.

Box 1.5 What's the Difference Between a Disinfectant, an Antiseptic and an Antibiotic?

All three of these chemical families can kill microbes, but each has its own particular use. A disinfectant is a chemical that can rapidly kill microbes but is so toxic to humans or other animals that it can't be used to treat an infectious disease – it can only be applied to inanimate surfaces. Disinfectants are, however, a very important means of reducing the spread of infectious diseases because they can be applied to surfaces that are likely to harbour dangerous microbes such as toilets and kitchens in the home, as well as the walls and floors of hospitals. Examples of effective disinfectants are bleach, ozone and glutaraldehyde.

Antiseptics are chemicals that can kill microbes, but, unlike disinfectants, they don't harm humans or other animals. They can, therefore, be used on external surfaces of the human body (such as the skin and mucous membranes) to kill pathogenic microbes that may be present there. Although they're safe to use on body surfaces, they're too toxic to be ingested or injected. Examples include chlorhexidine, iodine compounds, mercury compounds, alcohol and hydrogen peroxide.

Anti-viral, anti-fungal and anti-protozoal compounds, as well as antibiotics, are antimicrobial agents (drugs) that have such low toxicity for humans and other animals that they can be safely injected or swallowed and used to treat infections throughout the body.

However, the reality is that only a very small proportion of microbes are able to cause disease in healthy adults, although very young children, the elderly, the immunocompromised and those suffering from some chronic disease are more vulnerable. This vulnerability is because the immune system (the body's antimicrobial defence mechanisms) of infants is not fully developed while that of the elderly is in decline. Those on certain medications or suffering from some other underlying illness such as cancer, cystic fibrosis and

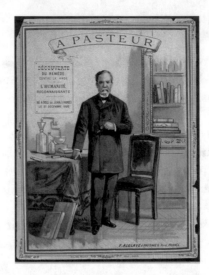

Fig. 1.11 Chromolithograph of Louis Pasteur. (Image courtesy of the Wellcome Collection. Attribution 4.0 International (CC BY 4.0))

diabetes often have immune defences that aren't fully functional. These groups of people are more vulnerable to many infectious diseases than the average healthy adult and are often referred to as being "at-risk" groups.

It's been estimated that the total number of microbes capable of causing infections in humans (we call these "pathogens") is no more than 1400 – this is an infinitesimal proportion of the 10^{12} species present on our planet. Pathogens are organisms that have learned to overcome the elaborate set of antimicrobial defence systems that have evolved in our bodies (Box 1.6).

Box 1.6 How Does the Human Body Defend Itself Against Microbes?

The many means by which our bodies defend us against microbial diseases can be broadly classified into two groups – innate and acquired immunity. Innate immune defences consist of a collection of physical and chemical/biochemical systems that operate continually and are effective against all microbes. The following are important components:

(a) The skin and mucous membranes (i.e. the moist inner surfaces of our body such as the mouth, respiratory tract, gut, urethra, vagina, etc.) constitute a physical barrier that prevents microbes gaining access to our sterile, inner tissues.

(b) The above surfaces are covered in antimicrobial compounds (more than 400 different types) produced by their constituent cells, and these can kill many pathogens.

(c) The cells that make up the skin and mucous membranes are continually being shed (and replaced by new ones), and this means that any microbes attached to them are continually being removed.

(d) Fluids such as urine, saliva and respiratory mucous continually flush away microbes from our internal body surfaces and deposit them into the stomach (in the case of saliva and respiratory mucous) where they are killed by the acidic gastric fluid or into the environment (in the case of urine).

Acquired immune defence systems, in contrast, may take days to weeks to become effective and involve particular types of cells (known as lymphocytes) as well as specific proteins (known as antibodies). An acquired immune response is very specific (in contrast to the generalised innate immune response) and targets only one particular microbe. The antibodies produced by lymphocytes bind to the target microbe which makes it easy to recognise by our phagocytic (i.e. cell-eating) cells which can then engulf and destroy the antibody-coated microbe. Vaccination against a particular pathogen primes the acquired immune system so that it reacts more rapidly than usual which can prevent symptoms developing.

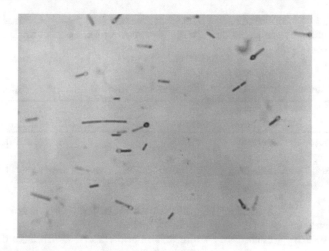

Fig. 1.12 *Clostridium tetani*, the bacterium that causes tetanus. Note the round spores at the end of some of the bacteria that are characteristic of this microbe. This bacterium's usual habitat is the soil (×956). (Dr. Holdeman, Centers for Disease Control and Prevention, USA)

So where do these pathogens come from? Basically there are two main types of pathogen – exogenous (originating from the environment) and endogenous (from our own microbiota). Exogenous pathogens are those microbes that don't ordinarily live on humans but live in the environment or on other animals. Examples include *Clostridium tetani* (Fig. 1.12), an organism that lives in soil and can cause the life-threatening disease, tetanus.

An example of an exogenous pathogen that lives mainly on other animals is *Brucella melitensis* (Fig. 1.13) which lives on sheep and goats and causes brucellosis.

Other examples are given in Table 1.2. These microbes are good at overcoming our antimicrobial defence systems, and if we come into contact with

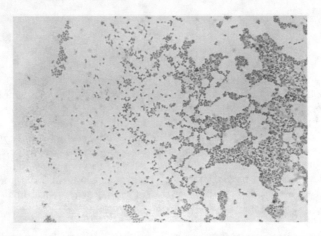

Fig. 1.13 *Brucella melitensis*, the bacterium responsible for brucellosis. This bacterium's usual habitats are goats and sheep. (Dr. W.A. Clark, Centers for Disease Control and Prevention, USA)

Table 1.2 Examples of exogenous pathogens of humans and their origins

Microbe	Type of microbe	Origin	Disease caused
Campylobacter jejuni	Bacterium	Poultry	Gastroenteritis
Salmonella typhimurium	Bacterium	Farm animals such as chickens, pigs and cattle	Gastroenteritis
Listeria monocytogenes	Bacterium	Domestic mammals, rodents, birds	Listeriosis
Bacillus anthracis	Bacterium	Animals; soil and water contaminated with animal faeces	Anthrax, a disease that usually affects the skin
Influenza A virus	Virus	Many animals including birds, pigs and horses	Influenza
Giardia lamblia	Protozoan	Many wild and domesticated animals; water contaminated with animal faeces	Giardiasis, a diarrhoeal disease
Acanthamoeba species	Protozoan	Freshwater and moist soil	Meningoencephalitis and eye infections
Rabies virus	Virus	Dogs, bats, monkeys and other animals	Rabies, a fatal disease of the nervous system
Severe acute respiratory syndrome coronavirus 2 (SARS-CoV-2)	Virus	Probably bats	Coronavirus disease 2019 (COVID-19)

them, then this can result in disease. Whether or not the outcome of such an encounter is disease depends on several factors such as the number of infecting microbes, the general health of the individual and whether or not the individual's defence systems are fully operative.

Our Treacherous Companions

The situation with regard to endogenous pathogens needs more detailed explanation. Over many millennia the types of microbes that inhabit our bodies have changed and have co-evolved with us so that a modern human is considered to be a symbiotic system consisting of a human **host** (composed of mammalian cells) and its microbiota (known as **symbionts**). **Symbiosis** means "living together", and there are three main forms this can take: commensal, when the host is unaffected but the symbiont benefits; mutualistic, both partners benefit; and parasitic, when the symbiont benefits while the host is harmed (Fig. 1.14).

For most of our lives, we have a mutualistic relationship with our microbial symbionts (Box 1.7). They protect us from pathogens in the environment, they provide us with nutrients and vitamins, and our immune system learns

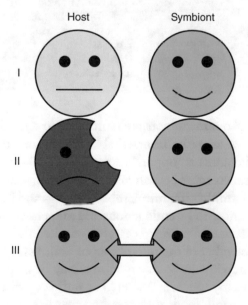

Fig. 1.14 Types of symbiotic relationships. I – commensal: the host is unaffected, while the symbiont benefits. II – parasitic: the host is harmed, while the symbiont benefits. III – mutualistic: both partners benefit.

to tolerate these beneficial microbes. However, this is a delicate balance that can be disrupted causing the relationship to become a parasitic one in which the human host suffers – we are then said to be in a **dysbiotic** state. What triggers such a change is often unknown. However, possibilities include some problem (temporary or long-term) with our antimicrobial defence systems (due to some trauma, stress, illness, malnutrition, etc.), a change in our diet, poor hygiene, hormonal changes and ageing.

Box 1.7 The Human-Microbe Symbiosis

The total number of genes in the microbes living in our gastrointestinal tract (GIT) is approximately 10 million which is about 500 times more than in the human genome. This means that our gut microbiota has a much greater capacity than us to make new molecules and break down existing ones. The relationship between *Homo sapiens* and its gut microbiota is an excellent example of a symbiosis. In the upper regions of the GIT (the stomach, duodenum and jejunum), a human provides an environment that does not encourage colonisation by microbes (it's acidic and dietary constituents pass through very rapidly), and it's in these regions that we extract many of the nutrients from our diet. In the colon, however, the environment is suitable for establishing a large and diverse microbiota – material passes through very slowly in this region. Here we take advantage of the enormous metabolic potential of our microbiota to break down those dietary constituents that we can't digest ourselves. Many of the resulting products are then absorbed by the colon, along with vitamins synthesised by some microbes. The microbially generated compounds that are absorbed supply us with 10% of our daily energy needs. The microbes, in turn, are supplied with an environment suitable for their growth and a constant supply of nutrients. As well as supplying nutrients, the indigenous microbiota also plays an important role in the development of our intestinal mucosa and immune system.

Some members of our microbiota are more likely to cause disease than others, and these are known as **pathobionts** (Table 1.3). Another, more descriptive, term used for pathobionts is "opportunistic pathogens" – this emphasises that these normally harmless microbes can take advantage of some problem (temporary or more long term) in our antimicrobial defences and initiate a disease.

As well as being infected by endogenous and exogenous pathogens, humans may succumb to disease from a third category of microbe. This type of microbe is one that isn't a member of the microbiota of healthy humans but is present only in infected individuals. Such organisms may be acquired directly from diseased individuals (by touch, saliva, etc.) or indirectly from an environmental source (such as water, food, surfaces, etc.) that has been contaminated by the infected individual. Such organisms are, therefore, neither members of the human microbiota nor true residents of the external environment – they can, however, be found in both locations. Examples of these organisms, and the diseases they cause, are given in Table 1.4.

Table 1.3 Important endogenous pathogens (i.e. pathobionts or opportunistic pathogens) of humans

Microbe	Type of microbe	Usual habitat in humans	Diseases caused
Staphylococcus aureus	Bacterium	Nostrils	Skin and wound infections, food poisoning, bone and joint infections
Streptococcus pyogenes	Bacterium	Throat	Pharyngitis, tonsillitis, skin infections
Streptococcus pneumoniae	Bacterium	Throat	Pneumonia, ear infections, meningitis
Neisseria meningitidis	Bacterium	Throat	Meningitis
Haemophilus influenzae	Bacterium	Throat	Ear infections, pneumonia, bronchitis, meningitis
Cutibacterium acnes	Bacterium	Skin	Acne
Gardnerella vaginalis	Bacterium	Vagina	Vaginosis
Streptococcus mutans	Bacterium	Mouth	Caries
Porphyromonas gingivalis	Bacterium	Mouth	Periodontitis
Bacteroides fragilis	Bacterium	Colon	Peritonitis
Clostridium perfringens	Bacterium	Colon	Gangrene
Escherichia coli	Bacterium	Colon	Urinary tract infections
Malassezia globosa	Fungus	Skin	Dandruff, dermatitis, pityriasis versicolor
Candida albicans	Fungus	Vagina, oral cavity, gastrointestinal tract	Oral and vaginal thrush
Entamoeba gingivalis	Protozoan	Mouth	Periodontitis
Herpes simplex virus-1	Virus	Mouth	Oral and genital ulcers

Table 1.4 Microbial pathogens that are transmitted (directly or indirectly) from a diseased to a healthy individual. Such pathogens aren't members of the human microbiota and aren't environmental organisms

Microbe	Type of microbe	Mode of transmission	Diseases caused
Mycobacterium tuberculosis	Bacterium	Directly via saliva (coughs, etc.); indirectly from surfaces contaminated with saliva	Tuberculosis
Shigella dysenteriae	Bacterium	Contaminated water	Shigellosis (bacillary dysentery)
Corynebacterium diphtheriae	Bacterium	Directly via saliva (coughs, etc.); indirectly from surfaces contaminated with saliva	Diphtheria
Neisseria gonorrhoeae	Bacterium	Sexual activities	Gonorrhoea
Treponema pallidum	Bacterium	Sexual activities	Syphilis
Trichophyton rubrum	Fungus	Direct contact or indirectly via contaminated towels, clothing, surfaces, etc.	Athlete's foot
Hepatitis A	Virus	Direct contact or via contaminated food or water	Hepatitis A

In Western countries, diseases due to all three types of pathogen occur, and in the chapters that follow, we'll discuss important examples of those we commonly and most frequently encounter. In each case, the symptoms of the disease will be described as well as the microbe that's responsible for it. The ways in which the causative microbe harms us and the methods available for treating and preventing the disease will also be covered.

Our Epithelium Is Our First-Line Defence Against Microbes

However, before going on to explore these various diseases, it's important to appreciate how the human body tries to fight off microbes to prevent an infectious disease and what happens if it starts to lose this battle. The outer boundary of our body is our epithelium (sometimes known as our "epithelial membrane") and beyond this lies the external environment. The epithelium itself and all that lies within it is "me". It's on the surface of this epithelium that we come into contact with the microbial world. The epithelium forms a continuous covering on all those surfaces of our body that make contact with the external environment. It's easy to appreciate that our skin is an important part of this covering. But it's less obvious that our throat, nostrils, lungs, stomach, intestinal tract, urethra, bladder and vagina are also covered in this epithelium. All of these various parts of the body are, in fact, in contact with the outside world. They are internal continuations of the surface of our body that we usually don't see – apart from our nostrils, mouth and throat which we can easily view in a mirror. These external and internal forms of the epithelium are anatomically similar but differ in that the external one (known as the epidermis) has a very dry surface while the internal ones (known as mucosae or mucous membranes) are moist. Regardless of these distinctions, both types of epithelia are composed of a particular type of cell known as an epithelial cell. Although there are several different types of epithelial cells, they all have one feature in common – they attach to one another to form resilient, cohesive sheets of cells (Fig. 1.15).

The epithelium is a formidable physical barrier that prevents microbes from gaining access to our internal tissues. Provided that it remains intact, few microbes can penetrate this barrier. The other defensive features of the epithelium depend on whether or not it's dry or moist. The dry epithelium of the skin has an outer layer consisting of dead epithelial cells known as the stratum corneum (Fig. 1.16). All microbes require moisture for their growth and survival so this dry surface is a very inhospitable environment for them. Furthermore, the

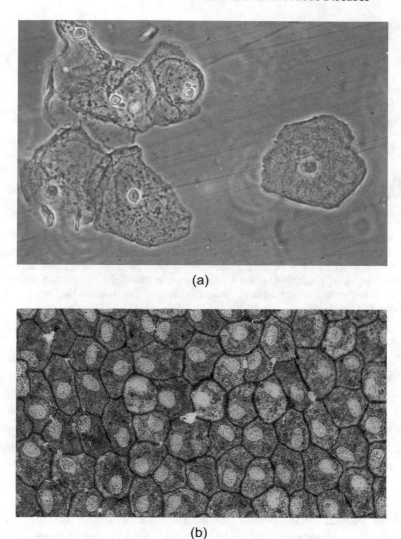

(a)

(b)

Fig. 1.15 Human epithelial cells and epithelium. **(a)** An individual human epithelial cell is shown on the right, and a group of cells (which are overlapping one another) is on the left. The round nucleus is clearly visible towards the centre of each cell. The cell is bounded by a flexible cytoplasmic membrane and so can change shape (magnification ×1000). (M. Rein, Centers for Disease Control and Prevention, USA). **(b)** A section of human epithelium showing tightly packed epithelial cells. The nucleus of each cell is visible as a pale blue-stained rounded structure. The cells are held together by strong adhesive forces between their cytoplasmic membranes (visible as dark, linear structures between the cells) so that they form a cohesive sheet of cells. Some of these sheets of cells form structures that are several layers thick (magnification ×400). (This file is made available under the Creative Commons CC0 1.0 Universal Public Domain Dedication via Wikimedia Commons)

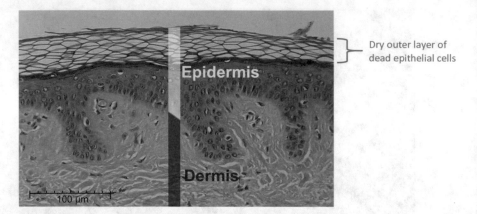

Fig. 1.16 Cross-section through human skin showing the epidermis and the outer layer of dead cells known as the stratum corneum. (I, the copyright holder of this work, release this work into the public domain. This applies worldwide. In some countries this may not be legally possible; if so: I grant anyone the right to use this work for any purpose, without any conditions, unless such conditions are required by law. Via Wikimedia Commons)

dying epithelial cells produce a lot of acids, and these accumulate to make the skin surface very acidic – few microbes can tolerate such an acidic environment.

The glands in our skin are continually producing sweat (0.5–1.0 litres per day) which has a high salt content. Evaporation of sweat from the skin's surface leaves behind high concentrations of salt, and most microbes can't grow in these salty conditions. Glands in the skin, as well as the epithelial cells themselves, continually produce a range of compounds that can kill microbes – these are known as **antimicrobial peptides** (AMPs). Currently, we know that humans produce approximately 400 different AMPs, each of these is able to kill only certain microbes and this is referred to as being the "antimicrobial spectrum" of the peptide. Some AMPs have a broad spectrum, whereas others have a narrow spectrum and can kill only a limited range of microbes. Collectively, however, this large range of AMPs constitute a formidable defence mechanism against those microbes that are out to do us harm.

Finally, a very important defence mechanism of epithelial surfaces, including the skin, is that the outer layer of epithelial cells is constantly being shed into the environment. Any microbes that have managed to survive on this inhospitable surface will, of course, be removed along with these cells (Fig. 1.17). It's been estimated that the average adult sheds 1 million skin particles each day which means that an enormous number of microbes are continually being removed from our bodies.

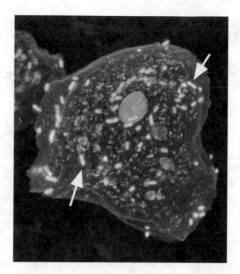

Fig. 1.17   Epithelial cell with pairs of bacteria (arrowed) and individual cells attached to it. (Image kindly supplied by Dr. Chris Hope, University College London)

Mucosal Surfaces of the Body: Slime Is Good

In contrast to the dry epithelium of the skin, mucosal surfaces don't have a layer of dead cells on their surface; they're covered in a layer of mucus that's produced continually (Fig. 1.18).

Mucus contains a variety of large molecules known as **mucins**, and these give mucus a gel-like consistency. Mucins are **glycoproteins** and can bind a large number of different microbes. They therefore trap microbes and stop them penetrating to the underlying epithelial cells. The mucus and the entrapped microbes are expelled from the body by mechanisms that vary between different body sites. In the respiratory tract, coughing and sneezing expel the mucus. This is beneficial to the coughing/sneezing individual but dangerous to others because it's an excellent means of spreading pathogens to other humans. In the urethra, the mucus layer is removed by urination. In the nasal cavity, the upper regions of the lung and the cervix, the mucus is removed by a specialised system known as the mucociliary escalator (Fig. 1.19). The epithelial cells of these regions are covered in hair-like structures (known as cilia) which move in a coordinated wave-like motion (like a Mexican wave) to propel the layer of mucus forwards – to the mouth in the case of the nasal cavity and upper regions of the lungs and to the vagina in the case of the cervix. In the mouth the mucus is then swallowed, and any microbes present are killed by the very acidic conditions of the stomach. In the vagina the acidic conditions there also kill many of the microbes in the mucus.

Fig. 1.18 The presence of a mucus layer (approximately 10 μm thick) on the surface of the nasal mucosa has been revealed by staining with a red dye. Scale bar = 10 μm. (Lee HJ, Yoo JE, Namkung W, Cho HJ, Kim K, Kang JW, Yoon JH, Choi JY. Thick airway surface liquid volume and weak mucin expression in pendrin-deficient human airway epithelia. *Physiol Rep.* 2015 Aug;3(8):pii: e12480

The mucus layer also contains high levels of AMPs produced by the underlying epithelial cells, and these are able to kill many of the microbes that are trapped there.

How Do We Manage to Co-exist with Our Microbiota?

Despite the impressive array of antimicrobial defence systems operating there, the epithelium is still covered in those microbes that are members of our microbiota. How is this possible? The short answer is that we don't really know. Microbiologists through the ages have usually focussed their attention on microbes that cause disease rather than those that live happily with us – our symbionts. However, during the last few decades, we've begun to appreciate the importance of our microbial symbionts and, consequently, have learnt far more about them. Our epithelial cells are able to distinguish between microbes that are our symbionts and those that aren't. In other words, it's as if we can recognise and make a distinction between a "microbial self" and a "microbial non-self". Although we don't know the whole story, it looks as though special sensors in epithelial cells (known as **Toll-like receptors** – TLRs) can distinguish between our symbionts and pathogenic microbes. A set of ten different TLRs have been found so far, and when these recognise that a particular microbe isn't one of our symbionts (and therefore may be harmful), then it sends a signal to our immune system that it should get ready to mount a defensive response. On the other hand, if the TLRs recognise that a microbe is a member of our microbiota, then no such signal is sent, and it's left alone.

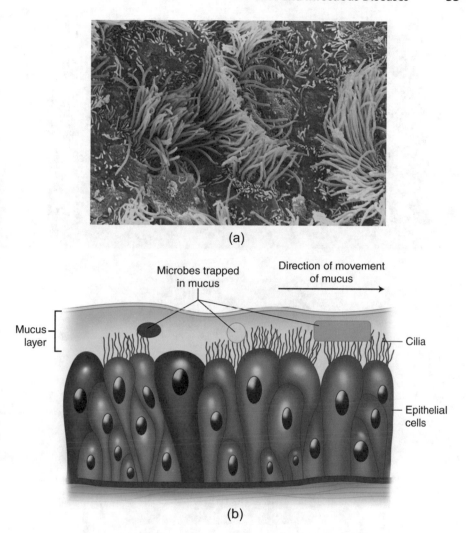

(a)

(b)

Fig. 1.19 The mucociliary escalator – a system for expelling mucus (and the microbes trapped in it) from the respiratory and female reproductive systems. (a) A section of the epithelium from the human respiratory tract showing cilia on the epithelial surface. (Credit: David Gregory and Debbie Marshall. CC BY 4.0). (b) Diagram showing the main features of the mucociliary escalator.. (c) The ciliated epithelium of the nasal mucosa (×400). (Gelardi M, Luigi Marseglia G, Licari A, Landi M, Dell'Albani I, Incorvaia C, Frati F, Quaranta N. Nasal cytology in children: recent advances. *Italian Journal of Pediatrics* 2012;38:51. (d) An individual ciliated cell (×2000). (Gelardi M, Luigi Marseglia G, Licari A, Landi M, Dell'Albani I, Incorvaia C, Frati F, Quaranta N. Nasal cytology in children: recent advances. *Italian Journal of Pediatrics* 2012;38:51

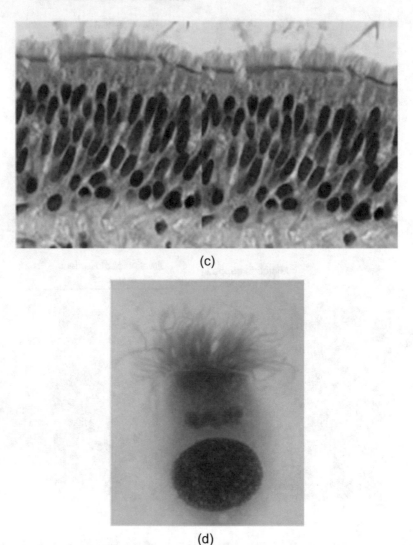

(c)

(d)

Fig. 1.19 (continued)

What Happens if We Get an Infection?

So, what happens if our epithelial defences fail to exclude a harmful microbe? This could happen if it becomes damaged in some way (such as by a physical injury or by some toxic chemical) or the microbe may be able to invade through the epithelium. The TLRs immediately alert the immune system by means of signalling molecules known as **cytokines**. These cytokines initiate what is known as an "inflammatory response" which attracts phagocytic cells such as **neutrophils** and **macrophages** (Fig. 1.20) to the site of the infection. These

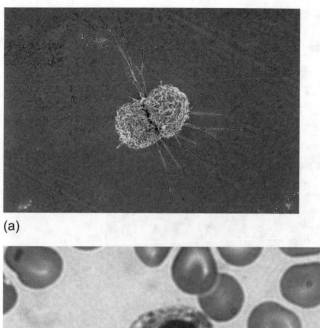

(a)

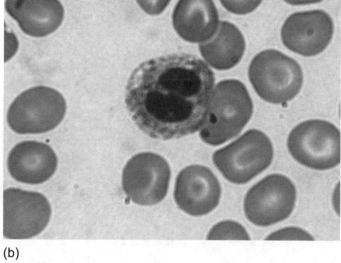

(b)

Fig. 1.20 Important cells of the human immune defence system. (a) A scanning electron micrograph of a macrophage (artificially colourised). (National Institute of Allergy and Infectious Diseases (NIAID)/CC BY (https://creativecommons.org/licenses/by/2.0)). (b) A polymorphonuclear leukocyte (also known as neutrophil or polymorph) with its characteristic multi-lobed nucleus – this one has two lobes, but several lobes may be present. It's surrounded by many red blood cells. (Ed Uthman from Houston, TX, USA/ CC BY (https://creativecommons.org/licenses/by/2.0)). (c) Electron micrograph (artificially colourised) showing a neutrophil (yellow) engulfing a large bacillus (orange). (Volker Brinkmann [CC BY 2.5 (https://creativecommons.org/licenses/by/2.5)]). (d) Two lymphocytes (stained blue) with characteristically large nuclei. Several red blood cells are also present (×1150). (Carl Flint and Dr. Volger, Emory University Hosp., Centers for Disease Control and Prevention)

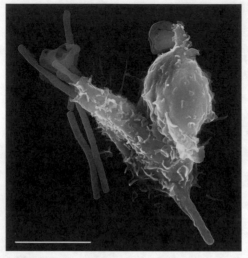

(c)

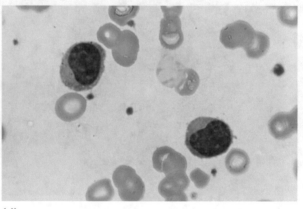

(d)

Fig. 1.20 (continued)

defence cells then engulf and kill the invading microbe. At the same time, other cells of the immune system (**lymphocytes**) are activated to produce antibodies which can bind to the invading microbe and stop it sticking to human tissues as well as helping the phagocytes to destroy the microbe. Inflammation is often accompanied by fever (Box 1.8) because one of the cytokines, interleukin-1, as well as being an important signalling molecule also affects the thermoregulatory centre in the brain, resulting in an increased body temperature.

The inflammatory response makes those blood vessels at the site of the infection increase in diameter (known as **vasodilation** – Fig. 1.21) and become more permeable due to the action of histamine which is released by immune cells.

Fig. 1.21   Vasodilation of blood vessels increases their diameter and their permeability. On the left is a normal blood vessel, while the one on the right shows the effects of inflammation.

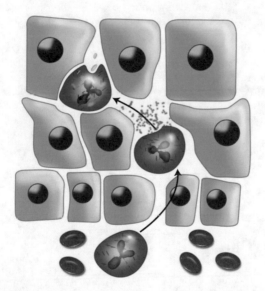

Fig. 1.22   The increased permeability of blood vessels during the inflammatory response enables neutrophils to migrate from the blood vessel into tissues where they can phagocytose any microbes that are present.

The vasodilation brings more blood (and therefore defence cells) to the infection site, while the increased permeability of the vessel walls makes it easier for immune defence cells and antibodies to leak out of the blood vessel and into the site of the infection to deal with the infecting microbe (Fig. 1.22).

These effects give rise to the four classic signs of inflammation: (i) redness, because there's more blood at the site; (iii) heat, again, because of the presence of larger quantities of blood; (iii) swelling (known as **oedema**), because large amounts of fluid have leaked out of the blood vessels; and (iv) pain, the histamine that makes the blood vessels more permeable also stimulates nerve cells resulting in pain. The oedema can also press against local nerve cells causing pain.

Box 1.8 What Is Fever?

Fever or pyrexia is a rise of your body temperature above normal which is usually around 37 °C or 98.6 °F. When your temperature is at or above 38 °C or 100.4 °F, most doctors, but not all, would say that you had a fever. It's hard to specify as there is variation of what is normal between individuals, according to the time of the day and the body site measured. There is also a diurnal fluctuation of normal body temperature by around 0.5°, with it being highest in the early evening.

Fever is a controlled change of the body temperature setting in response to a variety of triggers. This is commonly an infection but can also arise during other illnesses and processes including auto-immune diseases, trauma, cancer, vaccinations and drug side effects. A raised temperature is a symptom and is not a disease itself.

The body's temperature regulating centre is in the brain in an area known as the hypothalamus. When the hypothalamus is stimulated via chemical messengers (called pyrogens), the body responds in a variety of ways. Heat is generated by breaking down stored fat, and increased muscle contractions are experienced as shivering. Heat is conserved by constricting skin blood vessels which results in us feeling cold. Pyrogens are found in bacterial cell walls and can also be produced by a variety of human immune cells.

The role of fever is still debated; it may help your body to fight pathogens as it probably stimulates your immune system defences and can suppress some microbes which have a preference for particular temperatures.

In healthy people who are otherwise feeling well, a mild fever in itself is unlikely to cause serious or lasting damage, and, although it feels uncomfortable, it's not often dangerous. Obviously anyone with a fever should seek advice from a medical professional who will assess how unwell they are, identify the underlying cause and advise on the correct treatment.

Wearing loose clothing and using light bedding will help you to keep cool, and drinking plenty of fluids will reduce the dehydration that results from sweating. Fever can be treated with medication such as paracetamol and ibuprofen to make you more comfortable if required, but studies show that reducing a mild fever has little effect on recovery or reducing the risk of febrile convulsions. A fever that doesn't respond to medication doesn't necessarily imply a serious cause.

Fever is not always benign and there are some serious situations where urgent medical attention should be sought. These include anyone who has an immune deficiency, or other underlying conditions, which put them at risk of serious infections. Children aged under 3 months with a temperature of 38 °C or above and children aged 3–6 months with a temperature of 39 °C or above should consult a health professional without delay. A fairly common association in children up to the age of 5 years is a febrile seizure or convulsion; these are poorly understood but rarely serious.

The site of the infection soon becomes red, hot, painful and swollen. Large numbers of dead microbes, as well as phagocytes that have died during their battle with the invading microbes, accumulate in the remaining fluid, and this is known as pus. Pus is a thin, protein-rich fluid that contains dead phagocytic cells (mainly neutrophils) and microbes (both dead and live). The accumulation of pus results in what is known as an abscess; when pus accumulates near the skin surface, it's called a pustule or pimple. Eventually, in most cases, the invading microbe is defeated and the inflammatory response subsides and tissue repair restores the site to a healthy state. When this process takes no more than a few days, it's described as "acute inflammation". However, in some cases the microbe isn't totally eliminated, and the inflammatory response may continue for months or even years – this is known as "chronic inflammation" and can result in long-term damage to the body. Humans haven't been content to let nature takes its course and rely entirely on the ability of our immune response to combat an infectious disease. We've invented antibiotics to help the body ward off attack by pathogens.

What Drugs Can We Take to Fight Off an Infection?

A number of substances have been used over the centuries to combat infectious diseases. These have included compounds of arsenic, bismuth and mercury (Fig. 1.23) for the treatment of syphilis and gonorrhoea. However, although they're effective at killing microbes, they're also highly toxic to humans.

Effective antimicrobial chemotherapy, which uses drugs that can kill pathogens without harming their human host, arrived in 1907, thanks to the German scientist Paul Ehrlich. He produced the first synthetic antibacterial agent, an organoarsenic compound called salvarsan (or arsphenamine), and this was used in the successful treatment of syphilis (Fig. 1.24). For this reason, Ehrlich is regarded as the founding father of antimicrobial chemotherapy.

The next major breakthrough was the discovery of prontosil by the German bacteriologist Gerhard Domagk in the early 1930s. This was the first of a series of drugs known as the sulphonamides which were widely used, with great success, in the late 1930s. Meanwhile, in 1928, penicillin was discovered in London by Alexander Fleming, although it didn't come into clinical use until the early 1940s. It must be pointed out that penicillin was the first true

Fig. 1.23 Mercurous chloride was also known as calomel, a popular drug from the 1800s. It was prescribed for a number of infections – most notably syphilis. However, as it's highly toxic, it also acted as a slow poison as well as a cure. This packet contained calomel tablets to be taken orally and was supplied by the 11th Army Corps of the German Army to its medical personnel during World War I to combat sexually transmitted infections. (Credit: Science Museum, London. CC BY 4.0)

Fig. 1.24 Salvarsan was a synthetic drug developed by Paul Ehrlich to treat syphilis. Ehrlich coined the phrase "magic bullet" to describe this new wonder drug. The diluted yellow Salvarsan treatment was difficult and painful to inject. The drug in the kit was made by a German manufacturer and is stamped with the date "February 3, 1912". It was sold by a British chemist, W Martindale, who added all the equipment to prepare the injections. (Credit: Science Museum, London. CC BY 4.0)

Table 1.5 The main classes of antibiotics

Antibiotic class	How they work
Penicillins (e.g. amoxicillin)	Inhibit the synthesis of bacterial cell walls
Cephalosporins (e.g. cefotaxime)	Inhibit the synthesis of bacterial cell walls
Tetracyclines (e.g. doxycycline)	Inhibit protein synthesis
Glycopeptides (e.g. vancomycin)	Inhibit the synthesis of bacterial cell walls
Lincosamides (e.g. clindamycin)	Inhibit protein synthesis
Macrolides (e.g. erythromycin)	Inhibit protein synthesis
Polypeptides (e.g. polymyxin B)	Disrupt the cell wall of Gram-negative species
Quinolones/fluoroquinolones (e.g. ciprofloxacin)	Inhibit DNA replication
Oxazolidinones (e.g. linezolid)	Inhibit protein synthesis
Carbapenems (e.g. imipenem)	Inhibit the synthesis of bacterial cell walls
Aminoglycosides (e.g. gentamicin)	Inhibit protein synthesis

antibiotic – the previously discovered agents are classed as chemotherapeutic agents. The term antibiotic was first defined by Selman Waksman (the discoverer of streptomycin) in 1947 as "a chemical substance, produced by micro-organisms, which has the capacity to inhibit the growth of and even to destroy bacteria and other micro-organisms". Penicillin is produced by a fungus (belonging to the *Penicillium* genus) and so is a true antibiotic. Since then a number of different types (known as "classes") of antibiotic have been discovered, and these differ mainly in the way in which they kill, or inhibit the growth of, bacteria (Table 1.5). The widescale use of antibiotics largely put an end to the development of another promising approach to the treatment of infectious diseases – the use of bacteriophages (Box 1.9).

Box 1.9 Bacteriophage Therapy

Bacteriophages (literally "bacteria-eaters") are viruses that infect and destroy bacteria. Because they don't damage human cells, there was great interest in using them to treat bacterial infections during the early twentieth century. This idea was first suggested by the French-Canadian microbiologist Felix d'Herelle who, in 1919, used bacteriophages (often abbreviated to "phages") to successfully treat four children in Paris who had bacterial dysentery. A number of phage therapy centres were then established in Europe and India to treat a range of infections including dysentery, cholera and bubonic plague. With the introduction of sulphonamides and antibiotics, interest in phage therapy generally declined except in the former Soviet Union and Eastern Europe. Research into phage therapy continued in these countries and is currently used to treat a wide range of infections including those due to antibiotic-resistant organisms. The increasing problem of antibiotic resistance has regenerated interest in phage therapy in the West, although currently this approach has not been approved for human use in either the EU or USA.

Antibiotics are also classified on the basis of whether they kill, or only inhibit the growth of, bacteria – these are referred to as being **bactericidal** and **bacteriostatic**, respectively. Another classification system takes account of whether they're effective against only a few types of bacteria (known as "narrow spectrum") or a wide range of bacteria ("broad spectrum"). Great care must be taken to ensure that the correct antibiotic is used to treat a patient suffering from a bacterial infection. The main considerations are: (i) is the antibiotic effective against the particular species responsible for the infection and (ii) does the patient have an allergy or adverse reaction to that particular antibiotic? The best way to answer the first question is to isolate the bacterium from the patient and test it against a range of antibiotics to see which of these is the most effective – this is known as "antibiotic sensitivity testing" (Box 1.10). However, this can take several days, and seriously ill patients need to be given an antibiotic immediately. Consequently, the clinician often has to choose an antibiotic on the basis of which pathogen is most likely to be responsible for the infection and which antibiotic is most likely to be effective against it – this approach is known as "empirical therapy". Guidelines have been issued by responsible authorities in most countries to help the clinician choose the best antibiotic. With the growing problem of antibiotic resistance among pathogens, another factor to be taken into consideration is the likelihood of the pathogen being resistant to the particular antibiotic selected. Consequently, public health authorities carefully monitor the sensitivity of pathogens in their geographical area to various antibiotics. The clinician should, therefore, take this into account when trying to select an appropriate antibiotic for empirical therapy. Finally, once an antibiotic has been selected, it's important to ascertain if the patient is allergic to it. Allergy to penicillins and cephalosporins is a common problem – 7% and 10% of the UK and USA populations, respectively, are reportedly allergic to these antibiotics and allergies can overlap between drugs. For these, and many more reasons, we advise never to self-treat or use antibiotics without medical advice.

Box 1.10 How Do We Choose Which Antibiotic to Use to Treat an Infection?

The sensitivity (or susceptibility) of a bacterium to an antibiotic, or other antimicrobial agent, is a routine procedure carried out in hospital microbiology laboratories. Once the pathogen has been isolated from a sample taken from the patient, it's transferred (inoculated) onto an agar plate and spread over its surface. Small paper discs containing a known concentration of each of the antimicrobial agents are placed on the agar plate which is then incubated to allow the microbe to grow – this usually takes at least 16 hours (i.e. overnight). If the pathogen is resistant to a particular antibiotic, then it will grow right up to the edge of the disc containing that antibiotic. If the pathogen is sensitive to an antibiotic, then it won't be able to grow up to the edge of the disc, and this can be seen as a "zone of inhibition" around that disc. The microbiologist can then send a report to the clinician telling them which antibiotics are likely to be effective in treating the infection.

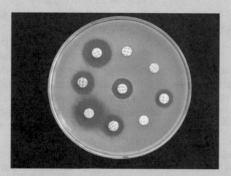

The pathogen has grown right up to the edge of three of the discs and therefore is resistant to the antibiotics contained in those discs. Zones of inhibition can be seen around the other six discs which means that any of these six antibiotics would be suitable for treating an infection caused by this pathogen. (Microrao, I, the copyright holder of this work, release this work into the *public domain*. This applies worldwide. In some countries this may not be legally possible; if so: *I grant anyone the right to use this work for any purpose, without any conditions, unless such conditions are required by law.*)

How Do Microbes Cause Damage to Our Body?

It's important to bear in mind that microbes aren't vindictive creatures. They don't intentionally set out to harm the animals they colonise, and, in many cases, it's in their best interest to prolong the life of their host because to move to a new host is a nuisance for them. All they're trying to do is survive in whatever environment they find themselves. So, when they land on a human body, all they're trying to do is ensure that they have food, warmth and shelter. However, they're immediately confronted with the body's defence systems, which they have to fight off. Therefore, they've evolved strategies to overcome these defence systems, and these are referred to as "virulence factors" (Table 1.6). Each microbe has a particular set of virulence factors that enable it to survive

Table 1.6 The main types of virulence factors produced by microbes

Virulence factor	Function
Capsule – a jelly-like coating surrounding the microbe	Prevents the microbe from being ingested and destroyed by the body's white blood cells
Enzymes secreted by the microbe into the environment	These can break down proteins, fats and polysaccharides in human tissue to small molecules that can then be absorbed by the microbe as nutrients
Adhesins – molecules on the surface of the microbe	Adhesins bind to complementary molecules (known as receptors) on the surface of human cells, thereby enabling the microbe to anchor itself and not be easily removed
Invasins – molecules that enable the microbe to invade human cells	Invasion of a human cell provides the microbe with an environment conducive to its growth and helps it to avoid the body's defence systems
Modulins	Molecules (usually on the outer surface of the microbe) that can interfere with our immune defence system
Exotoxins – molecules secreted by the microbe that can kill human cells	Exotoxins can kill the cells involved in the body's defence systems. The contents of the dead human cells can also be used as nutrients by the microbe
Endotoxins – complex molecules in the cell wall of Gram-negative bacteria that are made from a lipid and carbohydrates and are known as lipopolysaccharides (LPSs)	LPSs are important virulence factors of Gram-negative bacteria. They are highly toxic to human cells, induce the release of inflammatory cytokines, are antigenic, induce fever in mammals and can cause septic shock

on whichever part of the human body it prefers to colonise. The virulence factors employed by each of the microbes involved in the diseases covered in this book will be described in subsequent chapters. Once they've managed to overcome our antimicrobial defences, the next thing they have to do is find food, so that they can grow and reproduce. Often the only food sources available to them are the actual molecules that make up our tissues (proteins, polysaccharides, nucleic acids and fats), and in order to use these, they produce enzymes (which we often describe as being "toxins") that break down these molecules to smaller compounds that they can transport into their cells. The harm they cause to us by doing this is purely incidental – it's merely collateral damage.

The Drugs Don't Work: The Microbes Fight Back

As well as having developed strategies to circumvent our natural antimicrobial defences, microbes have also evolved methods of dealing with the antimicrobial agents we now employ to treat infections. Soon after the widespread use of penicillin, bacteria resistant to this antibiotic emerged. This same pattern of resistance development has followed the introduction of each new antibiotic into clinical practice. A large number of pathogens are now resistant to many of the antibiotics in current use, and this has become one of the most serious problems facing modern medicine. The main factors responsible for the growing resistance of bacteria to antibiotics are:

- Inappropriate use – such as being given to patients suffering from viral infections.
- Overuse – giving antibiotics when they aren't necessary.
- Widespread use in intensive farming – here they're used as growth stimulators.
- Non-prescription use – in some countries antibiotics can be obtained by the general public without a prescription and are often used unnecessarily.

How Can We Avoid Catching an Infectious Disease?

When it comes to infectious diseases, the thirteenth-century proverb "prevention is better than cure" is certainly true. Fortunately, thanks to the pioneering work of Edward Jenner in 1796 (Fig. 1.25), **vaccination** provides us with a highly

Fig. 1.25   Dr. Jenner performing his first vaccination, on James Phipps, a boy of 8. May 14, 1796. (Credit: Wellcome Collection. CC BY 4.0)

effective means of preventing a number of infectious diseases. The vaccine is delivered either orally or by injection and may consist of a dead microbe, the microbe in a weakened (attenuated) state, a component of the microbe (such as a protein) or a microbial product (such as a toxin). It may be effective against only one particular, or several different, strain(s) of the infecting microbe – the latter is known as a polyvalent vaccine. The vaccine stimulates the immune system in advance of being properly infected, so that if the pathogen is encountered, then a prompt and more efficient immune response will be generated before an infection can take hold. Within a short time (several weeks), the recipient develops what should be a long-term immunity to infection by that organism.

According to the WHO, effective vaccines are available for preventing the following infectious diseases: cholera, dengue fever, diphtheria, influenza, Japanese encephalitis, malaria, measles, mumps, pertussis, poliomyelitis, rabies, rubella, shingles, tetanus, tick-borne encephalitis, tuberculosis, typhoid fever and yellow fever. They are also available for preventing infections due to the following microbes: *Haemophilus influenzae* type b, hepatitis A, hepatitis B, hepatitis E, human papilloma virus, *Neisseria meningitidis*, *Streptococcus pneumoniae*, rotavirus and varicella.

The risk of transmitting a pathogen from an infected to an uninfected individual can be decreased by a number of simple hygiene measures such as handwashing (Box 1.11) and the use of antiseptics. Other means of reducing this risk depend on the route of transmission of the pathogen and are described in subsequent chapters.

Box 1.11 Handwashing, a Simple but Effective Way of Preventing the Transmission of Many Pathogens

Our hands can easily become contaminated with pathogenic microbes. Some of the many ways in which this can occur include:

- Putting our hands in front of our nose or mouth when we cough or sneeze
- Putting our fingers inside a nostril or into our mouth
- Touching the lesions produced by an infectious disease
- Touching the skin of individuals with an infectious disease
- Wiping our bottom after going to the toilet
- Touching surfaces, utensils, clothing, towels and bed linen that have pathogenic microbes on them
- Touching contaminated food

These microbes become attached to the outer layers of the stratum corneum and are, therefore, easily removed by washing with soap and water. Ordinary soap doesn't kill most microbes, but it enhances their removal from the hands.

(continued)

However, it can disrupt the outer membrane of some viruses and therefore inactivate them.

Important guidelines for handwashing as a means of reducing microbial contamination have been produced by the Centers for Disease Control and Prevention, and these are summarised below (https://www.cdc.gov/handwashing/show-me-the-science-handwashing.html).

1. *Wet your hands with clean, running water, turn off the tap, and apply soap.*

   Running water is more effective than standing water (in a bowl or sink) because hands could become re-contaminated if placed in standing water that has been contaminated through previous use.

   The temperature of the water doesn't affect microbe removal. However, warmer water may cause more skin irritation and is more environmentally costly.

   Using soap to wash hands is more effective than using water alone because the surfactants in soap remove dirt and microbes from the skin. Also, people tend to scrub hands more thoroughly when using soap, and this mechanical action helps to remove microbes.

   In the domestic setting, soaps containing antimicrobial agents are no more effective than ordinary soap in removing microbes.

2. *Lather your hands by rubbing them together with the soap, and make sure that this includes the backs of your hands, between your fingers and under your nails.*

   The mechanical action of lathering and scrubbing hands helps to remove dirt, grease and microbes from skin. Microbes are present on all surfaces of the hand, particularly under the nails, so the entire hand should be scrubbed.

3. *Scrub your hands for at least 20 seconds, preferably for 30 seconds.*

   Washing hands for about 15–30 seconds removes more microbes from hands than washing for shorter periods.

4. *Rinse your hands well under clean, running water.*

   Soap and mechanical action help to lift dirt, grease and microbes from the skin, but they need to be flushed away by running water. Rinsing the soap away also minimises skin irritation.

5. *Dry your hands using a clean towel or air-dry them.*

   Microbes are transferred more easily to and from wet hands; therefore, hands should be dried after washing. Using a clean towel and air-drying are the best ways of drying.

Want To Know More?

General Microbiology

American Society for Microbiology https://www.asm.org/browse-asm
Microbiology Society, UK https://microbiologyonline.org
Microbiology News, Science Daily https://www.sciencedaily.com/news/plants_animals/microbiology/
Micropia, Holland https://www.micropia.nl/en/discover/microbiology/
Todar's Online Textbook of Bacteriology http://textbookofbacteriology.net
Microbiology https://openstax.org/details/books/microbiology

Infectious Diseases

American Academy of Pediatrics https://www.healthychildren.org/English/health-issues/conditions/infections/Pages/Overview-of-Infectious-Diseases.aspx
Centers for Disease Control and Prevention, USA https://www.cdc.gov/diseasesconditions/index.html
Mayo Foundation for Medical Education and Research, USA https://www.mayoclinic.org/diseases-conditions/infectious-diseases/symptoms-causes/syc-20351173
National Foundation for Infectious Diseases, USA https://www.nfid.org/infectious-diseases/
Public Health Agency of Canada https://www.canada.ca/en/public-health/services/infectious-diseases.html
Public Health England https://www.gov.uk/topic/health-protection/infectious-diseases
World Health Organization https://www.who.int/topics/infectious_diseases/en/

Part II

Skin Infections

2

Acne

Abstract Acne vulgaris is a common skin disease in which the bacterium *Cutibacterium acnes* plays an important role although a number of other factors are involved including increased sebum production. Acne mainly affects adolescents, and, although it isn't a dangerous disease and has few complications, it can result in embarrassment, low self-esteem and social isolation. A variety of treatments are available for the condition including antimicrobial and anti-inflammatory agents as well as drugs that decrease sebum production.

Acne: What Is It?

Scourge of our adolescence. Endless hours spent in front of a mirror squeezing spots, pimples or zits. You know you shouldn't, but you can't help it. The constant searching and probing. Is it ready for squeezing? Weeks pass and it doesn't get any better. There are flare-ups and periods of relative calm, but it seems as though it will go on forever. The lonely nights spent skulking in your bedroom because you're too embarrassed to go out. As for the objects of one's affections, well, obviously, they're not going to look at old spotty face. The promised solutions followed by the invariable disappointments. The realisation that most adverts are just lies – this generates a hatred of capitalism and turns you into a socialist. Socialism wouldn't stand a chance if some entrepreneur came up with a spot remover, spot concealer or spot suppressor.

Spots (known by clinicians as "lesions") of all shapes, sizes and colours (white, black, yellow, deep ones, big ones, small ones, clusters of them, lonely outliers) are what we associate with this disease (Fig. 2.1).

M. Wilson, P. J. K. Wilson, *Close Encounters of the Microbial Kind*,
https://doi.org/10.1007/978-3-030-56978-5_2

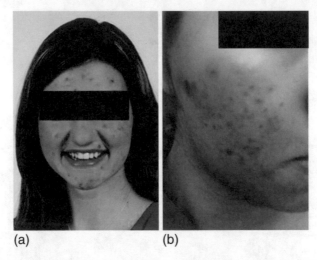

(a)   (b)

Fig. 2.1 Teenagers with acne. ((**a**) Psychosocial judgements and perceptions of adolescents with acne vulgaris: A blinded, controlled comparison of adult and peer evaluations. Ritvo E, Del Rosso JQ, Stillman MA, La Riche C. *Biopsychosoc Med.* 2011;5(1):11. This article is published under license to BioMed Central Ltd. (**b**) "aq`8" by RoyalSiamBeauty is licensed under CC BY 2.0)

Table 2.1 The main types of spots associated with acne

Type of spot	Description
Blackheads	Small black or yellowish bumps that develop on the skin. They are coloured because of the dark pigment (melanin) produced by cells of the inner lining of the hair follicle
Whiteheads	Have a similar appearance to blackheads but there is no pigment accumulation. They may be firmer than blackheads and will not empty when squeezed
Papules	Small red bumps that may feel tender or sore
Pustules	Similar to papules, but have a white tip in the centre due to the build-up of pus
Nodules	Deeper large hard lumps that build up beneath the surface of the skin and can be painful
Pseudo-cysts	The most severe type of spot caused by acne. They are large, deep, pus-filled lumps that look similar to boils and carry the greatest risk of causing permanent scarring

Clinicians group these lesions into six major types – blackheads (open comedones), whiteheads (closed comedones), papules, pustules, nodules and cysts (Table 2.1). A comedo (plural comedones) is a blocked hair follicle. The word comes from the Latin word "comedere" meaning to eat up. It was also a word used to refer to parasitic worms and then came to mean the worm-like appearance of the material squeezed out of a blackhead – very evocative isn't it? The presence and concentration of these lesions, as well as the degree of inflammation, are used to classify the severity of the disease.

Figure 2.2 shows examples of these spots.

These blemishes appear mainly on the face (in most patients), neck (in many patients), back (in 50% of patients) and chest (in 15% of patients). It's a disease that affects mainly adolescents, beginning at puberty and improving in the late teens or early twenties – so there is light at the end of the tunnel. About 80% of people aged 11–30 suffer from acne. Girls are affected at a younger age than boys – it's most common in girls from the ages of 14–17 and in boys from the ages of 16–19. The disease often disappears when a person reaches their mid-twenties. However, acne can continue into adult life with about 5% of women and 1% of men being affected over the age of 25 years.

Although the disease isn't life-threatening, it can have a huge psychological impact. The embarrassment and low self-esteem that come with the condition often exacerbate the challenges of adolescence and can leave the sufferer feeling vulnerable. As if this wasn't enough, a misunderstanding of the nature of the condition can lead to wasted time, effort and expense on inappropriate treatments. This can result in patients losing faith in being able to control the problem which can impact on the sufferer's mental heath.

What Causes Acne and Its Symptoms?

A number of factors are responsible for acne. First of all, the hormone changes that occur during puberty cause the oil-producing sebaceous glands next to the hair follicle (Fig. 2.3) to enlarge.

Sebum is an oily substance that waterproofs and lubricates the skin, but cells in the activated gland also produce other proteins and compounds that make inflammation more likely. Individuals suffering from acne may also have a defect in skin shedding so that the cells and keratin that line the hair follicle are not shed so easily. Inflammation and debris accumulating in the follicle result in the sebum being trapped. This creates the ideal environment for the proliferation of a microbe that's normally present on the skin, *Cutibacterium acnes*. This bacterium has a number of virulence factors that can cause further inflammation. Blockage of the hair follicles, accumulation of sebum and inflammation result in the characteristic spots of acne (Fig. 2.4).

The main long-term complication of acne that sometimes arises is scarring (Fig. 2.5). Scarring may occur when the most serious types of spots (nodules and cysts) burst deeper in the skin, the damage is repaired with collagen production which may be visible as a scar. It can also occur if spots are clumsily squeezed or

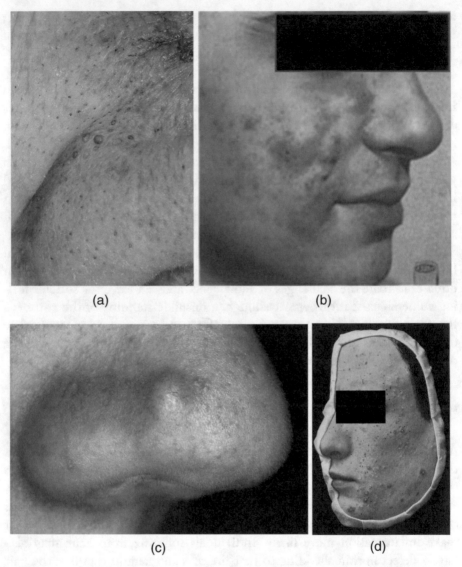

Fig. 2.2 Some of the characteristic skin lesions associated with acne. (**a**) Blackheads and whiteheads on the nose. (M. Sand, D. Sand, C. Thrandorf, V. Paech, P. Altmeyer, F. G. Bechara [CC BY 2.0 (https://creativecommons.org/licenses/by/2.0)]). (**b**) Cystic acne on the face. (This image is a work of a US military or Department of Defense employee, taken or made as part of that person's official duties. As a work of the US federal government, the image is in the public domain in the USA). (**c**) Papule on the nose. (Cutaneous lesions of the nose. Sand M, Sand D, Thrandorf C, Paech V, Altmeyer P, Bechara FG. Head Face Med. 2010;6:7. This article is published under license to BioMed Central Ltd). (**d**) This man has a variety of characteristic acne lesions – blackheads, whiteheads, papules and pustules. (Acne Vulgaris. Credit: Wellcome Collection. Attribution 4.0 International (CC BY 4.0))

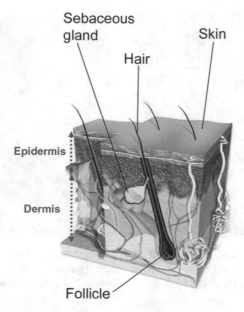

Fig. 2.3 Structure of a hair follicle in human skin. Most of the follicle is in the dermis, but it crosses the epidermis as well and opens out onto the surface of the skin. A single hair grows out of each follicle, and this has muscles attached which can make it lie flat on the skin or stick up from its surface. Each follicle has a sebaceous gland attached, and this secretes sebum which lubricates and moisturises both the hair and the skin surface near the opening of the follicle. (Modified from BruceBlaus [CC BY 3.0 (https://creativecommons.org/licenses/by/3.0)])

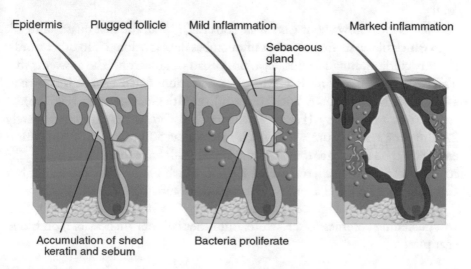

Fig. 2.4 Diagram showing the main stages involved in the development of acne. (OpenStax College [CC BY 3.0 (https://creativecommons.org/licenses/by/3.0)])

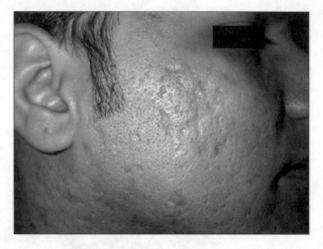

Fig. 2.5 Patient whose right cheek shows multiple scarring due to acne. (Novel Technology in the Treatment of Acne Scars: The Matrix-tunable Radiofrequency Technology. Ramesh M, Gopal M, Kumar S, Talwar A. J Cutan. Aesthet Surg. 2010 May-Aug; 3(2): 97–101

picked. Inflammation can also induce colour changes in the skin resulting in either darker or lighter pigmented areas, and these can persist for several months.

What Kind of Microbe Is *Cut. acnes*?

So what do we know about *Cut. acnes* (Box 2.1)? Well, the first question to ask is "to which of the six major groups of microbes does it belong?". In other words, is it a bacterium, fungus, virus, alga, protozoan or archaean? The answer is that it's a bacterium. *Cut. acnes* is a rod-shaped bacterium (such a shape is known as a "bacillus" which means a "stick" in Latin), and it's Gram-positive (Fig. 2.6).

Electron microscopy (Fig. 2.7) shows that *Cut. acnes* is approximately 2.0 μm long and 0.5 μm wide which means that 500 of them, if laid end to end, would be no longer than 1 mm. Some bacilli have long, slender, tail-like structures (similar to sperm) called flagella which enable them to swim, but *Cut. acnes* has no flagella and so isn't capable of moving by itself.

As can be seen in Fig. 2.8, in the laboratory *Cut. acnes* grows as small, convex, glistening colonies (often white, but other colours are possible) on blood agar plates.

Box 2.1 What's in a Name?

Acne comes from the Greek word "akme" which means an "eruption on the face". Vulgaris is a Latin adjective that means "common". Acne vulgaris is the most common type of acne.

Cut. acnes was first detected in acne lesions in 1893 by the German physician and dermatologist Paul Gerson Unna (figure). Because of its shape, he called it the "acne bacillus".

Paul Gerson Unna. (Unknown author [Public domain], via Wikimedia Commons. US National Library of Medicine, Digital Collections)

The bacterium was first grown in the laboratory in 1897 by Raymond Sabouraud, a French physician and bacteriologist. He named it "bacilli de seborrhea grasse" because he found it in sebaceous material squeezed out of the skin. It was subsequently named Bacillus acnes and then Corynebacterium acnes until in 1946 it was found to produce propionic acid and given the name Propionibacterium acnes, i.e. the rod-shaped bacterium that produces propionic acid and causes acne. It was re-named in 2016 as Cutibacterium acnes although many scientists and clinicians are likely to keep calling it Propionibacterium acnes for quite a while. Cutibacterium is derived from two Latin words – "cutis" meaning skin and "bacterium" meaning a rod.

Why a particular organism, such as *Cut. acnes*, is found only in a particular environment is an important topic in the branch of microbiology known as "microbial ecology". Although there are many millions of different microbial species on Earth, these aren't evenly, or randomly, distributed over the planet's surface. A particular microbe can only survive at a location that provides all of

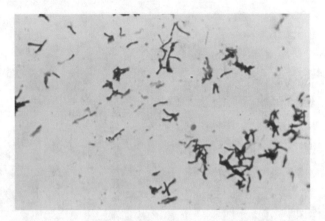

Fig. 2.6 Gram stain of *Cut. acnes* showing Gram-positive bacilli. Some have an irregular shape and some have formed short chains (×1150). (Image courtesy of D. Lucille K. Georg, Centers for Disease Control and Prevention, USA)

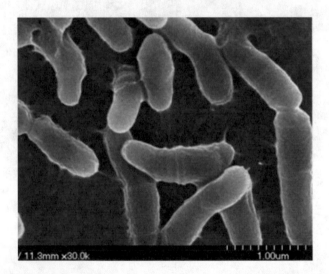

Fig. 2.7 Scanning electron micrograph of *Cut. acnes*. (Ryu S, Han HM, Song PI, Armstrong CA, Park Y (2015) Suppression of *Propionibacterium acnes* infection and the associated inflammatory response by the antimicrobial peptide P5 in mice. *PLoS ONE* 10(7): e0132619

the conditions it needs for its growth – we call this location its "habitat". Microbial ecology has given us a very good idea of what microbes we can expect to find in any particular environment on our planet. The surface of the

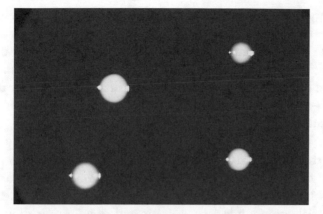

Fig. 2.8 *Cut. acnes* colonies on a blood agar plate. Each colony is approximately 2 mm in diameter. (Image courtesy of Bobby Strong, Centers for Disease Control and Prevention, USA)

skin is an environment with a particular set of conditions (known as "environmental determinants") that allow only certain organisms to survive and grow there. For example, it's very acidic, water isn't very abundant, and it's got a plentiful supply of fats and has regions with a low oxygen content (the hair follicles – Fig. 2.3) – such conditions are ideal for the growth of *Cut. acnes*. From Fig. 2.3 you can see why the long, narrow hair follicle is an oxygen-poor region – fresh air can't easily reach into the depths of this structure. Furthermore, the follicle usually has a sebaceous gland attached and from this pours a regular supply of fat-rich sebum which is a major source of nutrients for *Cut. acnes*. The constant outflow of sebum also helps to prevent oxygen getting into the follicle. However, a number of other bacterial species can also survive and grow under the conditions provided by the hair follicles, and in order to colonise this habitat, *Cut. acnes* produces a number of antimicrobial compounds that can kill such competitors. These substances include the waste products it excretes (propionic and acetic acids) and a range of antimicrobial peptides (known as bacteriocins) that can kill or inhibit other organisms. Because of differences in the concentration of hair follicles, variation in the environmental conditions in different skin regions and competition from other microbes, the concentration of *Cut. acnes* varies over the skin surface – it's most abundant on the scalp and the face where it's present at a concentration of more than one million bacteria per cm^2, whereas on the arms and legs it's fewer than 100 per cm^2.

For most young people, *Cut. acnes* lives in their hair follicles, happily consuming the fats produced there, and does no harm for most of the time. However, it's a pathobiont (see Chap. 1) which means that it can cause disease if the conditions are right. A combination of increased sebum production at puberty and blockage of hair follicles results in an environment that encourages growth of the organism to a level at which it can induce inflammation and this results in acne.

Teenagers have every reason to hate *Cut. acnes*, but, as already mentioned, we all have this bacterium on our skin, and it does have a number of beneficial effects. First of all it prevents the skin from being colonised by harmful bacteria that may be present in our environment. It does this by producing a variety of antibacterial compounds such as acetic and propionic acids as well as several bacteriocins that can kill a wide range of bacteria. This phenomenon is known as "colonisation resistance" and is an important function of our microbiota at all body sites. We certainly don't want to eliminate or disrupt our microbiota because this important defence mechanism is very effective at fighting off dangerous pathogens. Interestingly, *Cut. acnes* also produces an enzyme that converts linoleic acid (found on the skin) to 10,12-conjugated linoleic acid (CLA). CLA has a number of potential health-promoting properties as it contributes to the prevention of cancer, atherosclerosis (the accumulation of fat on arterial walls), diabetes, obesity and allergies.

How Is Acne Diagnosed?

Acne is diagnosed by examining the skin and noting the appearance and distribution of the lesions. The number and distribution of lesions, as well as the extent of inflammation, are used to classify the disease into four main types, and this is important for deciding on the type of treatment that's necessary:

- Grade 1 (mild) – mostly whiteheads and blackheads, with just a few papules and pustules. Total lesion count <30.
- Grade 2 (moderate) – multiple papules and pustules that are mostly confined to the face. Total lesion count 30–125.
- Grade 3 (moderately severe) – a large number of papules and pustules, as well as the occasional inflamed nodule. The back and chest are also affected.
- Grade 4 (severe) – a large number of large, painful pustules and nodules. Total lesion count >125.

How Is Acne Treated?

Because acne is due to a combination of a number of factors, it means that we can use a variety of approaches for treating it – either singly or, more powerfully, when they're used in combination. These include:

(i) Killing the bacteria involved with antibiotics such as tetracycline, erythromycin and clindamycin or an antiseptic such as benzoyl peroxide, hydrogen peroxide or azelaic acid.

(ii) Removing dead skin cells using retinoids – this stops them from building up inside, and blocking, hair follicles. These drugs also reduce inflammation. Topical retinoids such as tretinoin and adapalene are widely used.

(iii) Unplugging the follicles using a mild salicylic acid preparation.

(iv) In women we can reduce sebum production by using either the combined oral contraceptive pill or cyproterone.

(v) Using oral isotretinoin which probably acts against all of the factors that contribute to acne.

(vi) Using various types of light and laser therapy. These help in different ways by increasing healing, reducing the numbers of *Cut. acnes* on the skin and reducing inflammation. The effect of light can be increased by applying a special chemical (known as a photosensitiser) to the skin beforehand; this is an example of photodynamic therapy (see Box 3.4).

(vii) Reducing inflammation with either steroids or non-steroidal anti-inflammatory agents.

(viii) Physical therapies such as cryotherapy, cautery, microdermabrasion and diathermy can help in specific situations.

Which of these treatments is used depends on the severity of the condition, the type and location of the lesions, the sex of the patient, whether or not they're pregnant, patient preferences and the previous experiences of, and responses to, treatment. Medications do take time to work and so require perseverance - response is usually judged after at least 8 weeks. Treatment of scarring usually requires some form of cosmetic procedure or surgery, such as dermabrasion and laser therapy. Cosmetic camouflage can cover up scarring or pigment changes and can increase peoples' confidence in their appearance. This includes concealing products, the application of contouring and colour-correcting make-up.

How Can I Avoid Getting Acne?

Acne typically improves as people progress through adolescence, but it can persist into adulthood. You can prevent flare-ups of acne by using a variety of medications such as azelaic acid or adapalene.

A number of measures are helpful in caring for your skin during acne:

- Don't wash the affected areas of your skin more than twice a day – frequent washing can irritate the skin and make symptoms worse.
- Wash the affected area with a mild soap or antiseptic cleanser and use only lukewarm water – very hot or cold water can make acne worse.
- Don't squeeze blackheads or spots – this can make them worse and cause permanent scarring.
- Avoid using too much oily make-up and cosmetics – use water-based products that are described as non-comedogenic.
- Completely remove any make-up before going to bed.
- Wash your hair regularly and try to avoid letting hair fall across your face.
- Avoid humid environments like saunas.
- Stop smoking – nicotine makes comedones more likely.

Want to Know More?

American Academy of Dermatology https://www.aad.org/public/diseases/acne-and-rosacea/acne

British Association of Dermatologists http://www.bad.org.uk/for-the-public/patient-information-leaflets/acne/?showmore=1&returnlink=http%3A%2F%2Fwww.bad.org.uk%2Ffor-the-public%2Fpatient-information-leaflets

Canadian Dermatology Association https://dermatology.ca/public-patients/skin/acne/

Health Service Executive, Ireland https://www2.hse.ie/conditions/acne/acne-symptoms-and-diagnosis.html

Mayo Clinic, USA https://www.mayoclinic.org/diseases-conditions/acne/symptoms-causes/syc-20368047

MedlinePlus, US National Library of Medicine https://medlineplus.gov/acne.html

National Health Service, UK https://www.nhsinform.scot/illnesses-and-conditions/skin-hair-and-nails/acne

National Institute for Clinical Care and Excellence (NICE), UK. Acne vulgaris. 2019 https://cks.nice.org.uk/acne-vulgaris#!topicSummary

National Institute of Arthritis and Musculoskeletal and Skin Diseases, USA https://www.niams.nih.gov/health-topics/acne#tab-overview

Primary Care Dermatology Society, UK http://www.pcds.org.uk/clinical-guidance/acne-vulgaris

Dellavalle RP, Howland AR. Acne vulgaris. BMJ Best Practice, BMJ Publishing Group, 2018 https://bestpractice.bmj.com/topics/en-gb/101

Fox L, Csongradi C, Aucamp M, du Plessis J, Gerber M. Treatment modalities for acne. *Molecules.* 2016 Aug 13;21(8). pii: E1063. https://doi.org/10.3390/molecules21081063

Gebauer K. Acne in adolescents. *Australian Family Physician.* 2017 Dec;46(12):892–895.

Platsidaki E, Dessinioti C. Recent advances in understanding *Propionibacterium acnes* (*Cutibacterium acnes*) in acne. *F1000Res.* 2018 Dec 19;7. pii: F1000 Faculty Rev-1953. https://doi.org/10.12688/f1000research.15659.1. eCollection 2018.

Rao J, Chen J. Acne Vulgaris. Medscape from WebMD, 2020 https://emedicine.medscape.com/article/1069804-overview

Sutaria AH, Masood S, Schlessinger J. Acne Vulgaris. Treasure Island (FL): StatPearls Publishing LLC; 2020 https://www.ncbi.nlm.nih.gov/books/NBK459173/

3

Skin Abscesses

Abstract A boil is an accumulation of pus in a hair follicle that usually results from an infection by *Staphylococcus aureus*. This bacterium is present in the nostrils of about one third of the population. Complications are not common but the infection may spread into adjacent tissue resulting in cellulitis. Boils occur most frequently in adolescents and young adults. Most boils can be treated at home but occasionally antibiotics may be needed. *Staph. aureus* is also responsible for a variety of more serious infections, and some strains (e.g. MRSA) are resistant to a wide range of antibiotics.

What Are Boils and Abscesses?

Staphylococcus aureus is another skin-spoiler, a partner in crime for *Cut. acnes*. It's the dreaded boil-maker. A huge, red, tender lump filled with disgusting yellow pus (Fig. 3.1). The relief when the pus pops out between your probing finger nails, the horror when you can't tease it all out and it gets even sorer and starts to ache and throb. The fear that it will back-explode and, instead of being released, the pus will be forced into your bloodstream and you'll die of blood poisoning. Just like splinters of wood in your fingers that will eventually work their way into your heart and puncture it. As bad as apple seeds germinating into apple trees in your stomach. And, during your teenage years, why did the boil ALWAYS show itself on a Saturday night? All week your skin had been perfect, but you could guarantee that when you woke up on Saturday morning you were certain to have the beginnings of a boil. Then, by the evening, as you were about to go out and party, it had developed into a massive,

M. Wilson, P. J. K. Wilson, *Close Encounters of the Microbial Kind*,
https://doi.org/10.1007/978-3-030-56978-5_3

Fig. 3.1 On the left side of the face above the mouth, a boil can be seen showing clearly the presence of yellow pus. ("Eminem interlude" by Caitlinator is licensed under CC BY 2.0)

red, throbbing golf ball-sized monstrosity – and there was NOTHING you could do about it. Everyone looks at you and displays revulsion and pity – there was no way anyone was going to fall for you that night.

But you're not the only one to suffer from this dreadful affliction. Spare a thought for poor old Karl Marx who refers constantly to his recurrent boils and carbuncles in letters to his wife Jenny and to Friedrich Engels. They occurred all over his body – on his face, scalp, armpits, groin, penis, scrotum and inner thighs. He was often driven to lancing these boils with a razor and lamented that "the bourgeoisie will remember my carbuncles until their dying day".

The medical term for a boil is "furuncle" or "skin abscess". A boil is formed when a hair follicle (Fig. 2.3 in Chap. 2) becomes infected and pus accumulates resulting in a small swelling which is inflamed and tender. Pus is a white-yellow, yellow or yellow-brown fluid that collects at the site of an infection. It's made up mainly of dead white blood cells (known variously as polymorphonuclear neutrophils, polymorphs or neutrophils – Fig. 3.2) that have homed in on the infecting microbe to try to kill it. This is the response of the immune system which has detected the infecting microbe and has signalled to these phagocytic (i.e. cell-eating) cells that they should go to the danger area and kill off the microbial invader (Fig. 3.2).

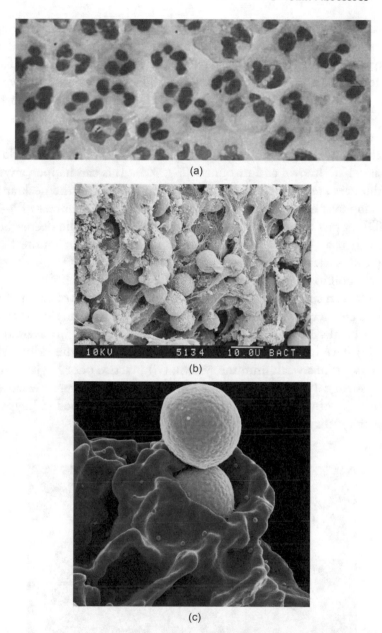

(a)

(b)

(c)

Fig. 3.2 Human polymorphonuclear neutrophils – also known simply as polymorphs or neutrophils. These are important phagocytic cells (12–15 μm in diameter) and comprise 55–70% of our white blood cells. They are the body's first line of defence against microbial tissue invaders. (a) A large number of human polymorphonuclear neutrophils showing their characteristic multi-lobed nuclei. (Image courtesy of Steven Glenn, Centers for Disease Control and Prevention, USA). (b) Scanning electron micrograph of pus showing the white blood cells (mainly neutrophils) that are present in the fluid. (Credit: David Gregory & Debbie Marshall. CC BY 4.0). (c) A neutrophil (purple) engulfing two cells of *Staph. aureus* (yellow). Following their internalisation by the neutrophil, the bacteria will be killed and digested. (Image courtesy of National Institutes of Health, USA)

The first stage in the formation of a boil is the appearance of redness at the infection site. This is followed by the formation of a small lump, and after 3 or 4 days, this lump turns yellowish due to the accumulation of pus. If a small cavity forms, then pus collects there, and this causes what is known as an abscess. Although boils can form anywhere on the body, they're most commonly found at hairy or sweaty regions such as the face, neck, armpits, shoulders, back and buttocks.

When a number of boils form close together and join beneath the skin, this results in what's known as a carbuncle (Fig. 3.3). This can happen anywhere, but carbuncles are most common on the back, thighs and the neck and usually develop over a few days. A carbuncle can grow to a diameter of 3–10 cm and will leak pus from a number of points – they tend to lie deeper beneath the skin than a boil and can take longer to heal. Men are more likely to develop carbuncles than women.

When a boil forms on the eyelid, it's called a "stye" (Fig. 3.4).

Anybody can suffer from boils, but you're at greater risk of them if you (i) have close contact with someone who has boils; (ii) have any condition that results in breaks in your skin such as acne, eczema and intravenous drug abuse; (iii) have been exposed to chemicals that damage the skin; (iv) have diabetes; (v) have a weak immune system; (vi) practice poor hygiene or regularly get sweaty; (vii) have poor nutrition; (viii) live in overcrowded conditions; (ix) use communal facilities; (x) are obese; and (xi) lived in Egypt in the fourteenth century BCE (see Box 3.1).

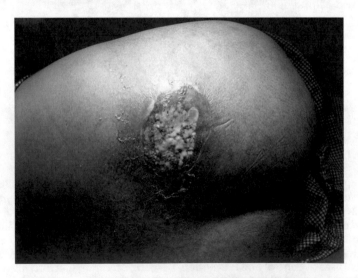

Fig. 3.3 A carbuncle on the buttock with multiple openings that discharge pus. (Drvgaikwad/CC BY (https://creativecommons.org/licenses/by/3.0))

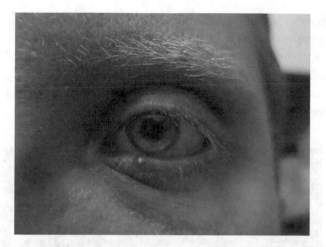

Fig. 3.4   A stye. (Image courtesy of Palosirkka. Public Domain via Wikimedia Commons)

Box 3.1 A Plague of Boils

In about the fourteenth century BCE, the Israelites were enslaved in Egypt. Because the pharaoh wouldn't set the Israelites free, the Book of Exodus tells us that God decided to punish him by inflicting ten plagues. The sixth of these was a plague of boils: Then the Lord said to Moses and Aaron, "Take handfuls of soot from a furnace and have Moses toss it into the air in the presence of Pharaoh. It will become fine dust over the whole land of Egypt, and festering boils will break out on men and animals throughout the land" (Exodus 9:8–9). Consequently, the pharaoh, his servants, all Egyptians and even their animals developed painful boils all over their bodies.

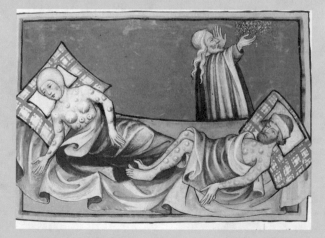

The plague of boils inflicted on the Egyptians as shown in the Toggenburg Bible (Switzerland) of 1411 CE. (Kupferstichkabinett, Staatliche Museen, Germany. Public Domain via Wikimedia Commons)

Although most boils and carbuncles don't cause further problems, complications do occasionally arise. In some individuals the infection can spread from the follicle into the surrounding dermis (the deeper layer of the skin) – this is known as cellulitis (Fig. 3.5) and can be very painful. More rarely, invasion of the bloodstream (i.e. septicaemia – Box 3.2) can occur, and this is life-threatening. Larger boils and carbuncles can also lead to scarring.

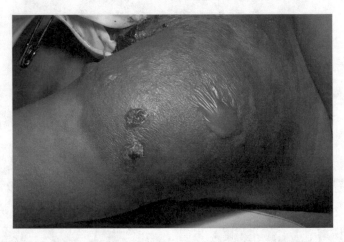

Fig. 3.5 This shows cellulitis in the left shoulder of a child. The child had developed a secondary staphylococcal infection at the site of a vaccination. Note the signs of cellulitis – the spreading erythema (redness) around the vaccination site, swelling and pus formation. (Allen W. Mathies, MD/California Emergency Preparedness Office (Calif/EPO), Immunization Branch, Centers for Disease Control and Prevention, USA)

Box 3.2 Septicaemia and Sepsis

These are two life-threatening conditions that can arise during the course of any infection. Septicaemia (often called "blood poisoning") means invasion of the bloodstream by a microbe. It's sometimes classified on the basis of what type of microbe is involved – bacteraemia, when the invader is a bacterium, viraemia when it's a virus and fungaemia when it's a fungus. Many different microbes can invade the bloodstream during an infection. The most frequent infections that result in a septicaemia are those involving the urinary tract, lungs, kidneys and the abdomen. Septicaemia sometimes results in sepsis.

(continued)

Box 3.2 (continued)

Sepsis is a complex multifactorial condition in which many organs are damaged due to the body's disordered response to an infection. The damage results from the actions of the body's immune system rather than the direct effects of the infecting microbe. Among other things, the resulting inflammation causes blood clots which stop oxygen from reaching vital organs, and these then fail. Bacteria are responsible for about 90% of cases. The source of the infection responsible for most cases of sepsis is the respiratory tract (60%). Other sources include the abdomen (26%), bloodstream (20%), skin (14%) and urinary system (12%). In the USA there are over 1 million cases of sepsis each year, and the mortality rate is 28–50%. In England there are about 190,000 cases of sepsis each year resulting in about 23,000 deaths.

How frequently boils occur in the general population is difficult to estimate, but in the UK in 2010, 0.5% of the population consulted their doctor because of boils. However, this is only the tip of the iceberg as most of us don't bother to go to the doctor if we have a boil. Boils are generally rare in children and occur most commonly in adolescents and young adults. 10% of those of us who have boils will be affected again within 12 months.

Which Microbe Causes Boils and How Does It Produce Them?

Staphylococcus aureus (Box 3.3), the most important cause of boils, can get into a hair follicle directly or when the skin is damaged by a cut or graze or sharp object. This bacterium is a Gram-positive coccus, and its cells usually group together to form small bunches (Fig. 3.6).

Box 3.3 What's in a Name?

Staph. aureus was first identified and named in 1884 by the German physician and microbiologist Friedrich Rosenbach. The name *Staphylococcus* for this genus of bacteria (which contains more than 40 different species) was coined in 1882 by the British surgeon Sir Alexander Ogston.

Sir Alexander Ogston. (R.M. Morgan Ltd. [CC BY 4.0 (https://creativecommons. org/licenses/by/4.0)])
  The name is derived from the microscopic appearance of the bacterium and is a combination of the Greek "staphyle", which means a bunch of grapes, and "kok-kos", which means a grain or seed. The species name, *aureus*, was used because of the golden colour of colonies of the organism when it is grown on an agar plate.

(continued)

Box 3.3 (continued)

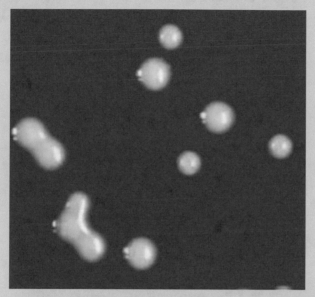

Colonies of *Staph. aureus* growing on a blood agar plate showing their golden colour. (Image courtesy of: *PLoS Pathog.* 2016 Nov;12(11):e1006024. Dynamic in vivo mutations within the ica operon during persistence of *Staphylococcus aureus* in the airways of cystic fibrosis patients. Schwartbeck B et al.

But where does this bacterium come from? *Staph. aureus* is found mainly in the respiratory tract (particularly the nostrils – Fig. 3.7) of many healthy humans. It's also present at other body sites, but generally less frequently (Table 3.1).

From crime novels and movies, we know that genetic fingerprinting (analysis of the DNA present in human cells) can identify individual human beings. The same technique can also be used to identify, and distinguish between, individual bacteria belonging to the same species – such microbes are known as "strains" of that particular species. Genetic fingerprinting has shown that a boil, or other infection due to *Staph. aureus*, is usually caused by the strain present in the nose of that individual. In other words, it's an infection caused by a member of our own microbiota, i.e. *Staph. aureus* is a pathobiont or opportunistic pathogen (see Chap. 1). This finding has important implications for preventing hospital-acquired infections – see Box 3.4.

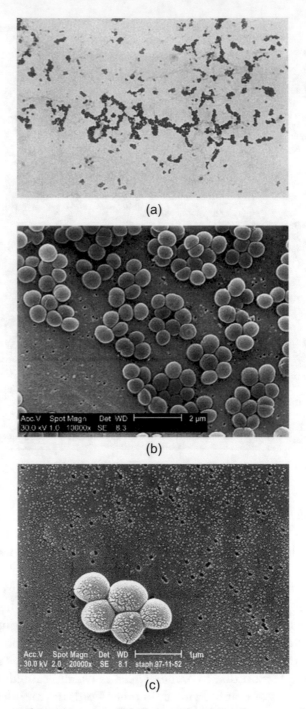

Fig. 3.6 Images of *Staph. aureus*. (**a**) Gram stain of *Staph. aureus* showing Gram-positive cocci in clusters. Magnification ×250. (Image courtesy of Dr. Richard Facklam, Centers for Disease Control and Prevention, USA). (**b**) A digitally colourised scanning electron micrograph of *Staph. aureus*. Magnification ×10,000. (Image courtesy of Matthew J. Arduino, Centers for Disease Control and Prevention, USA). (**c**) Scanning electron micrograph of a group of *Staph. aureus* cells. Magnification ×20,000. (Jim Biddle, Centers for Disease Control and Prevention, USA)

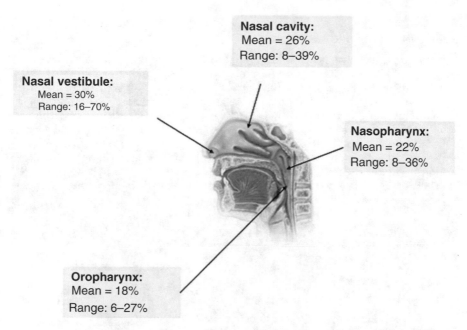

Nasal cavity:
Mean = 26%
Range: 8–39%

Nasal vestibule:
Mean = 30%
Range: 16–70%

Nasopharynx:
Mean = 22%
Range: 8–36%

Oropharynx:
Mean = 18%
Range: 6–27%

Fig. 3.7 Frequency of detection of *Staph. aureus* in various regions of the respiratory tract of healthy adults. For each site the mean value is given as well as the range of values that have been detected in various studies

Table 3.1 Frequency of detection of *Staph. aureus* at various body sites in healthy adults

Body site	Frequency of detection (%)
Armpits	8
Chest/abdomen	15
Perineum	22
Intestinal tract	17–31
Vagina	5
Urethra of men	8

Box 3.4 Preventing Hospital-Acquired Infections Due to *Staph. aureus*

Patients are in hospital because they're ill, which generally means that they're more susceptible to contracting an infectious disease. It's in hospital that our indigenous microbiota (see Chap. 1) can become treacherous and turn against us. Studies have shown that *Staph. aureus* living in our nose is able to cause not only boils but can also infect the wounds that are left after surgery resulting in what are known as "surgical site infections (SSIs)". Up to 5% of surgical sites have been shown to be infected within the first month following surgery. In

(continued)

Box 3.4 (continued)

order to prevent such hospital-acquired infections, patients are usually screened for the presence of *Staph. aureus* in their nostrils, and, if they're found to carry this organism (and approximately 30% of us do – see Fig. 3.7), then they're treated to eliminate it. This usually involves applying mupirocin cream, an antimicrobial agent, to the nostrils several times a day for 5 days. However, recently a simpler and more rapid approach has been developed which uses a light-activated antimicrobial agent or "photosensitiser". In this procedure (known as "antimicrobial photodynamic therapy"), the photosensitiser is applied to the nostrils, and this is followed by a 4 minutes exposure to red light (see figure below). Studies have shown this to be very effective at eliminating *Staph. aureus* from the nostrils and in reducing SSIs.

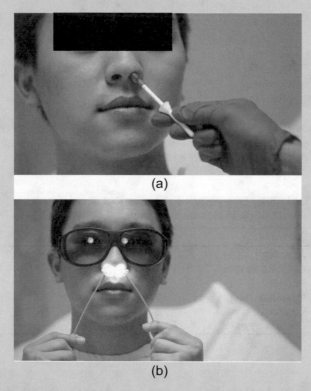

(a)

(b)

Figure. Nasal decontamination using a light-activated antimicrobial agent (photosensitiser). In (a) the photosensitiser is being applied to the nostrils of the patient. In (b) the nostrils are being irradiated with red light from a laser via optical fibres. (Images courtesy of Ondine Biomedical Inc., Vancouver, Canada)

We've known about *Staph. aureus* for more than 150 years and so have found out a lot about it. For a start it's a facultative anaerobe which means that it's not bothered whether or not there's any oxygen in its environment – it can grow with or without it. Also, although (like all life forms) it needs water to grow, it can cope with relatively dry conditions. Furthermore, it's resistant to acids and so can survive and grow on the acidic skin surface. As for what it needs as food, it can use many of the large molecules present in skin (such as elastin, hyaluronic acid and nucleic acids) as well as many of the compounds produced by the sweat and sebaceous glands such as sugars, amino acids, glycerol and fatty acids (see Box 3.5). Breakdown of the large molecules present in skin will, of course, damage the host tissues that contain these macromolecules, but so what? *Staph. aureus* doesn't care.

These properties mean that *Staph. aureus* is very versatile and can grow in a variety of environments found on humans and in their surroundings.

Box 3.5 Macromolecules in Humans That Can Act as Food Sources for Microbes

The skin consists of a number of huge molecules (known as macromolecules) such as proteins, glycosaminoglycans, nucleic acids and lipids. These molecules are too large to enter microbial cells and so can't be used as nutrients. However, many species produce enzymes that can break down macromolecules to smaller molecules (sugars, amino acids, fatty acids) that can then be transported into a microbial cell and used as nutrients. The table below shows some of the most important macromolecules found in human skin and how microbes can use them as nutrients, once they've been broken down.

Type of macromolecule	Macromolecule	Molecules produced by the action of microbial enzymes on the macromolecule
Protein	Keratin Collagen Elastin Fibrillin	Amino acids
Glycosaminoglycan	Hyaluronic acid Chondroitin sulphate Dermatan sulphate	Sugars
Lipid	Many different types	Fatty acids and glycerol
Nucleic acid	Many different types	Sugars, nucleotides

Staph. aureus has many virulence factors (Table 3.2) that enable it to cause boils and other diseases. First of all, it protects itself from our phagocytic cells by producing a thick capsule (Fig. 3.8) which makes it difficult for

Table 3.2 Virulence factors of *Staph. aureus*

Virulence factor	Function
Capsule	Protects against phagocytic cells of the immune defence system. Also enables the bacterium to stick to surfaces such as host tissues, medical devices, etc.
Enzymes such as proteases, glycosaminoglycanases, nucleases and lipases	These break down macromolecules so releasing nutrients for the microbe; they also damage host tissues
Cytotoxins	These kill human cells and so interfere with our antimicrobial defence system; they also release nutrients from the dead cells
Enterotoxins	Cause diarrhoea
Toxic shock syndrome toxin	Causes toxic shock syndrome

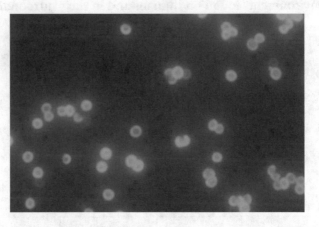

Fig. 3.8 In this photomicrograph, a thick jelly-like capsule (coloured light green) can be seen surrounding many of the cells of *Staph. aureus* which are grey in colour. Those bacteria without a capsule are coloured blue. (Image courtesy of: *BMC Microbiol*. 2013 Mar 22;13:65. Production of capsular polysaccharide does not influence *Staphylococcus aureus* vancomycin susceptibility. Jansen A, Szekat C, Schröder W, Wolz C, Goerke C, Lee JC, Türck M, Bierbaum G

neutrophils to engulf and destroy it. This capsule also helps the bacterium to attach itself to surfaces such as medical devices and body tissues.

Staph. aureus also produces **cytotoxins** and these kill some of the cells of our immune system and so reduces its ability to get rid of the organism. These dying human cells release nutrients that can be used by the microbe. Some strains of *Staph. aureus* produce **enterotoxins** that can cause diarrhoea, while others release toxic shock syndrome toxin which results in toxic shock syndrome, a potentially fatal infection.

As well as being responsible for boils, *Staph. aureus* can cause a number of other diseases, some of which are life-threatening, and these are summarised in Table 3.3. It's important to bear in mind that the more serious diseases are all much rarer than boils and it's highly unlikely that you'll get one of them.

Table 3.3   Diseases, other than boils, caused by *Staph. aureus*

Disease	Comments
Folliculitis	Characterised by clusters of white-headed pimples around hair follicles
Infective endocarditis	A serious, life-threatening condition arising from infection of the inner lining (endocardium) of the heart
Surgical site infections	Wound infections that arise following surgery
Toxic shock syndrome	A serious, life-threatening condition usually associated with tampon use
Pneumonia	Usually occurs in hospitalised patients
Osteomyelitis	*Staph. aureus* is the most frequent cause of this disease
Prosthetic device-associated infections	For example, artificial heart valves, replacement joints, intravascular catheters, breast implants
Food poisoning	Can be contracted from meats, poultry and egg products, salads, cream-filled pastries, sandwich fillings, milk and dairy products

Box 3.6 MRSA: The Dreaded Superbug

MRSA, the great killer, fills us with horror. This is the bug that we worry about catching if we go into hospital for a routine operation. So, what is MRSA? It's an abbreviation for "methicillin-resistant *Staphylococcus aureus*", and this implies that the bacterium is resistant only to an antibiotic called methicillin. If that was the case, then there wouldn't be that much to worry about. But there's more to it than that.

In the 1940s penicillin became readily available and revolutionised the treatment of infectious diseases. Nowadays it's hard to appreciate the huge impact this had at the time – penicillin was regarded as a wonder drug. However, the human race was then given a perfect example of evolution and natural selection in action – exposure of *Staph. aureus* to penicillin encouraged the emergence of strains of the organism that were resistant to the antibiotic. To get round this problem a new antibiotic, methicillin, was introduced in 1959. This was very effective until in 1961 the first methicillin-resistant strain of the organism emerged – this marked the birth of MRSA. Fortunately, other antibiotics were developed and were used to treat infections due to MRSA, but, not surprisingly, MRSA then developed resistance to them as well. Currently, there are strains of MRSA that are resistant to most of the known antibiotics. The abbreviation

(continued)

Box 3.6 (continued)

MRSA can, therefore, also mean multiply-resistant *Staph. aureus*. At present the only antibiotic effective against some strains of MRSA is vancomycin, but there are reports of emergence of resistance to this "last-line" antibiotic. *Be afraid, be very afraid*.

However, scientists are actively engaged in a search for new antibiotics. In addition, alternative approaches are being developed to killing antibiotic-resistant microbes that may be useful for the treatment of the infections they cause. One such approach is antimicrobial photodynamic therapy (aPDT) which can be used to prevent and/or treat infections on the surface of the body (see Box 3.4). Another approach involves the use of bacteriophages (figure). These are viruses that can kill bacteria, and some of them are able to target only one particular bacterial species. An appropriate bacteriophage could, therefore, be administered to a patient to deal with a particular bacterium (once it's been identified) that's responsible for an infection in any part of the body.

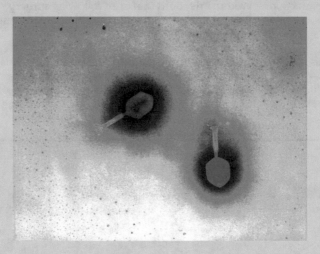

Figure. Electron micrograph of two bacteriophages (known as T4) that can kill the bacterium *Escherichia coli*. Each phage is approximately 100 nm long and 90 nm wide. In comparison, a cell of *E. coli* is approximately 2000 nm long and 1000 nm wide. (Credit: David Gregory & Debbie Marshall. Attribution 4.0 International (CC BY 4.0))

How Can We Treat Boils and Prevent Them from Occurring?

You can easily treat most boils at home by regularly applying a warm, moist compress to the affected area three or four times a day. This soothes the pain and draws the pus to the surface, and the boil eventually bursts. You should then wash the skin with an antiseptic and cover with a plaster or bandage until the wound heals. You should never squeeze a boil because this can spread the pathogen to other parts of your skin; any pus that comes out should be cleaned up carefully. Other treatments depend on the severity and location of the problem; options include antiseptic soaps and ointments, antibiotic tablets or creams. Most small boils will drain on their own and clear up within a few weeks without treatment, but if it doesn't heal, you may have to have it drained by a medical practitioner. This involves numbing the area and piercing the boil with a sterile needle or scalpel.

Antibiotics are sometimes necessary and may be used:

- In cases of carbuncles or abscesses.
- If you have a high temperature or are unwell otherwise.
- If a secondary or spreading infection, such as cellulitis, develops (Fig. 3.9).
- When the boil is on your face – facial boils have a higher risk of causing complications.
- If you're in severe pain and discomfort.
- If you're immune-impaired.
- If the boil is in a challenging or difficult location – especially if you have poor circulation.
- When the boil isn't clearing up on its own.

The *Staph. aureus* (or other pathogen) in boils and carbuncles can spread to other parts of the body as well as to other people. The risk of developing boils or allowing them to spread can be reduced by simple precautions including:

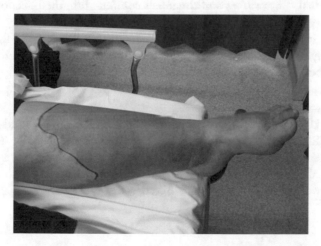

Fig. 3.9 Cellulitis of the leg. ("Cellulitis leg" by jlcampbell104 is licensed under CC0 1.0)

- Maintaining good personal hygiene especially handwashing after touching affected areas
- Using a separate face cloth and towel
- Washing underwear, bed linen and towels at a high temperature
- Cleaning and covering wounds with a dressing until they heal
- Careful disposal of used dressings

Complications aren't common but you may develop scarring especially if the boil is large. Recurrent infections can be prevented using antiseptic treatments and avoiding sweaty activities. Cellulitis and sepsis are prevented by correct and timely treatment, especially targeting those who are most at risk.

Want to Know More?

American Academy of Dermatology https://www.aad.org/public/skin-hair-nails/skin-care/boils-and-styes

British Association of Dermatologists http://www.bad.org.uk/for-the-public/patient-information-leaflets/boils/?showmore=1&returnlink=http%3A%2F%2Fwww.bad.org.uk%2Ffor-the-public%2Fpatient-information-leaflets

DermNet, New Zealand https://www.dermnetnz.org/topics/staphylococcal-skin-infection

Health Service Executive, Ireland https://www.hse.ie/eng/health/az/b/boils/

Mayo Clinic, USA https://www.mayoclinic.org/diseases-conditions/boils-and-carbuncles/symptoms-causes/syc-20353770

MedicineNet https://www.medicinenet.com/boils/article.htm

MSD Manual https://www.msdmanuals.com/en-gb/home/infections/bacterial-infections-gram-positive-bacteria/staphylococcus-aureus-infections

National Institute for Clinical Care and Excellence (NICE), UK. Boils, carbuncles, and staphylococcal carriage. 2017 https://cks.nice.org.uk/boils-carbuncles-and-staphylococcal-carriage#!topicSummary

Northern Ireland Direct, Government Services https://www.nidirect.gov.uk/conditions/boils-and-carbuncles

Patient Info, UK https://patient.info/skin-conditions/boils-carbuncles-and-furunculosis

Herchline TE. Staphylococcal Infections. Medscape from WebMD, 2019 https://emedicine.medscape.com/article/228816-overview

Troxell T, Hall CA. Carbuncle. Treasure Island (FL): StatPearls Publishing LLC; 2020 https://www.ncbi.nlm.nih.gov/books/NBK554459/

4

Dandruff

Abstract Dandruff is a disease of the scalp in which fungi belonging to the genus *Malassezia* are thought to play an important role. These fungi are also invariably present on the scalp of dandruff-free individuals. The disease results in flaking of the scalp and affects about 50% of adults worldwide. The disease is rarely eliminated but can be controlled using shampoos containing anti-fungal agents. *Malassezia* species are responsible for a number of other diseases (pityriasis versicolor, seborrheic dermatitis and atopic eczema), but these are less prevalent than dandruff.

Dandruff: What Is It?

You're going out on a date, you've been told "you look great in dark colours" so you put on your best grey suit or your famous "little black dress". You look in the mirror, and your eyes are immediately drawn to the deep layer of snow-flakes on your shoulders. Off with the suit or dress – no more dark clothes for me, ever. Your night is ruined. And then there are the formal occasions. You've passed your final exams and have attained a First Class Honours Degree – everyone is thrilled. Then comes the degree awards ceremony, and you MUST wear your black graduation gown. You stand on the podium waiting to receive your certificate from the vice chancellor and your shoulders are just COVERED in those horrible white flakes. The vice chancellor shakes your hand, and this dislodges an avalanche of skin scales all down your gown, and yes, they can be seen on the photo your proud parents have taken. Oh the shame!

© Springer Nature Switzerland AG 2021
M. Wilson, P. J. K. Wilson, *Close Encounters of the Microbial Kind*,
https://doi.org/10.1007/978-3-030-56978-5_4

But even the Beatles suffered from dandruff. On arriving in the USA at the start of their 1964 tour, they were asked "What is the biggest threat to your careers, the atom bomb or dandruff". Ringo replied "The atom bomb. We've already got dandruff". In 2007 a photograph of George Harrison was sold for $13,000 – he had signed it "George 'Dandruff' Harrison". When asked "Do you wear wigs?", John replied "If we do, they must be the only ones with real dandruff".

Dandruff (also known as "pityriasis capitis", "seborrheic dermatitis confined to the scalp", pityriasis simplex, furfuracea or capitis) is a condition that affects the scalp and results in the production of skin flakes (Figs. 4.1 and 4.2). In some cases, this is accompanied by redness and itching of the scalp. The first recorded use of the word dandruff was in 1545, but its origins aren't very clear. The second part of the word is probably derived from the Northumbrian or East Anglian term "huff", "hruff" or "hurf" which means a scab.

Although dandruff isn't contagious and doesn't have any serious medical consequences, it's embarrassing and often results in low self-esteem. Also, it can be more severe in people with immune problems or those on immune-supressing medication. Since as long ago as 1873, a type of fungus belonging to the genus *Malassezia* has been associated with the disease but, as in the case of acne (Chap. 2), the disease is multifactorial and so the story is more complicated – it's not a simple matter of becoming infected with this fungus. In order to understand the nature of the disease, you need to know something

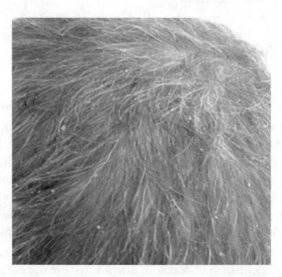

Fig. 4.1 Flakes of skin are clearly visible on this adult male afflicted by dandruff. (Peripheral neuropathy in Sézary syndrome: coincidence or a part of the syndrome?. Karadağ YS, Gülünay A, Oztekin N, Ak F, Kılıçkap S. Turk J. Haematol. 2013 Dec;30(4):420–1. doi: https://doi.org/10.4274/Tjh.2012.0163

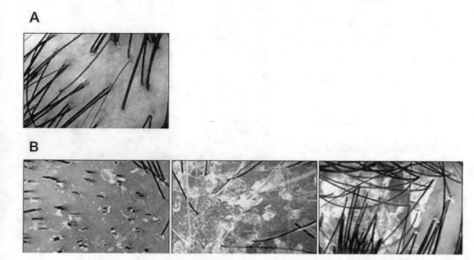

Fig. 4.2 Close-up view of the scalp of a dandruff-free individual (a) and of three people with dandruff (b). Flaking of the scalp is clearly visible in the three dandruff-afflicted people. (Park HK, Ha M-H, Park S-G, Kim MN, Kim BJ, Kim W (2012) Characterization of the Fungal Microbiota (Mycobiome) in Healthy and Dandruff-Afflicted Human Scalps. PLoS ONE 7(2): e32847. https://doi.org/10.1371/journal.pone.0032847

about the structure of the skin that's found on the scalp, particularly the outer layers as these are the ones affected by dandruff.

The scalp is, for most of us for most of our lives, a hairy region, and this has a profound effect on its environment and, consequently, the types of microbe that can grow there. The structure of the skin is shown in Fig. 2.3 of Chap. 2, and a cross-section through the epidermis showing its layered structure is shown in Fig. 4.3.

The outer region, the epidermis, consists of five layers of cells. New skin cells (known as keratinocytes or corneocytes) are continually being produced in the innermost layer, the stratum basale, and pushed towards the surface. As the new cells are pushed upwards, they undergo a process known as keratinisation. This involves the production of a protein called keratin (Fig. 4.4) and is accompanied by the death of the cell and loss of its contents.

Keratin is an insoluble, fibrous protein, and when it accumulates on the skin surface in the dead cells of the stratum corneum (Fig. 4.5), it protects the underlying tissues from heat, microbes, and environmental chemicals. It's therefore a very useful and important protein. The dead, keratinised cells that make up the stratum corneum are shed into the environment (Fig. 4.5) and take with them any microbes from the external environment that have attached themselves to the skin surface.

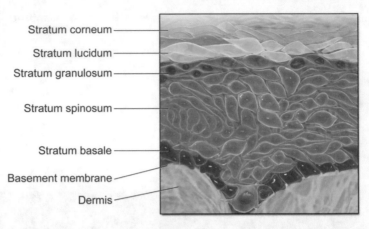

Fig. 4.3 Diagram of a cross-section through the epidermis showing the main layers. The epidermis is separated from the underlying dermis by the stratum basale and the basement membrane. (Blausen.com staff (2014). "Medical gallery of Blausen Medical 2014" WikiJournal of Medicine 1 (2). https://doi.org/10.15347/wjm/2014.010. ISSN 2002-4436. [CC BY 3.0])

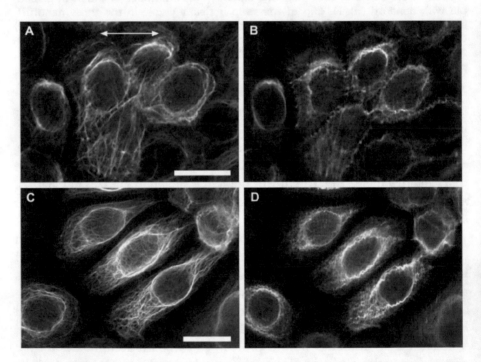

Fig. 4.4 Human keratinocytes (between 6 and 10 cells in each of the four images) showing a dense network of intracellular filaments (the white strands inside each cell) composed of keratin which constitutes between 10% and 70% of the proteins in the cell, depending on its stage of development. Bars = 25 µm. (The intermediate filament network in cultured human keratinocytes is remarkably extensible and resilient. Fudge D, Russell D, Beriault D, Moore W, Lane EB, Vogl AW. PLOS ONE (2008). 3:e2327. https://doi.org/10.1371/journal.pone.0002327. Published under CC BY 4.0)

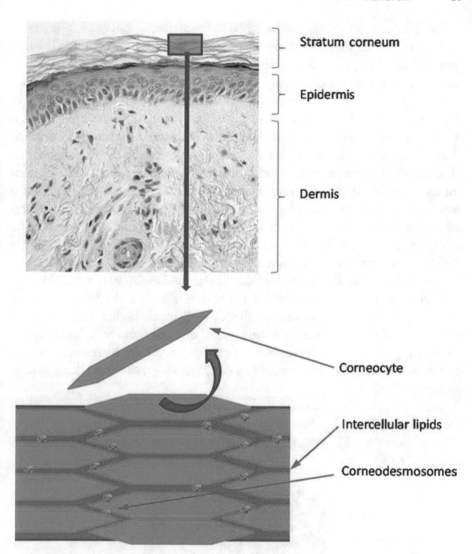

Fig. 4.5 Cross-section through human skin showing the various layers – dermis, epidermis, and the stratum corneum. The stratum corneum consists of layers of lipids sandwiched between dead keratinocytes (empty-looking cells), also known as corneocytes, that contain keratin. Corneocytes are anchored to one another by means of protein complexes known as corneodesmosomes. The corneocytes in the outermost layer of the stratum corneum are continually being shed into the environment. (Glycan distribution and density in native skin's stratum corneum. J. Danzberger, M. Donovan, C. Rankl, R. Zhu, S. Vicic, C. Baltenneck, R. Enea, P. Hinterdorfer. *Skin Res Technol.* 2018 Aug; 24(3): 450–458

This is a very important antimicrobial defence mechanism because it means that any dangerous microbes that attach to the skin surface are continually being removed and dumped back into the environment. Their attachment to the surface of our body is, therefore, only temporary and limits the time available for them to penetrate into our inner tissues. The gaps between the corneocytes in the stratum corneum are filled with oily substances (known as lipids) which come from the dead keratinocytes and are also produced by sebaceous and sweat glands. This makes the outer layer very waterproof. It's also very acidic because of the acids that are produced during the keratinisation process. The stratum corneum, therefore, is a very inhospitable and, because of its shedding nature, a very temporary environment for microbes. What a brilliant system for protecting us against environmental microbes! The transformation of a new cell in the stratum basale to a shed corneocyte takes 2–4 weeks, and the rate of shedding is such that the whole of the stratum corneum is renewed every 15 days.

In a healthy adult the number of skin scales shed each day is approximately 10^8, with the vast majority of these consisting of only a single cell. In individuals with dandruff, the rate of shedding is much higher with a greater proportion of scales consisting of more than one cell – many scales comprise several hundreds, or even thousands, of cells (Fig. 4.6). In the severest cases of dandruff, the weight of scales shed can be 20 times greater than that from an unaffected individual.

The severity of the condition varies among patients and can be assessed by a trained observer using the scale shown in Table 4.1.

In addition to skin flaking, the other symptoms of dandruff include an itchy, red or scaly scalp.

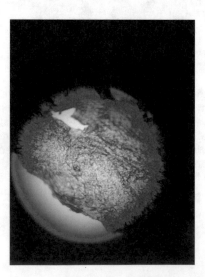

Fig. 4.6 A flake of dandruff viewed through a microscope. (Angelo Popovic/CC BY (https://creativecommons.org/licenses/by/4.0))

Table 4.1   Classification of severity grades and scores for adherent scalp flaking

Severity description	Description	Severity grade score
O	Healthy scalp with no dryness or dandruff	0
A	Fine dryness on scalp surface	1
B	Small powdery flakes partially adhering to scalp	2
C	Moderately flaky scales loosely attached to scalp	3
D	Large pronounced crusty scaling adhering to scalp	4
E	Very large crusty scaling congealed into plates adhering to scalp	5

Complications

Symptoms, especially when they are ongoing, can impact on people's self-esteem. The damaged skin in affected areas can become infected by other microbes such as bacteria. In susceptible people the rash can become widespread and severe cases can be seen in special situations such as immunosuppression. Rarely a potentially serious condition called erythroderma can result which needs specialist treatment and may require admission to hospital.

Box 4.1 Flake News: Dinosaurs Had Dandruff

A small feathered dinosaur (*Confuciusornis*) that lived in China about 125 million years ago has recently been shown to have suffered from dandruff. This is, at present, the oldest known case of the disease. The fossilised remains of the meat-eating feathered dinosaur, which is about the size of a modern crow, included very small flakes of fossilised skin (Figure).

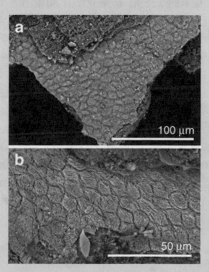

Figure. Flakes of skin from *Confuciusornis* at low (a) and high (b) magnifications. Note the sheets of groups of hexagonal-shaped corneocytes in the

(continued)

Box 4.1 (continued)

flakes. (Fossilized skin reveals coevolution with feathers and metabolism in feathered dinosaurs and early birds. McNamara ME, Zhang F, Kearns SL, Orr PJ, Toulouse A, Foley T, Hone DWE, Rogers CS, Benton MJ, Johnson D, Xu X, Zhou Z. Nat Commun. 2018 May 25;9(1):2072. This article is licenced under a Creative Commons Attribution 4.0 International License, which permits use, sharing, adaptation, distribution and reproduction in any medium or format, as long as you give appropriate credit to the original author(s) and the source, provide a link to the Creative Commons licence and indicate if changes were made)

The prehistoric skin flakes are the only evidence scientists have of how dinosaurs shed their skin. The material shows that rather than losing their outer layer in one piece, or in large sheets, as is common with modern reptiles, the feathered dinosaurs adapted to shed their skin in tiny flakes.

How Common Is Dandruff?

Dandruff is a very common condition and affects about 50% of the adult population worldwide. It's estimated that at least 50 million people in the USA and 11 million in the UK suffer from dandruff. It also means big business – in the USA $300 million is spent annually on over-the-counter products to treat the condition.

The problem generally begins at puberty, reaches peak incidence and severity at the age of about 20 years and becomes less prevalent among those over 50 years of age. Despite its general absence in children, a similar condition is found in neonates and is often called "cradle cap" (Fig. 4.7). The peak prevalence of this occurs in 3-month-old infants of which 70% are affected. Only 10% of 1-month-old infants and 7% of 1 year olds have the condition.

Dandruff is more prevalent in men than women and varies between different ethnic groups. In a recent study dandruff prevalence was found to be 81–95% in African-Americans, 66–82% in Caucasians and 30–42% in Chinese. The severity of the condition is generally worse in the winter.

Surprisingly, very few studies have been carried out on the epidemiology of dandruff. However, in a recent study in France, dandruff was found in 17% of 1703 individuals over the age of 15 years and was more frequent in men (21%) than women (13%). The prevalence decreased with age (Fig. 4.8).

Fig. 4.7 A baby with cradle cap. (Diagnosis of Atopic Dermatitis: Mimics, Overlaps, and Complications. Siegfried EC, Hebert AA. J Clin Med. 2015 May; 4(5): 884–917. This article is an open-access article distributed under the terms and conditions of the Creative Commons Attribution License (http://creativecommons.org/licenses/by/4.0/))

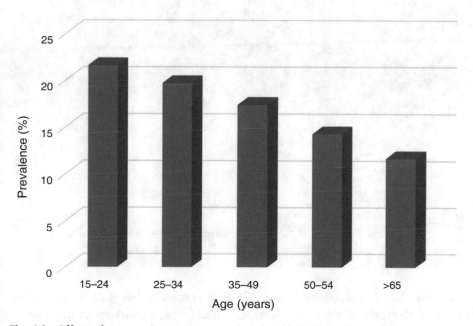

Fig. 4.8 Effect of age on the prevalence of dandruff in a French population

What Causes Dandruff?

Unfortunately, there's no easy answer to this question, but since 1873 certain fungi have been thought to be involved in some way. The main culprits for many years have been considered to be species belonging to the genus *Malassezia* (Box 4.2).

Although species belonging to the genus *Malassezia* are fungi, they're rather

---

Box 4.2 What's in a Name?

Malassezia species have been known since 1846 when Karl Ferdinand Eichstedt, a German physician, detected their presence in the skin of a patient suffering from tinea versicolor, a skin disease common in hot, humid climates. In 1873 Sebastiano Rivolta suggested that yeast-like fungi were also present in dandruff and in 1874 Louis-Charles Malassez, a French anatomist and physician, subsequently identified the fungus in scalp scales from patients with dandruff. The fungus was eventually grown and fully described, and the genus to which it belonged was named *Malassezia* in 1889 in recognition of Malassez's discovery.

M. le Dr MALASSEZ,
Membre de l'Académie de Médecine,
Directeur adjoint du Laboratoire d'Histologie
au Collège de France.

Drawing of Louis-Charles Malassez. (Bibliothèque interuniversitaire de Santé/ Licence Ouverte)

---

peculiar. The vast majority of the 120,000 species of fungi so far identified grow in the form of microscopic, thread-like (filamentous) structures known as hyphae. However, a few grow as individual cells and so are classified as "yeasts". A typical example of this type of fungus is the yeast used in bread-making and brewing – *Saccharomyces cerevisiae*. *Malassezia* species are peculiar because they produce hyphae but can also exist in the form of an oval, yeast-like single cell (Fig. 4.9). Because of this ability to exist in two distinct forms, *Malassezia* species are known as "dimorphic fungi".

Fourteen different species of the genus have been identified of which eleven have been found on the skin of humans. Just like the situation with *Cut. acnes* and acne (Chap. 2), the mere presence of *Malassezia* on your skin doesn't mean that you'll be afflicted by dandruff – other factors are involved. Nevertheless, many studies have found greater numbers of *Malassezia* species on the scalps of dandruff sufferers and have also shown that the proportions of these fungi on the skin increases with the severity of the condition. Furthermore, *Malassezia* species are known to be associated with several other skin diseases including pityriasis versicolor, atopic dermatitis and folliculitis. The two species of *Malassezia* that are most closely associated with dandruff are *Mal. globosa* and *Mal. restricta*.

How Do These Fungi Cause Dandruff?

Like acne, dandruff is a disease associated with puberty, and this is a time of hormonal changes and increased sebum production by sebaceous glands of the skin. The scalp has a greater density of sebaceous glands (between 400 and 900 per cm^2) than many other skin regions such as the abdomen, forearm and leg (approximately 100 per cm^2), and they produce approximately 650–700 mg of sebum per day. Like most *Malassezia* species, *Mal. globosa* and *Mal. restricta* can't grow in the absence of lipids which are their main nutrients. Sebum is a very rich source of lipids so it's tempting to argue that the increased production of sebum during puberty stimulates an increase in the proportion of *Malassezia* species on the scalp and this results in dandruff.

Malassezia species have a number of virulence factors that could contribute to skin damage and, ultimately, to the increased release of skin flakes. For example they produce a large variety of lipases – these are enzymes that can break down lipids to glycerol and fatty acids. *Mal. globosa* produces 14 such enzymes, while *Mal. restricta* produces 12 of them. Fatty acids are compounds that cause inflammation and some of them (e.g., oleic acid) stimulate the release of skin flakes from the scalp. Furthermore, *Malassezia* species can convert squalene (a major component of sebum) into compounds that stimulate the multiplication of keratinocytes, and this encourages skin flaking. The lipases also break down the lipids that hold together the corneocytes in the stratum corneum – flakes of corneocytes can therefore more easily leave the surface of the skin.

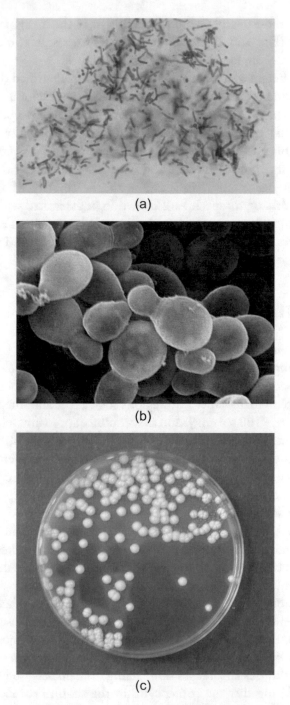

(a)

(b)

(c)

Fig. 4.9 Images of *Malassezia* species. (a) Photomicrograph of a *Malassezia* species showing both forms of this dimorphic fungus – single cells and thread-like hyphae. (Image courtesy of Dr. Lucille K. Georg, Centers for Disease Control and Prevention, USA).

If dandruff is a consequence of increased sebum production during puberty leading to higher proportions of skin-damaging *Malassezia* species, then surely all teenagers should develop dandruff? Obviously this isn't the case which means that individuals vary in their susceptibility to developing the condition. This suggests that additional factors in susceptible individuals make their skin vulnerable to damage by *Malassezia*. These factors could include one or more of the following: (i) the immune response of the individual to *Malassezia*, (ii) genetic factors that affect the structure of the stratum corneum, (iii) personal habits such as excessive shampooing and/or hair combing and (iv) variation in the quantity and/or composition of sebum produced at puberty.

Box 4.3 Skin Shedders and Transmission of the Important Pathogen *Staph. aureus*

It should come as no surprise that indoor air and indoor floor dust contain high proportions of microbes that originate from humans. Microbes that live on humans comprise 17% and 20% of the microbes found in indoor air and floor dust, respectively. Many of these microbes are members of the skin microbiota and include species belonging to the genera *Staphylococcus*, *Propionibacterium* and *Corynebacterium*. These bacteria are dumped into the environment on the skin scales (squames) we are continually shedding. While shedding occurs spontaneously and is an important antimicrobial defence mechanism (see Chap. 1), friction between clothing and the skin also generates many squames. The number of squames is greatly increased during exercising and undressing. A single act of undressing releases as many as 500,000 squames, and each of these carries about 5 live microbes.

Not all regions of the skin contribute equally to the shedding process – the nose and the perineal area are particularly active in this respect. Because these two areas are the preferred habitats of *Staph. aureus*, the shedding of skin scales is an effective means of dispersing this organism – particularly by men, who've been shown to shed more staphylococci than women. As discussed in Chap. 3 *Staph. aureus* is an important pathogen (see Table 3.3), being responsible for a variety of diseases. The number of staphylococci in the air can be increased tenfold (to approximately 360 cells/m^3) by a "good" skin shedder. The scales are dispersed into the air when the individual moves or undresses and when contaminated articles such as clothes or bedding are disturbed. In hospitals, transmission of *Staph. aureus* via skin scales is a major route by which other patients become infected. The number of skin scales in the air can be reduced by dusting surfaces with damp cloths.

← ——————————————————————————————

Fig. 4.9 (continued) **(b)** Scanning electron micrograph of the yeast form of a *Malassezia* sp. (Image courtesy of Janice Haney Carr, Centers for Disease Control and Prevention, USA). **(c)** Colonies of a *Malassezia* species growing on an agar plate. (Cabañes FJ (2014) Malassezia Yeasts: How Many Species Infect Humans and Animals? PLoS Pathog 10(2): e1003892. https://doi.org/10.1371/journal.ppat.1003892

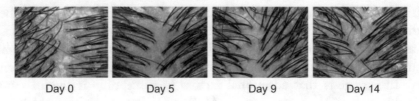

Day 0   Day 5   Day 9   Day 14

Fig. 4.10 Effect of a zinc pyrithione-containing shampoo on the scalp of a dandruff-afflicted individual. The shampoo (applied once every 2 days) achieved a 70% reduction in the degree of scalp flaking after 14 days. (Evaluation of therapeutic potential of VB-001, a leave-on formulation, for the treatment of moderate adherent dandruff. Bhattacharyya A, Jain N, Prasad S, Jain S, Yadav V, Ghosh S, Sengupta S. BMC Dermatol. 2017 May 3;17(1):5. https://doi.org/10.1186/s12895-017-0058-5. This article is distributed under the terms of the Creative Commons Attribution 4.0 International License (http://creativecommons.org/licenses/by/4.0/), which permits unrestricted use, distribution and reproduction in any medium, provided you give appropriate credit to the original author(s) and the source, provide a link to the Creative Commons licence and indicate if changes were made)

How Can Dandruff Be Treated?

A major clue suggesting that *Malassezia* fungi do have an important role in causing dandruff is that the condition can be treated successfully with anti-fungal shampoos. The most effective anti-fungal agents used in shampoos include zinc pyrithione, selenium sulphide, sulphur, ketoconazole, coal tar, tea tree oil, imidazoles (such as climbazole) and hydroxypyridones (such as piroctone olamine and ciclopirox). Of these, the most widely used and effective agent is zinc pyrithione (Fig. 4.10). Some of these anti-fungal agents have additional activities that may be beneficial. For example, coal tar is also anti-inflammatory and reduces sebum production while selenium sulphide stops keratinocyte proliferation.

Salicylic acid and coconut oil preparations reduce scale and so improve symptoms and can help other treatments become more effective. Steroid scalp treatments can also be useful in reducing itchiness. Oral antifungal agents are rarely used but may be necessary for severe cases – particularly in people with a weak immune system or if the rash has become widespread.

Treatment usually controls the problem in the short term rather than cures it, and recurrence of the condition is common. Long-term treatment is usually needed if recurrence becomes a problem.

How Can I Avoid Getting Dandruff?

There isn't much you can do to prevent dandruff, but the following may help:

- Try not to scratch your scalp when using shampoo.
- Brush your hair daily and wash it at least three times a week. After washing your hair, rinse it thoroughly to get all the shampoo out. Using a shampoo that contains tea tree oil daily may help reduce dandruff.
- Avoid using chemicals on your scalp, such as those in hair colouring products. These disrupt the microbiota and can kill the bacteria that control *Malassezia*.
- Using hair gels and hair sprays can irritate the scalp in some people and so are best avoided.
- Spending time outdoors can help reduce dandruff. However, ultraviolet light from the sun can damage your skin, and sunscreen protection for exposed areas is advisable. The time spent outdoors should be limited.
- Stress or tiredness can trigger a flare up so should be avoided.
- Using an anti-fungal shampoo once a week.

While dandruff is the most prevalent of the diseases caused by *Malassezia*, these fungi are also associated with a number of other diseases, although these are much less common. The main features of these conditions and their prevalence are summarised in Table 4.2. It must be emphasised that all of these diseases are multifactorial in nature, and the exact role played by *Malassezia* is often unclear.

Table 4.2 Other diseases in which *Malassezia* species play a role

Disease	Main features	Prevalence (%)
Pityriasis versicolor	Discoloured patches of skin covered by fine scales Found mainly on the back, chest and neck Affects all ages but more frequent in third and fourth decades of life Unsightly but harmless	<8% in the USA; <4% in the UK
Seborrheic dermatitis	A relapsing condition characterised by redness, itching and greasy, large scales Affects mainly the scalp, eyebrows, paranasal folds, chest, back, axillae and genitals Most common in infants and those in third and fourth decades of life	<12% in the USA; <4% in the UK
Atopic Eczema	Itchy, red, dry and cracked skin Most commonly affected areas in children are the back of the knees and front of the elbows, whereas in adults the hands and feet are usually affected Starts in childhood but then often disappears during adolescence	15–20% of children and 1–3% of adults worldwide
Folliculitis	Inflammation of the hair follicles mainly on the shoulders, back and chest Characterised by itching and red bumps or white-headed pimples around hair follicles May be caused by a variety of microbes other than *Malassezia* including a variety of bacteria, viruses and mites	Uncertain

Want to Know More?

American Academy of Dermatology https://www.aad.org/public/diseases/hair-and-scalp-problems/dandruff-how-to-treat

Health Service Executive, Ireland https://www.hse.ie/eng/health/az/d/dandruff/

Indian Journal of Dermatology https://www.ncbi.nlm.nih.gov/pmc/articles/PMC2887514/

Mayo Clinic, Mayo Foundation for Medical Education and Research, USA https://www.mayoclinic.org/diseases-conditions/dandruff/symptoms-causes/syc-20353850

Medical News Today https://www.medicalnewstoday.com/articles/152844.php

National Health Service, UK https://www.nhs.uk/conditions/dandruff/

National Institute for Clinical Care and Excellence (NICE), UK. Seborrhoeic dermatitis. 2019 https://cks.nice.org.uk/seborrhoeic-dermatitis#!topicSummary

Handler MZ. Seborrheic Dermatitis. Medscape from WebMD, 2019 https://emedicine.medscape.com/article/1108312-overview

Janniger CK, Schwartz RA. Seborrhoeic dermatitis. BMJ Best Practice, BMJ Publishing Group, 2018 https://bestpractice.bmj.com/topics/en-gb/89

Saunders CW, Scheynius A, Heitman J. *Malassezia* fungi are specialized to live on skin and associated with dandruff, eczema, and other skin diseases. *PLoS Pathogens.* 2012;8(6):e1002701. https://doi.org/10.1371/journal.ppat.1002701. Epub 2012 Jun 21.

Tucker D, Masood S. Seborrheic Dermatitis. Treasure Island (FL): StatPearls Publishing LLC; 2020 https://www.ncbi.nlm.nih.gov/books/NBK551707/

Turner GA, Hoptroff M, Harding CR. Stratum corneum dysfunction in dandruff. *International Journal of Cosmetic Science.* 2012 Aug;34(4):298–306. https://doi.org/10.1111/j.1468-2494.2012.00723.x. Epub 2012 May 17.

5

Fungal Nail Infections

Abstract Infections of the nail by fungi is a common problem throughout the world. The fungi responsible vary with climate, and *Trichophyton rubrum* is responsible for 90% of cases of the disease in temperate climates. The fungus invades and digests the protein (keratin) that makes up the nail and can, ultimately, result in the nail becoming detached. The disease is treated with anti-fungal drugs but this is a long, slow process because the drugs don't penetrate into the nail very easily. Treatment failure and re-infection are very common.

One of us (MW) has had a recent episode of onychomycosis of the toenails, and what a nuisance that was. It wasn't painful but my toenails looked dreadful – brown and crumbly – and I was too embarrassed to go to my doctor. I couldn't, of course, wear sandals at the beach and so I had to walk around in socks and shoes when it was 30 °C. That just made things a lot worse because the fungus was given a lovely warm and moist environment in which to grow and proliferate. What I did was to paint on anti-fungal nail varnish every day, but that didn't seem to make much difference and wasn't even a nice colour! Also, whenever there was nobody around, I used to take off my socks and shoes when I was in the garden and expose my feet to the sunshine hoping that a dry environment and ultraviolet light would do the trick. It took more than 2 years to clear up, and during that time was an ever-present source of concern and unease.

© Springer Nature Switzerland AG 2021
M. Wilson, P. J. K. Wilson, *Close Encounters of the Microbial Kind*,
https://doi.org/10.1007/978-3-030-56978-5_5

What Is Onychomycosis?

Fungal infection of the nails (onychomycosis) is a common complaint which affects 23% of Europeans, 20% of East Asians and 14% of Americans. The various parts of the nail and how it's embedded in the finger (or toe) are shown in Fig. 5.1.

The nail has three main regions – the free edge, nail plate (or body) and nail root (that part of the nail that can't be seen). It consists of several layers of tightly packed dead, toughened skin cells (Fig. 5.2) and has a mostly pink colour because of the blood vessels in the underlying tissue known as the nail bed. The nail bed is the skin beneath the nail plate and consists of the usual epidermis and dermis (see Chap. 4). The nail is embedded in the skin at three nail folds, one of which, the eponychium (or cuticle), forms a protective seal above the nail and its root. The nail matrix is an area underneath the cuticle that produces new nail. The nail grows (by about 0.1 mm per day) because new cells produced by the matrix gradually push out the old, keratinised cells

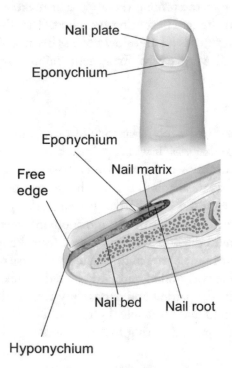

Fig. 5.1 Diagram showing the nail and its associated structures. (Modified from: Blausen.com staff (2014). "Medical gallery of Blausen Medical 2014". *WikiJournal of Medicine* 1 (2). https://doi.org/10.15347/wjm/2014.010. ISSN 2002-4436)

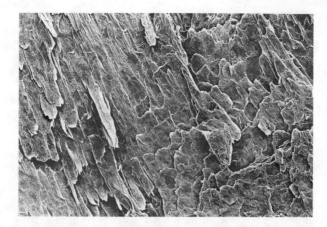

Fig. 5.2 The nail plate as seen through an electron microscope. Note the tightly packed layers of keratin that are derived from dead skin cells (keratinocytes). (Credit: Dr Thanuja Perera. CC BY 4.0. Via Wellcome Image Library)

that form the nail plate. The hyponychium forms a seal that protects the nail bed where the free edge joins the finger/toe.

The main symptoms of onychomycosis are that the nail becomes discoloured (white, black green or yellow), thickens, becomes crumbly and starts to detach from the underlying nail bed (Fig. 5.3). As the disease progresses, pieces of the nail may break off, and the underlying tissue can become inflamed and painful (Fig. 5.3). It can affect a single nail or, unusually, all nails. Toenails are seven times more likely to be affected than fingernails with the first and fifth toenails being the most commonly affected.

Although the disease is hardly life-threatening, it's common and can affect the sufferer in a number of ways. It can cause embarrassment because of its appearance and so can result in social withdrawal and low self-esteem. In some cases it can cause occupational discomfort or interfere with everyday activities. It can also permanently damage the nail and can lead to serious complications in at-risk individuals such as those with diabetes or those with impaired immune systems due to AIDS, leukaemia or cancer therapy. Complications include cellulitis due to secondary infection with bacteria (see Chap. 3) and, more rarely, osteomyelitis. Its high treatment costs make it an important public health problem.

Age is a major risk factor for the infection with one study reporting that 60–79 year olds are 26 times more likely to have the infection that those under 19 years of age. The disease is seen three times more often in men than in women. Other predisposing factors include diabetes (one third of patients with diabetes are affected), HIV infection, psoriasis (patients with psoriasis have a 50% greater chance of onychomycosis), smoking, cancer therapy and

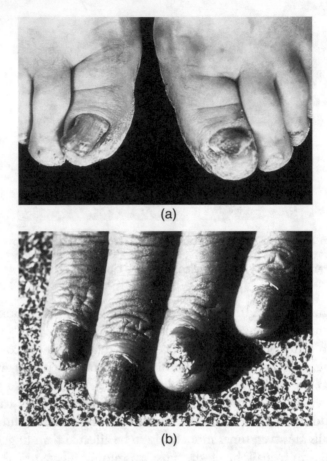

(a)

(b)

Fig. 5.3 Images of infected nails of **(a)** the toes and **(b)** the fingers. ((a) Dr. Lucille K. Georg, Centers for Disease Prevention and Control, USA. **(b)** Dr. Libero Ajello, Centers for Disease Prevention and Control, USA)

peripheral arterial disease. The condition isn't common in tropical climates which suggests that wearing socks and shoes, which create a warm and moist environment thereby encouraging fungal growth, is an important risk factor. Other risk factors include participating in physical activities, exposure to wet work, wearing ill-fitting shoes, using commercial swimming pools, working with chemicals, walking barefoot and nail biting.

What Causes the Disease?

The main fungi responsible for the disease in temperate climates are known as dermatophytes (literally "skin plants" – this is because fungi were originally classified as being plants), a term that includes three fungal genera – *Microsporum, Epidermophyton,* and *Trichophyton.* However, in tropical and sub-tropical regions, other fungi may be responsible such as *Candida albicans* (see Chap. 23) and species belonging to the genera *Scopulariopsis, Fusarium, Aspergillus* and *Acremonium.* It's important to identify the fungus responsible prior to any treatment, as nail trauma can cause a similar appearance and this is done by examining a sample of the infected nail under the microscope. The sample is immersed in potassium hydroxide solution to dissolve the nail, and this makes it easier to identify the fungus. If the fungus can't be identified in this way, then a nail sample can be grown in the laboratory to obtain colonies of the organism for further tests.

Box 5.1 What's in a Name?

The word onychomycosis is derived from the Greek for nail (onyx), fungus (mykes) and condition (osis), i.e. a condition that affects the nail and is caused by a fungus. The disease is also known as "tinea unguium" from the Latin for worm (tinea) and nail (unguis).

The name of the genus, *Trichophyton,* comes from two Greek words – "thrix" and "phuton" meaning "hair" and "plant", respectively. The name "hairy plant" refers to the fluffy appearance of a colony of the fungus when it's grown on an agar plate – it appears to be covered in hairs. *T. rubrum* was first identified in 1845 by Professor Pehr Henrik Malmsten at the Karolinska Institute in Sweden, but it wasn't until 1909 that the Italian bacteriologist Aldo Castellani grew it in the laboratory. The species name is related to the red colour of the underside of a colony of the organism when it's grown on an agar plate. The upper surface of the colony is usually white.

T. rubrum (Fig. 5.4) is an important dermatophyte that is found almost exclusively in humans. In temperate climates it's responsible for 90% of cases of onychomycosis.

The organism is a typical fungus and so grows as long, filamentous strands with a diameter of 1–2 um. It produces two types of spore-containing structures (known as conidia): (a) microconidia which are small (2–3 × 2–4 μm), teardrop-shaped (Fig. 5.5a, b macroconidia which are larger structures (4–8 × 8–50 μm) and contain several spores (Fig. 5.5b). These structures are important in its identification.

It grows both in the presence or absence of oxygen and produces enzymes that can break down a number of different proteins including keratin which is the main component of the nail. The production of a keratin-degrading enzyme (known as a keratinase) is its main virulence factor. The organism uses this enzyme to break down the keratin to smaller molecules (amino acids) which it can use as nutrients. In the most common form of onychomycosis,

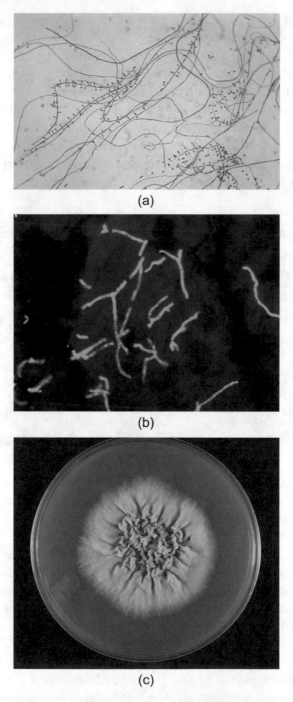

(a)

(b)

(c)

Fig. 5.4 Images of *T. rubrum*. (**a**) *T. rubrum* as seen through a microscope at a magnification of ×475. This shows long, filamentous hyphae attached to which are small teardrop-shaped spores. (Dr. Lucille K. Georg, Centers for Disease Prevention and Control, USA). (**b**) *T. rubrum* observed in an infected nail and viewed at a higher magnification than that used in (**a**). (*J Fungi (Basel)*. 2019 Mar; 5(1): 20. Recent Findings in

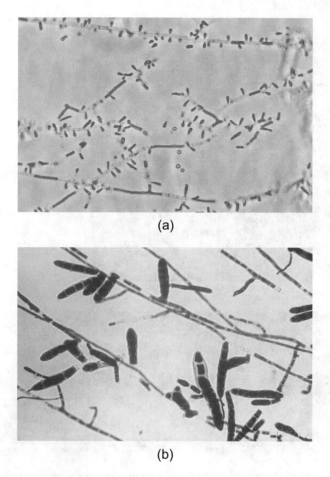

(a)

(b)

Fig. 5.5 Microconidia and macroconidia of *T. rubrum*. (**a**) Hyphae of *T. rubrum* with small microconidia attached (×500). (Dr Libero Ajello, Centers for Disease Prevention and Control, USA). (**b**) Hyphae of *T. rubrum* with macroconidia. Note the presence of several spores in each macroconidium (×474). (Dr Lucille K. Georg, Centers for Disease Prevention and Control, USA)

Fig. 5.4 (continued) Onychomycosis and Their Application for Appropriate Treatment. Michel Monod and Bruno Méhul. This article is an open-access article distributed under the terms and conditions of the Creative Commons Attribution (CC BY) licence (http://creativecommons.org/licenses/by/4.0/)). (**c**) A colony of *T. rubrum* growing on an agar plate. The appearance of a colony of an organism (whether a fungus or bacterium) is very helpful in trying to identify it. (Dr. Libero Ajello, Centers for Disease Prevention and Control, USA)

(a)

(b)

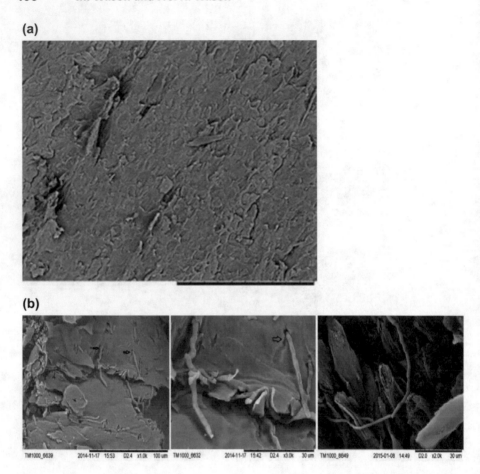

TM1000_6639 2014-11-17 15:53 D2.4 x1.0k 100 um TM1000_6632 2014-11-17 15:42 D2.4 x3.0k 30 um TM1000_8849 2015-01-08 14:49 D2.0 x2.0k 30 um

Fig. 5.6 Invasion of the nail plate by *T. rubrum* in a patient with onychomycosis. (**a**) A healthy intact nail with minimal damage (magnification ×400). (**b**) Images of nails infected with the fungus. In all images the nail plate shows significant damage, and hyphae can be seen piercing through the nail plate (arrows). Image magnifications: left = ×1000, central = ×3000, right = ×2000. (*BMC Infect Dis*. 2015 Nov 17;15:532. An ultrastructural study of *Trichophyton rubrum* induced onychomycosis. Yue X, Li Q, Wang H, Sun Y, Wang A, Zhang Q, Zhang C. This article is distributed under the terms of the Creative Commons Attribution 4.0 International License (http://creativecommons.org/licenses/by/4.0/), which permits unrestricted use, distribution and reproduction in any medium, provided you give appropriate credit to the original author(s) and the source, provide a link to the Creative Commons licence and indicate if changes were made)

the fungus invades the nail bed (Fig. 5.6) and the underside of the nail plate beginning at the free edge. It migrates through the underlying nail bed and matrix and causes mild inflammation. This results in detachment of the nail plate from the nail bed (a process known as onycholysis) and thickening of the region underneath the nail. Other possible routes of invasion are via the nail plate itself and via the cuticle, but these occur less frequently. Nails that have been damaged in some way are more easily infected.

T. rubrum is also a major cause of athlete's foot, jock itch and ringworm (see Chap. 6).

How Can It Be Treated?

The main goal of treatment is to get rid of the fungus from the nail. This is often difficult to achieve and can take a long time and needs prolonged or repeated treatments. An option of not treating at all is appropriate if you aren't troubled by the nail changes or you don't suffer from any other symptoms or aren't at risk of complications. If you don't choose treatment, then you should take care to avoid spreading the problem to others or to different parts of your own body.

The fact that the fungus doesn't penetrate deeply into the skin suggests that it should be possible to treat mild cases by applying an anti-fungal agent to the surface of the nail, i.e. topical therapy. Unfortunately, the drugs don't penetrate into the nail very readily, and you have to file down the thickened nail; this is helpful because it also reduces the amount of abnormal nail that remains to be treated. This approach generally has a low success rate, which is further reduced when people lose faith in the slow response and get tired of carrying out the fiddly treatment process.

The most widely used agents are amorolfine and ciclopirox. Amorolfine is applied as a 5% lacquer once or twice weekly for 6–12 months, but the success rate is less than 15%. Ciclopirox is used as an 8% lacquer and is applied once daily for 24 weeks on fingernails and for 48 weeks on toenails – cure rates are generally less than 10%. Some newer, more effective topically applied agents include tioconazole and efinaconazole which have success rates of 22% and 18%, respectively. Some studies have shown success rates are improved using a combination of oral and topical treatments.

More severe cases require the use of oral anti-fungal agents such as terbinafine, griseofulvin or itraconazole. Terbinafine is the anti-fungal agent of choice and kills fungi by preventing the production of new cell walls and membranes. Unfortunately, long-term treatment is required for a successful outcome, and the drug must be taken for at least 6 weeks for infections of fingernails and for 12–16 weeks for toenails. Treatment sometimes lasts much longer. Failure rates are surprisingly high with 25% of treatments being unsuccessful. Complete removal of the nail is sometimes offered as an option, if other treatments have failed.

Newer developments in treatment include drilling tiny holes in the nail surface to aid penetration of the topical treatments, using laser or UV light or photodynamic therapy. Research is also being carried out into the use of ultrasound or electrical current to deliver drugs to underneath the nail.

Relapses are common, even after successful treatment, and occur in 25–30% of patients. This may be due to reinfection from the environment including contaminated footwear due to the lengthy survival of the spores or hyphae, a genetic predisposition to the disease, predisposing factors such as diabetes, or conditions that weaken the nail such as psoriasis or trauma. However, given the ubiquitous nature of *T. rubrum*, re-infection isn't surprising; often whole households are affected by the same strain of the fungus.

How Can I Avoid Getting the Disease?

The fungus is described as being anthropophilic which means that it prefers humans to other animals. Nevertheless, it's also found on pets and farm animals. You can reduce the chances of getting the infection by avoiding infected humans and other animals as well as **fomites** (such as clothing or nail clippers) that have been in contact with infected individuals. The fungus can survive in the environment and survive best in warm, humid areas. Therefore you should avoid walking barefoot in those communal facilities where the environment is conducive to the growth of the fungus, such as public bathing areas and changing rooms. Drying well after bathing and changing socks regularly are also important preventive measures as are treating shoes with antifungals and replacing old footwear which may be contaminated. Keeping your nails short and wearing well-fitting shoes will help to reduce damage to, or weakening of, the nails. Treating fungal skin infections promptly helps to reduce spread between the nails and the skin. If your fingernails are affected, then it's important to avoid repeated hand washing or immersion of your hands in water because this makes your skin more susceptible to infection.

Thickened misshapen nails can injure the surrounding skin or cause problems with shoe fitting and walking. Rarely the infection may cause a difficult-to-treat dermatophytoma which is a dense collection of fungal filaments in, or under, the nail plate. Infection in the nails can spread to the skin or even further to other locations on the body, especially if there's a lot of scratching. Bacterial infection can take hold in damaged nails, especially in at-risk individuals, and this can sometimes lead to cellulitis of the foot or leg.

Want to Know More?

American Academy of Dermatology https://www.aad.org/public/diseases/contagious-skin-diseases/nail-fungus

American Family Physician https://www.aafp.org/afp/2001/0215/p663.html

British Association of Dermatologists http://www.bad.org.uk/shared/get-file.ashx?id=205&itemtype=document

British Skin Foundation https://www.britishskinfoundation.org.uk/fungal-infections-of-the-nails

Centers for Disease Prevention and Control, USA https://www.cdc.gov/fungal/nail-infections.html

DermNet New Zealand Trust https://www.dermnetnz.org/topics/fungal-nail-infections/

Doctor Fungus - Educational and Community Resource for the Mycoses Study Group Education and Research Consortium https://drfungus.org/knowledge-base/onychomycosis/

Mayo Clinic, Mayo Foundation for Medical Education and Research https://www.mayoclinic.org/diseases-conditions/nail-fungus/symptoms-causes/syc-20353294

National Health Service, UK https://www.nhs.uk/conditions/fungal-nail-infection/

National Institute for Clinical Care and Excellence (NICE), UK. Fungal nail infection, 2018 https://cks.nice.org.uk/fungal-nail-infection#!topicSummary

Bodman MA, Krishnamurthy K. Onychomycosis. Treasure Island (FL): StatPearls Publishing LLC; 2020 https://www.ncbi.nlm.nih.gov/books/NBK441853/

Gupta AK, Mays RR. The impact of onychomycosis on quality of life: a systematic review of the available literature. *Skin Appendage Disorders*. 2018 Oct;4(4):208–216. https://doi.org/10.1159/000485632. Epub 2018 Feb 13.

Gupta AK, Stec N. Recent advances in therapies for onychomycosis and its management. *F1000Research*. 2019 Jun 25;8. pii: F1000 Faculty Rev-968. https://doi.org/10.12688/f1000research.18646.1. eCollection 2019.

Reinecke JK, Hinshaw MA. Nail Health in Women. *International Journal of Women's Dermatology* 2020 Feb 5;6(2):73–79.

Tosti A. Onychomycosis. Medscape from WebMD, 2018 https://emedicine.medscape.com/article/1105828-overview

6

Fungal Infections of the Skin

Abstract A number of fungi, known as dermatophytes, are able to cause skin infections (often called ringworm or tinea) at various body sites including the scalp, face, hands, feet, groin, and trunk. They're the most common infections of humans. These fungi belong to three main genera – *Epidermophyton*, *Microsporum* and *Trichophyton*. Infection is accompanied by an itchy, inflamed characteristic rash. The disease can be transmitted from person to person or via contaminated personal care items or other objects. The fungi produce enzymes that break down skin components such as keratin and collagen to small molecules that are used as nutrients. The infection can be treated using anti-fungal agents either applied to the skin or taken orally. However, treatment failure is common often because of the prolonged treatment period required.

What Is a Dermatophyte Infection?

Dermatophytes are fungi that cause infections at a number of body sites. Most commonly these are on the surface of the skin, but very rarely they can cause infections in deeper areas of the skin and body in at-risk individuals. Such infections are also known as tinea or ringworm and are named on the basis of which part of the body is affected (Table 6.1).

On flat areas of skin, the infection spreads outwards from a central point which usually results in a ring-like, red or silvery rash with an inflamed edge which is scaly and itchy. The disease symptoms depend on which part of your body is affected.

Table 6.1   Infections caused by dermatophyte fungi

Disease	Body site affected
Tinea capitis	Scalp
Tinea barbae	Beard and moustache areas
Tinea faciale or faciei	Face
Tinea corporis	Trunk, arms and legs
Tinea cruris (also known as "jock itch")	Inner thighs and groin
Tinea manuum	Hands
Tinea pedis (also known as "athlete's foot")	Feet
Tinea unguium	Toenails and fingernails

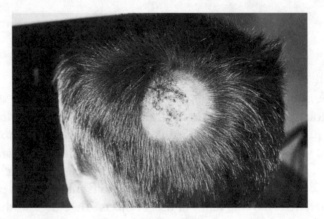

Fig. 6.1   Tinea capitis. (Centers for Disease Control and Prevention, USA)

Infection of the scalp (Fig. 6.1) results in small patches of scaly skin which may be sore. This may be accompanied by patchy hair loss and itchiness of the scalp. In more severe cases small, pus-filled sores may appear on the scalp as well as crusting and bald patches. Occasionally a large inflamed sore that oozes pus forms on the scalp – this is known as a kerion. The disease is seen mostly in young children – particularly those between 5 and 10 years of age and those who are of Afro-Caribbean origin. In the USA 3–8% of children are affected, and it's more common in men than in women. The disease is rare in adults although occasionally it's found in the elderly.

Infection of the foot (Fig. 6.2) causes an itchy, dry, red and flaky rash, and this usually appears in the spaces between the toes. In severe cases other symptoms may develop including cracking of the skin in the affected area, blister formation, skin swelling, a burning or stinging sensation in the skin and scaling around the sole, between the toes and on the side of the foot. Tinea pedis is the most common of the fungal infections of the skin, and it's been estimated that 70% of the world's population will suffer from this infection at some time during their life. It's worth noting there are non-fungal causes of

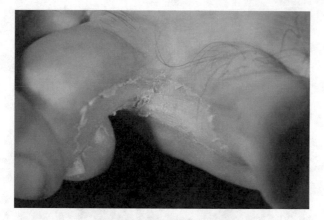

Fig. 6.2 Athlete's foot. (Dr. Lucille K. Georg, Centers for Disease Control and Prevention, USA)

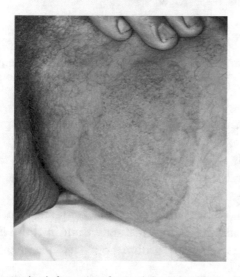

Fig. 6.3 Tinea cruris on the left groin of an adult male. (Dr. Lucille K. Georg, Centers for Disease Control and Prevention, USA)

rashes between the toes, and these include bacterial infections and soft corns. Children are rarely affected and most cases occur after puberty – the prevalence then increases with age. Men are 2–4 times more likely to be affected than women. In developed countries between 30% and 40% of the population at any one time suffer from the condition.

The symptoms of tinea cruris (Fig. 6.3) include scaly, flaky skin on the inner thighs, itchiness and redness around the groin area, inner thighs and anus and red-brown sores, which may have blisters or pus-filled sores around

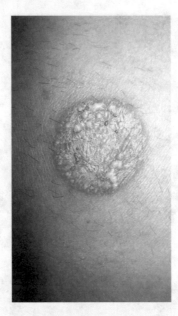

Fig. 6.4 Tinea corporis. (This work has been released into the public domain by its author, Corina G. This applies worldwide. Via Wikimedia Commons)

the edge. It's three times more common in men than in women, and adults are affected much more commonly than are children.

In tinea corporis (Fig. 6.4), the lesions are usually on the exposed skin of the trunk, arms and legs and consist of scaly flat patches with raised edges; there may be vesicles and pustules. Occasionally the lesions appear as overlapping or concentric circles. It occurs more frequently in those who live in countries with a warm and humid climate than in those living in temperate climates. It affects all age groups but its prevalence is highest in pre-adolescents.

In tinea barbae, a red, scaling rash is present, and pustules form in the follicles (Fig. 6.5). It's not a common disease in Western countries and affects only adolescent and adult men.

Tinea manuum usually affects only one hand and scaling, and redness are prominent. It often causes the skin to become thicker on the palm and between the fingers. It sometimes accompanies tinea pedis and is probably the result of scratching affected feet, but it occurs less frequently.

Tinea faciale (Fig. 6.6) usually appears as red scaling patches with borders that have papules, vesicles and/or crusts. It occurs most commonly on the cheeks, followed by the nose, around the eyes, chin and forehead. The condition occurs worldwide but is more common in tropical regions with high

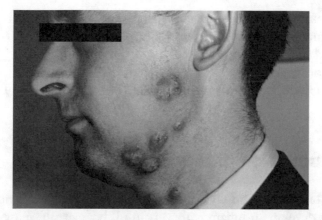

Fig. 6.5   Tinea barbae. (Image courtesy of Centers for Disease Control and Prevention, USA)

Fig. 6.6   Tinea faciale. (Dr. Libero Ajello, Centers for Disease Control and Prevention, USA)

temperatures and humidity. It occurs most frequently in children and in adults aged 20–40 years.

Who Are Most Likely to Get a Dermatophyte Infection?

People most at risk of developing a dermatophyte infection are those who:

- Have diabetes
- Are obese

- Are immunocompromised such as HIV patients or those being treated with immunosuppressants for cancer or other conditions
- Have had fungal infections in the past
- Have poor hygiene
- Have pets which are infected or come into close contact with animals and soil

What About Complications?

Complications of dermatophyte infection aren't common. Patients with tinea capitis may develop an inflammatory, painful, boggy scalp mass known as a kerion. This can be treated with a long course of an oral anti-fungal agent because the infection doesn't respond to surface treatments. The lesions of tinea pedis may become infected with bacteria, particularly *Streptococcus pyogenes* or *Staphylococcus aureus*. If this occurs, topical or systemic antibiotics may be necessary.

If the rash is incorrectly treated with steroid creams instead of anti-fungal agents, then the symptoms may improve, but the infection remains untreated and spreads; this is called "tinea incognita" as the appearance of the rash will have been changed by the treatment which makes it more difficult to identify.

Rarely the infection invades the epidermal layers into deeper areas of the skin and further into the body. This is more likely to occur in immunosuppressed people or when the skin is broken. When the deeper layers of the skin are affected, it's called "deep dermatophytosis", when it has spread further, it's called "invasive dermatophytosis". One example of the infection penetrating underneath the epidermis and into the dermis is called Majocchi's granuloma (Fig. 6.7) where hair follicles become inflamed. It causes small, sore and red lumps in the skin such as nodules, papules and pustules, most commonly on the limbs.

Which Microbes Are Responsible?

It's often stated that dermatophyte infections are not only the most common fungal infections but are the most common infections of humans in the world. This is mainly because tinea pedis is such a common problem in the developed world while tinea capitis is very common in the developing world. However, it's also important to remember that dermatophytes colonise 30–70% of humans without causing any disease. Approximately 40 different

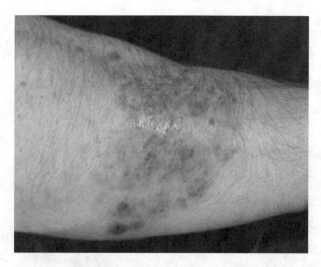

Fig. 6.7   A patient with Majocchi's Granuloma on the arm. (Tinea Corporis Gladiatorum Presenting as a Majocchi Granuloma. Kurian A, Haber RM. *ISRN Dermatol*. 2011; 2011: 767589. PMID: 22363858

species are responsible for infecting humans, and most of these belong to three fungal genera *Epidermophyton* (Fig. 6.8a, b), *Microsporum* (Fig. 6.8c, d) and *Trichophyton* (see Chap. 5).

Box 6.1 What's in a Name?

Epidermophyton derives its name from the Greek words "epi", "derm" and "phyton" which mean "upon", "skin" and "plant", respectively. Hence it means "a plant that lives on the skin". Fungi were once considered as belonging to the plant kingdom before they were classified as a type of microbe. The fungus was discovered by the Hungarian physician and mycologist David Gruby in 1842 who isolated it from a patient with tinea barbae.

Microsporum is derived from the Greek words "micros" and "sporos" which mean small and seed, respectively. This microbe was also discovered by David Gruby who isolated it from a patient with tinea in 1843.

Dermatophytes can invade and grow in dead keratin which is found in the stratum corneum, the outermost layer of the skin. They rarely invade the inner layers of the epidermis. They tend to grow outwards on the skin from a central point to produce a ring-like pattern – hence the term "ringworm". Some dermatophytes are spread from person to person (anthropophilic organisms), others are transmitted to humans from soil (geophilic organisms), while some are

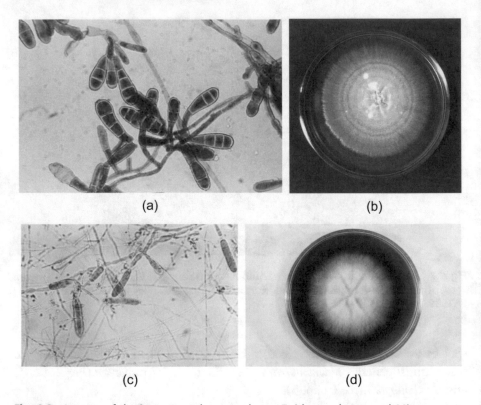

(a)   (b)

(c)   (d)

Fig. 6.8 Images of the important dermatophytes *Epidermophyton* and *Microsporum*. Images of the other important dermatophyte, *Trichophyton*, are shown in Chap. 5.
(a) Photomicrograph of an *Epidermophyton* species showing its large bulbous reproductive structures (known as macroconidia) and filamentous hyphae (×475). (Dr. Lucille K. Georg, Centers for Disease Control and prevention, USA). (b) A colony of an *Epidermophyton* species growing on an agar plate. (Dr. Lucille K. Georg, Centers for Disease Control and prevention, USA). (c) Photomicrograph of a *Microsporum* species showing its large cigar-shaped reproductive structures (known as macroconidia) and filamentous hyphae (×475). (Dr. Lucille K. Georg, Centers for Disease Control and prevention, USA). (d) A colony of a *Microsporum* species growing on an agar plate. (Dr. Lucille K. Georg, Centers for Disease Control and prevention, USA)

spread to humans from other animals (zoophilic organisms). The fungi can also be transmitted to humans via objects such as hats, upholstery, hairbrushes and combs. Most skin infections are caused by anthropophilic organisms, although zoophilic organisms are common in developing countries. The outermost layer of the skin, the stratum corneum (see Chap. 2), consists of dead cells containing high levels of keratin which is a protein that dermatophytes use as their main nutrient. While the normal skin defence mechanisms (see Chap. 1) are effective at combating these fungi most of the time, they can be disrupted

Table 6.2 Dermatophytes responsible for the various forms of tinea

Infection	Main dermatophytes responsible (other fungi may also be responsible)
Tinea pedis	*Trich. rubrum, Trich. mentagrophytes, Trich. tonsurans, Epiderm. floccosum*
Tinea capitis	*Micro. canis, Micro. gypseum, Trich. equinum, Trich. verrucosum, Trich. tonsurans, Trich. violaceum*
Tinea barbae	*Trich. rubrum*
Tinea corporis	*Trich. rubrum*
Tinea cruris	*Trich. rubrum, Epiderm. floccosum, Trich. mentagrophytes, Trich. verrucosum*
Tinea faciale	*Micro. canis, Trich. mentagrophytes, Trich. rubrum*
Tinea manuum	*Trich. rubrum, Trich. mentagrophytes, Epiderm. floccosum*

leaving the skin vulnerable to infection. The dry surface of the skin, for example, prevents microbes from growing there, but continual exposure to moist conditions (such as humid climates, wet clothing, excessive sweating) makes it suitable for fungal colonisation and growth. Dermatophytes can attach themselves to the stratum corneum, and, by secreting enzymes that can breakdown keratin (known as keratinases), they obtain the nutrients they need for their growth. They can then invade the outer skin layers and become established there. As well as producing keratinases, many dermatophytes also secrete other enzymes such as lipases which can break down the fats that are plentiful on the skin surface. Collagenases are also produced, and these break down collagen which is a major constituent of human tissue. These enzymes produce small molecules (glycerol, fatty acids and amino acids) which are used as nutrients for the dermatophytes, but, in doing so, they also damage human tissue. The fungi start growing in the stratum corneum within 24 hours of arriving on the skin surface. Damage to the skin tissues induces the release of cytokines, and this results in an inflammatory response which, ultimately, disposes of the fungi.

The main dermatophytes responsible for the various types of infection are listed in Table 6.2.

How Are These Diseases Diagnosed?

The diseases are diagnosed on the basis of the appearance and location of the skin rash. Skin scrapings can also be examined by microscopy (after immersion in KOH to dissolve the skin) and are used to inoculate culture media to try to grow and identify the fungi responsible. PCR (see Chap. 1) of skin scrapings is increasingly being used to identify the DNA of the fungi responsible for the infection.

How Is the Infection Treated?

The main goal of treatment is to eradiate the fungi responsible for the infection, and this can be achieved by using topical or systemic anti-fungal agents, depending on the site and severity of the infection. Some of the available agents are listed in Table 6.3.

How Can I Avoid Becoming Infected?

The fungi responsible for tinea can survive on furniture, hairbrushes, clothing, bed sheets and towels and so can be spread through contact with these kinds of items. Therefore, if someone in your household has a fungal infection, it's important to avoid sharing their personal care items. Wearing something on your feet in gym and swimming pool changing rooms reduces the chances of picking up dermatophytes from the floor surfaces. Pets may be a source of infection; therefore, you should treat any animal displaying symptoms (such as patches of missing fur) as soon as possible. You should avoid wearing shoes, such as trainers, that trap sweat and create a moist environment. Prompt treatment helps to prevent the spread of the infection to others and also reduces the chances of complications in at-risk people.

Table 6.3   Anti-fungal agents used in the treatment of dermatophyte infections

Infection	Anti-fungal agents used
Tinea capitis	Oral griseofulvin for *Microsporum* infections and oral terbinafine for *Trichophyton* infections Anti-fungal shampoos (such as selenium sulphide and ketoconazole) may be used to reduce fungal spread and risk of transfer to others
Tinea barbae, tinea manuum	Oral griseofulvin, terbinafine or itraconazole
Tinea faciale, tinea corporis, tinea cruris, tinea pedis	These are treated with topical anti-fungal agents. The most effective are those from the allylamine group – naftifine, terbinafine and butenafine. Other possible agents include clotrimazole, ketoconazole and miconazole

Want to Know More?

American Academy of Dermatology https://www.aad.org/public/diseases/contagious-skin-diseases/ringworm

American Family Physician https://www.aafp.org/afp/2014/1115/p702.html

British Association of Dermatologists http://www.bad.org.uk/shared/get-file.ashx?id=1998&itemtype=document

British Journal of Family Medicine https://www.bjfm.co.uk/tinea-infection-an-illustrated-guide

Centers for Disease Control and Prevention, USA https://www.cdc.gov/fungal/diseases/ringworm/index.html

Doctor Fungus, Mycoses Study Group Education and Research Consortium https://drfungus.org/knowledge-base/tinea-corporis-tinea-cruris-tinea-pedis/

Mayo Clinic, Mayo Foundation for Medical Education and Research https://www.mayoclinic.org/diseases-conditions/ringworm-body/symptoms-causes/syc-20353780

MedlinePlus, U.S. National Library of Medicine https://medlineplus.gov/tineainfections.html

National Health Service, UK https://www.nhsinform.scot/illnesses-and-conditions/infections-and-poisoning/ringworm-and-other-fungal-infections

National Institute for Clinical Care and Excellence (NICE), UK. Fungal skin infections, 2018 https://cks.nice.org.uk/fungal-skin-infection-body-and-groin#!topicSummary

Primary Care Dermatology Society http://www.pcds.org.uk/clinical-guidance/tinea

Baltazar LM, Ray A, Santos DA, Cisalpino PS, Friedman AJ, Nosanchuk JD Antimicrobial photodynamic therapy: an effective alternative approach to control fungal infections.. *Frontiers in Microbiology.* 2015 Mar 13;6:202. doi: https://doi.org/10.3389/fmicb.2015.00202. eCollection 2015.

Baumgardner DJ. Fungal infections from human and animal contact. *Journal of Patient-Centred Research and Reviews.* 2017 Apr 25;4(2):78–89. doi: https://doi.org/10.17294/2330-0698.1418. eCollection 2017 Spring.

Handler MZ. Tinea Capitis. Medscape from WebMD, 2020 https://emedicine.medscape.com/article/1091351-overview

Hay R. Therapy of skin, hair and nail fungal Infections. *Journal of Fungi (Basel).* 2018 Aug 20;4(3). pii: E99. doi: https://doi.org/10.3390/jof4030099.

Lesher JL. Tinea Corporis. Medscape from WebMD, 2018 https://emedicine.medscape.com/article/1091473-overview

Nigam PK, Saleh D. Tinea pedis. Treasure Island (FL): StatPearls Publishing LLC; 2020 https://www.ncbi.nlm.nih.gov/books/NBK470421/

Robbins CM. Tinea pedis. Medscape from WebMD, 2018 https://emedicine.medscape.com/article/1091684-overview

Tosti A. Dermatophyte infections. BMJ Best Practice, BMJ Publishing Group, 2019 https://bestpractice.bmj.com/topics/en-gb/119

Yee G, Al Aboud AM. Tinea Corporis. Treasure Island (FL): StatPearls Publishing LLC; 2020 https://www.ncbi.nlm.nih.gov/books/NBK544360/

7

Chickenpox

Abstract Chickenpox (varicella) is an infection due to the varicella-zoster virus that mainly affects children under the age of 10 years. The main symptom is a rash of red itchy spots that turn to blisters – it's often painful and accompanied by flu-like symptoms. It's usually mild and self-limiting and lasts about 1 week. In adults the symptoms are usually more severe and can lead to pneumonia in up to 14% of cases. Infection with the virus during pregnancy can cause birth defects and, if it occurs a few days before delivery, can result in a life-threatening infection in the newborn. It's highly contagious, and in temperate climates outbreaks are more common in late winter and spring. It's spread by touching infected people and contaminated surfaces and in droplets in the air produced by coughing or sneezing. Anti-viral agents aren't usually needed in most cases of childhood infection, but other types of medication should be given to relieve pain and itching. However, anti-viral agents such as acyclovir should be considered in adults and offered to those with a moderate or high risk of developing a severe infection or complications. Spread of the disease can be reduced by keeping infected children away from nursery or school for at least 5 days, and adults shouldn't return to work until 5–6 days after the first appearance of the rash or until it has dried up. Vaccination against the disease is extremely effective and is practised in many, but not all, countries. At-risk individuals can also be protected against the infection by giving them antibodies against the virus – this is known as "passive immunoprophylaxis".

© Springer Nature Switzerland AG 2021
M. Wilson, P. J. K. Wilson, *Close Encounters of the Microbial Kind*,
https://doi.org/10.1007/978-3-030-56978-5_7

What Is Chickenpox?

Chickenpox (also known as varicella) is a disease that most commonly affects children under the age of 10 years. It's a viral infection and is caused by the varicella zoster virus (VZV) Thankfully you normally only catch it once as the immunity that develops afterwards is usually long lasting. It's a mild, self-limiting and very common disease that most children will catch in those countries that don't vaccinate against it. In the USA children are vaccinated against the disease, but in many European countries, including the UK, they aren't.

Box 7.1 What's in a Name?

Varicella is a word derived from "Variola" which is the medical term for the deadly disease smallpox. Variola comes from the Latin word "varius" which means speckled or spotted – a characteristic feature of smallpox. Varicella means "little variola" and probably refers to the fact that it is a much milder disease than the deadly smallpox. Zoster is a Greek word that means a girdle or warrior's belt (see Chap. 8). How the disease became known as chickenpox isn't certain, and it's been suggested that (i) it refers to the fact that it's a less deadly disease than smallpox, which has similar symptoms; (ii) the spots resemble the peck marks made by a chicken; and (iii) the spots resemble chick peas. In 1875 Rudolf Steiner discovered that chickenpox was due to an infection by showing that volunteers who were exposed to fluid from the blister of a patient with varicella developed the disease. However, it wasn't until 1954 that the VZV was isolated from the blister fluid of patients with varicella – this was achieved by the Nobel prize-winning American virologist Thomas Weller.

What Are the Symptoms of Chickenpox?

The main symptom of the infection is a rash of red itchy spots that can cover the entire body (Fig. 7.1). The rash usually first appears on the face, scalp or trunk and then spreads to the limbs. It's often accompanied by pain (90% of patients) and flu-like symptoms (12% of patients).

The incubation period for the disease is about 14 days, and, prior to the appearance of the spots, other symptoms are often experienced including pain (41% of patients), itchiness (27% of patients), tingling (12% of patients), nausea, fever, sore throat, muscle aches, headache and loss of appetite. In adults these symptoms occur more frequently and are generally more severe than in children. The red spots soon turn into blisters (known as the "eruptive stage") and become very itchy after about 12 hours (Fig. 7.2).

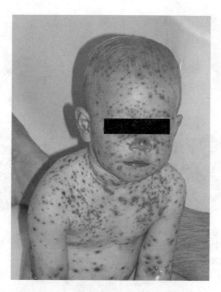

Fig. 7.1 A child suffering from chickenpox showing the characteristic rash. (Posterjack. I, the copyright holder of this work, release this work into the public domain. This applies worldwide. In some countries this may not be legally possible; if so: I grant anyone the right to use this work for any purpose, without any conditions, unless such conditions are required by law. Via Wikimedia Commons)

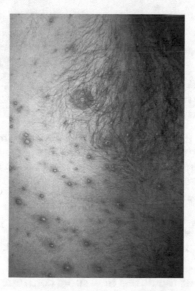

Fig. 7.2 View of the surface of the skin of the right chest of a patient with chickenpox. The rash is in the eruptive stage and consists mainly of fluid-filled blisters. (K.L. Herrmann Centers for Disease Control and Prevention, USA)

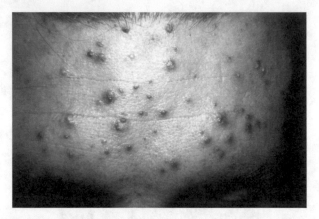

Fig. 7.3 Forehead of a patient with chickenpox. Some of the blisters have ruptured and begun to form their characteristic crusty surface. (K.L. Herrmann Centers for Disease Control and Prevention, USA)

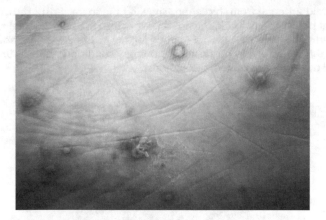

Fig. 7.4 A close view of the sole of the foot of a patient with chickenpox. The lesions are in various stages of development, some are in their fluid-filled, vesicular form, while others having entered the crusty stage and are covered by a scab. (K.L. Herrmann Centers for Disease Control and Prevention, USA)

After 1–2 days the fluid in the blisters becomes cloudy, and then the blisters dry out and become crusty within 7–10 days (Fig. 7.3). The crusts fall off after 1–2 weeks.

However, throughout the illness new spots continue to appear, and so later on the patient will have spots at all stages of development (Fig. 7.4). Healthy children usually have 200–500 spots in 2–4 successive crops.

Are There Any Complications of Chickenpox?

Serious complications in healthy people don't occur very often. The most common complication in children is a secondary infection of the skin with either *Staph. aureus* or *Strep. pyogenes*, and these may require antibiotic treatment. Very rarely, varicella in children can lead to more serious infections including Reye's syndrome, appendicitis, myocarditis, pancreatitis, arthritis, eye infections, meningitis and encephalitis.

Pneumonia may affect 5–14% of adults with varicella, and those most at risk are pregnant women, immunosuppressed people and smokers. It usually occurs 1–6 days after the rash first appears. Other complications in adults include hepatitis and encephalitis, and generally adults are more unwell with the infection and more likely to be admitted to hospital. Later reactivation of VZV causes shingles (see Chap. 8).

Pregnant women who develop varicella after 36 weeks gestation have the highest risk of having a baby affected by chickenpox. Infection with varicella in the newborn is serious and can be life-threatening – the mortality rate can be as high as 30%. Pregnant women themselves are at an increased risk of developing a severe illness and complications if infected with the virus.

Infection during the first 28 weeks of pregnancy can lead to a condition known as foetal varicella syndrome (FVS). However, FVS is rare, and the risk of it occurring in the first 12 weeks of pregnancy is less than 1%, between weeks 13 and 20 the risk is 2%, while between 20 and 28 weeks, it is <1%. FVS can result in scarring, eye defects (such as cataracts), shortened limbs and brain damage.

Infection between 28–36 weeks of pregnancy will not cause the baby any symptoms. However, the child may develop shingles as an infant due to reactivation of the infection they contracted in the womb.

In healthy children the disease is generally mild and self-limiting and lasts approximately 1 week. People with a moderate risk of severe infection or complications include those who (i) are more than 13 years of age, (ii) have a pre-existing skin disease such as atopic dermatitis, (iii) have an underlying lung disease, (iv) are taking salicylate medications and (v) are taking short courses of oral corticosteroids. People with a high risk of severe disease include those who (i) are immunocompromised (such as those who have an organ transplant, are receiving chemotherapy or have HIV infection), (ii) are less than 4 weeks old, (iii) are taking long-term oral corticosteroids or high-dose systemic immunosuppressants and (iv) are pregnant.

How Common is Chickenpox?

VZV is highly contagious, and more than 90% of those who haven't been vaccinated against it will become infected with the virus if exposed to it. In the USA, the UK and Japan, more than 80% of people become infected with the VZV by the age of 10 years. The virus is more prevalent in temperate climates, and outbreaks are more common in late winter and spring – epidemics occur every 2–3 years. In temperate climates the greatest incidence of varicella is in children 1–9 years of age, but in tropical climates the disease is more often acquired in adulthood.

Globally, it's been estimated that the number of cases each year is 140 million, and this results in 4.2 million severe complications and 4200 deaths each year. In developed countries mortality rates are very low but are highest in adults and the immunocompromised. In the USA, prior to the introduction of a vaccination programme in 1995, there were approximately four million cases of chickenpox each year, and this resulted in 11,000–13,500 hospital admissions and 100–150 deaths. Following the introduction of vaccination, the incidence of the disease decreased by 90% over the following decade. In the UK, which doesn't have a vaccination programme, the number of new cases each year is approximately 300 per 100,000, and the mortality rate is 0.2–0.4 per one million. Approximately 80% of those who die are adults.

What Causes Chickenpox?

The virus responsible for chickenpox, the varicella zoster virus, belongs to the same family as the herpes simplex virus (HSV – discussed in Chap. 23) and has a similar structure (Fig. 7.5).

The virion has a diameter of 150–200 nm and has a nucleocapsid (100 nm in diameter) containing DNA. It's described as being an "enveloped virus" because its outer coat (envelope) is formed from the cytoplasmic membrane of the human cell in which it was produced. The envelope of the virus has a number of glycoprotein spikes that are involved in adhering to, and invading, human cells.

Viruses are fascinating creatures. For a start, they're extremely fortunate as they don't even have to reproduce themselves – they get other creatures to do this for them. Amazing! Surely then they must be the most complex of creatures? But no, the very opposite. Structurally they're the simplest forms of life on earth, and, like some molecules, they can even be crystallised. They're so

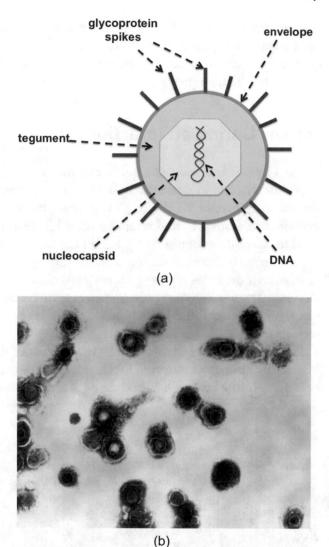

Fig. 7.5 The varicella zoster virus. (a) Diagram showing the structure of the varicella zoster virus. (b) The varicella zoster virus as seen through an electron microscope. (Dr. Erskine Palmer, Centers for Disease Control and Prevention, USA)

simple that they don't even consist of a cell – we call an individual virus a "particle" or "virion". We're taught at school that the general characteristics of all living creatures are that they feed, breathe, grow, reproduce and respond to stimuli. But viruses do none of these things – so are they alive?

As described in Chap. 1, viruses often consist of no more than a molecule of nucleic acid (which can be either DNA or RNA) surrounded by a protein coat. But the great success of this life form is that it can hijack other, more complex, cells to make many copies of itself. The unfortunate cell, after becoming no more than a virus factory, then usually dies and releases large numbers of new viruses. That's a pretty smart way of existing.

What Happens During an Infection?

VZV can be spread by touch or via the air either as virions (shed from rup-tured skin vesicles) or attached to skin cells or in droplets from the respiratory tract. It enters the body mainly via the respiratory tract but can also enter through the conjunctiva. It invades, and is replicated within, epithelial cells of the pharynx and tonsils. The glycoprotein spikes on the surface of the virion bind to specific molecules (known as receptors) on the surface of the epithelial cells. Binding of the virion to its receptor on the epithelial cell triggers the ingestion of the virion by the cell by a process known as **endocytosis** (Fig. 7.6). The virion is now within a small fluid-filled "balloon" inside the cell – this is known as a vacuole (Fig. 7.6).

The virion then releases its DNA which enters the nucleus of the host cell (Fig. 7.7). The nucleus produces more viral DNA and proteins which start to migrate to the surface of the host cell. When they get there, these molecules

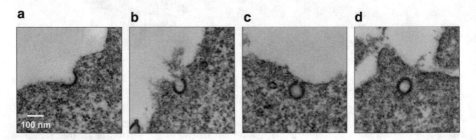

Fig. 7.6 Series of electron micrographs showing the uptake of a virus by a human cell and the subsequent formation of a virus-containing vacuole inside the cell. **(a)** The cell membrane begins to form an indentation when the virus binds to it. **(b)** and **(c)** The membrane of the cell progressively envelops the virus **(d)** eventually the virus is enclosed within a vacuole inside the cell. (Vesicular stomatitis virus enters cells through vesicles incompletely coated with clathrin that depend upon actin for internalization. Cureton DK, Massol RH, Saffarian S, Kirchhausen TL, Whelan SP. PLoS Pathog. 2009 Apr;5(4):e1000394. doi: https://doi.org/10.1371/journal.ppat.1000394. Epub 2009 Apr 24

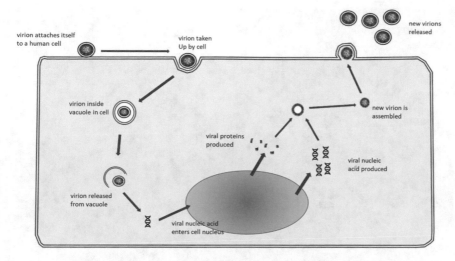

Fig. 7.7 Diagram showing the stages involved in the replication of an enveloped virus such as VZV in a human cell. This gives a general outline of how many enveloped viruses replicate within human cells – the exact details vary from virus to virus

assemble to form a new virion which then causes the membrane to produce a bud that closes round it. The bud then separates from the host cell as a new virion. After 4–6 days the VZV enters the bloodstream – this is known as a viraemia. It's carried by the blood to organs such as the liver, spleen and nerve cell ganglia (Box 7.2) where it once again replicates in cells of these organs.

Box 7.2 Neurons and Ganglia

A neuron (nerve cell) is a rather oddly shaped cell that has a large star-shaped central "cell body" containing the nucleus. The cell body has many thin branches known as dendrites which receive incoming signals from other nerve cells via junctions known as synapses. The cell body also has a long, thin projection called an axon (protected by a waxy sheath) which ends in several thin terminals which connect with another neuron, a muscle or a gland. A signal is received by the dendrites, passes along the axon and is then transmitted to another neuron (or a muscle or gland) via the axon terminals.

(continued)

Box 7.2 (continued)

Illustration showing the main features of a nerve cell. A line of three nerve cells is shown, and these connect to one another other via a synapse. (Credit: Bill McConkey. Attribution 4.0 International (CC BY 4.0))
    The cell bodies of neurons are often clustered together to form a structure known as a ganglion which is usually located outside of the brain and spinal cord.

While in the ganglia the virus can enter a dormant (latent) state and remain like that indefinitely – virions have been detected in as many as 4% of the body's nerve cells. The dormant phase of VZV occurs mainly in the dorsal root ganglia (situated along the spine) and the cranial nerve ganglia – particularly the trigeminal ganglion which is inside the skull.

After approximately 9 days, a second viraemia then occurs, and the virus is transported to the skin where it invades and kills epithelial cells. The virus causes inflammation of blood vessels in the skin, and this increases their permeability which means that blood plasma can leak out and so deliver antibodies and phagocytic cells to deal with the virus. This inflammatory response, together with the large-scale destruction of skin cells, results in the rash and blister formation that are characteristic of varicella.

How Is Chickenpox Diagnosed?

In children, the infection is usually diagnosed on the basis of the appearance and distribution of the spots and blisters, together with the occurrence of itchiness and a mild fever. Laboratory tests aren't usually necessary. However in adults, pregnant women and high-risk patients, confirmation of the disease and testing to see their level of immunity may be necessary. PCR (see Chap. 1) can be used to detect VZV DNA in samples of blister fluid, and the level of antibodies to VZV can be determined by analysis of a blood sample.

How Is Chickenpox Treated?

Anti-viral agents aren't usually needed for treating children with the disease. They should be given paracetamol to relieve pain but not non-steroidal anti-inflammatory drugs such as ibuprofen as these increase the risk of complications. To relieve itching, the child can be given antihistamines, or calamine lotion can be applied to the lesions. It's important that the child refrains from scratching so as to avoid secondary infection and possible scarring of the skin. The fingernails, therefore, should be kept short and clean, and the hands should be covered at night to prevent scratching during sleep.

Anti-viral agents such as acyclovir should be considered for those patients who have severe symptoms, complications or a moderate or high risk of developing a severe infection – these at-risk groups have been listed previously.

How Can I Avoid Getting Chickenpox?

Although a child is most infectious 1–2 days before developing the rash, the risk of spreading the disease can be reduced by keeping infected children away from nursery or school for at least 5 days. Adults shouldn't go back to work until after the last blister has burst, crusted over and the fluid has dried up, which is usually about 5–6 days after the appearance of the rash. If you have chickenpox, you should stay away from vulnerable people including pregnant women, newborn babies and anyone who has a weak immune system such as those who are having cancer therapy or who are taking steroids. VZV isn't a particularly hardy virus, but it can survive in the environment, especially in body fluids and other secretions, and so can be transmitted to others via contaminated surfaces, objects, clothing, bedding, etc. These should be disinfected where possible (with 1% sodium hypochlorite or 70% ethanol) or washed before being used by others.

Vaccination of children against the disease has been shown to be extremely effective, and a large number of countries, including the USA, have vaccination programmes. In those countries such as the UK, and many other European countries, who don't routinely vaccinate children, the vaccine is given to protect people who are most at risk of a serious chickenpox infection as well as to those who could pass the infection on to someone who is at risk. These include (i) healthcare workers who aren't immune such as a nurse who has never had chickenpox, (ii) those who are living with someone who has a weakened immune system such as the child of a parent receiving chemotherapy and (iii) those with occupational exposure to children or military recruits. The vaccine isn't suitable for pregnant women, neonates and those who are immunocompromised.

For those at risk of serious complications who are exposed to the infection and who haven't previously had varicella themselves, an alternative treatment is available. It involves giving them antibodies against the virus by an injection into a muscle – this is known as "passive immunoprophylaxis". This can be given following exposure to an infected person or during an epidemic. It provides immediate, but only short term, immunity and so reduces the risk of developing an infection after recent exposure to the virus. The patient is injected with human plasma collected from those blood donors who have high levels of antibodies against VZV. This is useful for a variety of situations including pregnant women who have no immunity, immunocompromised patients and young babies.

Want to Know More?

American Academy of Pediatrics https://www.healthychildren.org/English/health-issues/vaccine-preventable-diseases/Pages/VaricellaChickenPox.aspx?gclid=EAIaI QobChMIkaPO1rXa5QIVzbTtCh0fZgWKEAAYAiAAEgJdmvD_BwE

Centers for Disease Control and Prevention, USA https://www.cdc.gov/chickenpox/index.html

DermNet, New Zealand https://www.dermnetnz.org/topics/chickenpox/

Family Doctor, USA https://familydoctor.org/condition/chickenpox/

Patient Info, UK https://patient.info/doctor/chickenpox-pro

Royal College of Obstetricians and Gynaecologists, UK https://www.rcog.org.uk/globalassets/documents/patients/patient-information-leaflets/pregnancy/pi-chickenpox-and-pregnancy.pdf

KidsHealth, USA https://kidshealth.org/en/parents/chicken-pox.html

Mayo Clinic, USA https://www.mayoclinic.org/diseases-conditions/chickenpox/symptoms-causes/syc-20351282

National Foundation for Infectious Diseases, USA https://www.nfid.org/infectious-diseases/chickenpox-varicella-2/

National Health Service, UK https://www.nhs.uk/conditions/chickenpox/ https://www.nhsinform.scot/illnesses-and-conditions/infections-and-poisoning/chickenpox

National Institute for Clinical Care and Excellence (NICE), UK. Chickenpox, 2018 https://cks.nice.org.uk/chickenpox#!topicSummary

Ayoade F, Kumar S. Varicella Zoster (Chickenpox). Treasure Island (FL): StatPearls Publishing LLC; 2019 https://www.ncbi.nlm.nih.gov/books/NBK448191/

Bridger NA. School, child care and camp exclusion policies for chickenpox: A rational approach. *Paediatrics and Child Health.* 2018 Sep;23(6):420–427. https://doi.org/10.1093/pch/pxy096. Epub 2018 Aug 16. PMID: 30919830

Cohen J, Breuer J. Chickenpox: treatment. *BMJ Clinical Evidence.* 2015 Jun 15;2015. pii: 0912. PMID: 26077272

Gershon AA. Is chickenpox so bad, what do we know about immunity to varicella zoster virus, and what does it tell us about the future? *Journal of Infection.* 2017 Jun;74 Suppl 1:S27-S33. https://doi.org/10.1016/S0163-4453(17)30188-3. PMID: 28646959

Papadopoulos AJ. Chickenpox. Medscape from WebMD, 2018 https://emedicine.medscape.com/article/1131785-overview

Riordan A. Acute varicella-zoster. BMJ Best Practice, BMJ Publishing Group, 2020 https://bestpractice.bmj.com/topics/en-gb/603

Wutzler P, Bonanni P, Burgess M, Gershon A, Sáfadi MA, Casabona G. Varicella vaccination - the global experience. *Expert Review of Vaccines.* 2017 Aug;16(8):833–843. https://doi.org/10.1080/14760584.2017.1343669. Epub 2017 Jul 13.

8

Shingles

Abstract Shingles (herpes zoster) is a condition that arises as a result of the reactivation of the varicella zoster virus (VZV) following a previous chicken-pox infection. The re-activated virus then infects the region of the skin that is supplied by the nerve it's been inhabiting resulting in a painful rash and itchy blisters. Pain during, and prior to, the rash is common, and other symptoms can include fever, headache, malaise or fatigue. The whole episode usually lasts 2–4 weeks. Complications of shingles can arise, and these are more likely in the elderly or in those who are immunocompromised. The most common is postherpetic neuralgia which affects 20–70% of shingles sufferers over the age of 50 years and can last for months or years. Up to 20% of cases affect the eye resulting in a condition known as "herpes zoster ophthalmicus" that can cause ulceration and permanent scarring of the cornea and glaucoma. Treatment of shingles is usually focussed on relieving the pain and itching that accompany the disease. However, anti-viral agents such as acyclovir can be given to sufferers in at-risk groups, including those over the age of 50 years, or those with significant symptoms. Anti-viral drugs are most effective if given within 72 h of the rash starting. Those with shingles can transmit VZV to others, and this can cause chickenpox in those without immunity to the virus. Effective vaccines against shingles are available and are administered to those over 50 years of age in many countries.

© Springer Nature Switzerland AG 2021
M. Wilson, P. J. K. Wilson, *Close Encounters of the Microbial Kind*,
https://doi.org/10.1007/978-3-030-56978-5_8

What Is Shingles?

Shingles, also known as herpes zoster (HZ), is a disease caused by reactivation of the varicella zoster virus (VZV) following a previous episode of chickenpox which may have occurred many years earlier. Shingles comes from the Latin word "cingulum" which means a girdle – this refers to the typical pattern of the rash which often forms a band around one side of the abdomen (Fig. 8.1). Zoster is a Greek word that also means a girdle or warrior's belt.

Following chickenpox infection, the VZV can reside for a long time in some nerve cells in a dormant state, when it is reactivated, it gives rise to an infection in the nerve and the skin this nerve supplies (Fig. 8.2). The location of the skin rash will depend on which particular nerve has been affected.

The characteristic symptoms are a painful rash that develops into itchy blisters, similar to chickenpox (Fig. 8.3). The blisters become yellowish after a few days then flatten and dry out. Scabs form where the blisters were, and this can result in slight scarring. New blisters may continue to develop for up to a week after the first appearance of the rash.

Before the rash appears, other symptoms you may experience include (i) a headache, (ii) burning, tingling, numbness or itchiness of the skin in the

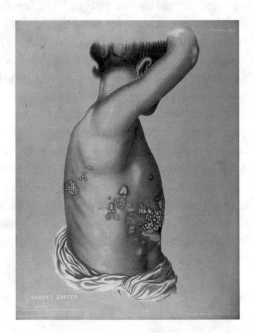

Fig. 8.1 Diseased skin and blisters on the torso of a young man suffering from shingles showing a characteristic band of spots on the abdomen. Chromolithograph by E. Burgess, 1850/1880. (Credit: Wellcome Collection. CC BY 4.0)

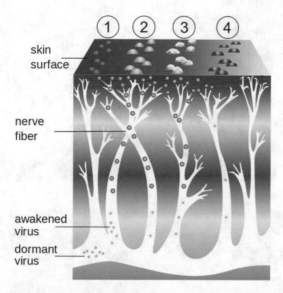

Fig. 8.2 Shingles is due to the reactivation of the varicella zoster virus which has been in a latent state within a nerve ganglion. The disease progresses through several stages. A cluster of small bumps (1) turns into blisters (2) that resemble chickenpox lesions. The blisters fill with pus, break open (3), crust over (4) and finally disappear. (Renee Gordon. Unless otherwise noted, the contents of the Food and Drug Administration website (www.fda.gov) – both text and graphics – are public domain in the USA. [1] (August 18, 2005, last updated July 14, 2015). Via Wikimedia Commons)

affected area, (iii) a feeling of being generally unwell and (iv) a fever. The rash can affect any part of your body (depending on which nerve has been infected) although the chest and abdomen are the most common areas. It's usually accompanied by a pain that is described as burning, stabbing or throbbing. The intensity of the pain can vary from mild to severe, and the affected area of skin is usually tender. Pain is less common in young healthy patients and is rare in children. It usually starts a few days before the rash appears and can remain for a few days or weeks after the rash has healed. In approximately 20% of patients, additional symptoms occur such as fever, headache, malaise or fatigue. An episode of shingles usually lasts 2–4 weeks. Although 6–14% of patients can suffer from a second episode of shingles, it's not common to experience it more than twice.

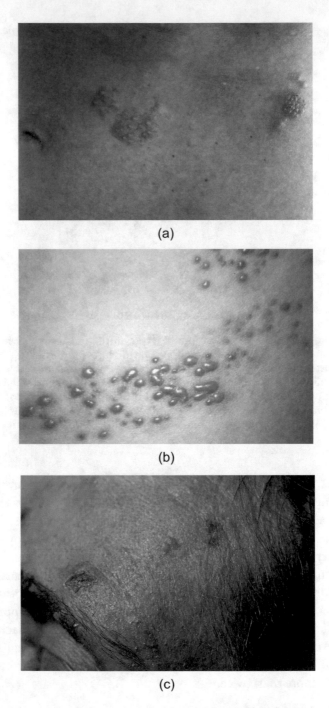

(a)

(b)

(c)

Fig. 8.3 Development of the rash associated with shingles. (a) Early shingles rash showing the start of blistering. (NIAID / CC BY (https://creativecommons.org/licenses/by/2.0)). (b) At this stage the rash has developed into blisters. (K.L. Herrmann, Centers for Disease Control and Prevention, USA). (c) Shingles rash on the forehead. This illustrates the late stage of the rash in which most of the blisters have burst and scabs have formed. (Heinz F. Eichenwald, MD, Centers for Disease Control and Prevention, USA)

What Are the Complications?

Complications of shingles arise most often in the elderly or in those who are immunocompromised. The most common complication is severe nerve pain (neuralgia) and intense itching affecting the area where the rash has been. This condition is known as "postherpetic neuralgia" and affects approximately 20% of shingles patients who are over the age of 50 years and as many as 70% of those over 60 years of age. The types of pain experienced include (i) constant or intermittent burning, aching, throbbing, stabbing or shooting pain; (ii) allodynia, pain triggered by something that would not usually be painful such as changes in temperature or light touch; and (iii) hyperalgesia, an enhanced response to pain. It may resolve after 3–6 months but can last for several years – 50% of patients experience it for more than 1 year. It can be treated with conventional painkillers, but these can be less effective for severe symptoms, and sometimes specialist nerve painkillers (neuropathic analgesics) are required. Sometimes specialist advice from Pain Teams or Pain Management Clinics is necessary.

Another possible complication is an eye condition known as "herpes zoster ophthalmicus" (Fig. 8.4). This affects 10–20% of patients with shingles and can cause ulceration and permanent scarring of the cornea, inflammation of the eye and optic nerve and glaucoma, i.e. an increase in pressure inside the eye. Anti-viral agents such as acyclovir are essential for the treatment of this condition.

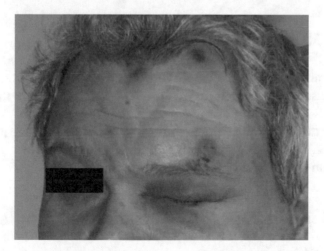

Fig. 8.4 A patient suffering from herpes zoster ophthalmicus. (StromBer 22:24, 15. Mär. 2008 (CET) / Public domain via Wikimedia Commons)

Infection of the rash by bacteria such as *Staph. aureus* and *Strep. pyogenes* may occur (in approximately 1.1% of cases), and these should usually be treated with antibiotics. Rarely, serious complications can develop including pneumonia, hepatitis, meningitis or encephalitis.

What Are the Risk Factors for Shingles?

The main risk factors for shingles are (i) being more than 50 years of age; (ii) suffering from HIV infection, the incidence of shingles is 15 times greater in HIV positive patients; and (iii) immunosuppression associated with various conditions such as long-term steroid treatment, chemotherapy, organ transplantation and physical and emotional stress.

How Common Is Shingles?

In the UK the annual incidence of shingles is 3.4 cases per 1000 population although this varies significantly with age. In people less than 50 years of age, the incidence is <2 cases per 1000, whereas in those >80 years it's 11 cases per 1000. Approximately 1 in 4 people in the UK will suffer from shingles during their lifetime. In the USA the annual incidence is 4 per 1000 population, and this results in approximately one million cases of shingles each year. Approximately 1 in 3 people in the USA will suffer from shingles during their lifetime.

How Is Shingles Diagnosed?

Shingles is almost always diagnosed on the basis of the characteristic symptoms. Because of the way that nerves are distributed in the body, the rash only affects one side of the body at any one time, this is a very helpful pattern when looking for causes of a blistering rash. Further laboratory tests are rarely necessary, but VZV DNA can be detected by PCR in samples taken from blister fluid or from affected skin.

What Happens During an Episode of Shingles?

Although it's known that shingles is due to the reactivation of dormant VZV in the nerve ganglia, the triggers are poorly understood. The most likely factors are a decrease in the effectiveness of the immune system with age or the immunosuppression associated with various conditions as listed previously. Once the virus has been reactivated, it spreads from the ganglion (a cluster of nerve cells) to the associated nervous tissue and skin. This results in inflammation and destruction of the nerve cells and surrounding tissue and also of the skin tissue of the region (known as a "dermatome") that's served by those nerves.

How Is Shingles Treated?

Shingles is usually a self-limiting condition, and treatment is normally focussed on relieving the symptoms of the disease. For mild pain, you can take analgesics such as paracetamol and ibuprofen, but for severe pain you may need opioid analgesics such as codeine. Sometimes steroids are used but only in combination with antivirals. Anti-inflammatory drugs should be avoided in children as they increase complications of the rash.

Other helpful measures include (i) keeping the rash as clean and dry as possible, this reduces the risk of secondary bacterial infections; (ii) wearing loose-fitting clothing, this reduces discomfort; (iii) using a non-adherent dressing to cover the blisters, this reduces the risk of virus transmission to others; (iv) applying calamine lotion, this has a soothing, cooling effect and relieves itching; and (v) applying a cool compress to weeping blisters for 20 minutes several times a day, this soothes the skin and keeps blisters clean.

Anti-viral agents such as aciclovir, valaciclovir or famciclovir are a useful treatment option especially for those who (i) are over 50 years of age, (ii) have a weakened immune system, (iii) experience moderate to severe pain, (iv) have a moderate to severe rash, (iv) have a rash that isn't on the trunk or (v) experience eye symptoms. They are most effective if started as early as possible, within 72 h of the rash appearing. These agents stop viral multiplication and reduce the severity and duration of the infection. They also help to prevent complications.

How Can I Avoid Getting Shingles?

You can only develop shingles when the VSV in your body, from a previous chickenpox infection, is re-activated. So both avoiding chickenpox infection and keeping your immune system as healthy as possible will reduce the risk of developing shingles. Vaccines are available to prevent both chickenpox and shingles, and long-term anti-viral medication can be used to prevent recurrences in at-risk people. Transmission of VZV from patients with shingles to those who aren't immune to the virus doesn't give rise to shingles but can cause chickenpox. The virus can be transmitted via fluid from the shingles rash blisters, and therefore direct body contact should be avoided. Covering the lesions also decreases the risk of transmission. To help prevent the spread of the virus, patients with shingles shouldn't share towels or flannels and should avoid swimming and contact sports. They should also avoid work or college while they have oozing blisters that can't be covered.

A vaccine consisting of a live, attenuated VZV (Zostavax®) has been shown to be effective in preventing shingles (51% efficacy) as well as postherpetic neuralgia in those >50 years of age. However, because it's a live vaccine, it shouldn't be used in immunosuppressed people, pregnant women, or children. More recently (in 2017) a new VZV vaccine (Shingrix®) was approved by the US Food and Drug Administration for adults aged ≥50 years. This isn't a live vaccine and is based on one of the proteins of VZV. It's been shown to be 97% effective at preventing shingles.

Want to Know More?

American Academy of Dermatology https://www.aad.org/diseases/a-z/shingles-treatment

British Skin Foundation https://www.britishskinfoundation.org.uk/shingles-herpes-zoster?gclid=EAIaIQobChMI7Yu6jsXa5QIVh7HtCh04oggsEAAYASAAEgIPHfD_BwE

Centers for Disease Control and Prevention, USA https://www.cdc.gov/shingles/index.html

DermNet, New Zealand https://www.dermnetnz.org/topics/herpes-zoster/

Family Doctor, USA https://familydoctor.org/condition/shingles/

Health Service Executive, Ireland https://www.hse.ie/eng/health/az/s/shingles/

Herpes Viruses Association, UK https://herpes.org.uk/shingles-support-society/

Mayo Clinic, USA https://www.mayoclinic.org/diseases-conditions/shingles/symptoms-causes/syc-20353054

National Health Service, UK https://www.nhs.uk/conditions/shingles/ https://www.nhsinform.scot/illnesses-and-conditions/infections-and-poisoning/shingles

National Foundation for Infectious Diseases, USA https://www.nfid.org/infectious-diseases/shingles/

National Institute for Clinical Care and Excellence (NICE), UK. Shingles, 2019 https://cks.nice.org.uk/shingles#!topicSummary

National Institute on Aging, USA https://www.nia.nih.gov/health/shingles

Patient Info, UK https://patient.info/skin-conditions/shingles-herpes-zoster-leaflet

Canaday DH. Herpes Zoster in the older adult. John AR, *Infectious Disease Clinics of North America*. 2017 Dec;31(4):811–826. https://doi.org/10.1016/j.idc.2017.07.016.

Gater A, Uhart M, McCool R, Préaud E. The humanistic, economic and societal burden of herpes zoster in Europe: a critical review.. *BMC Public Health*. 2015 Feb 27;15:193. https://doi.org/10.1186/s12889-015-1514-y.

Janniger CK. Herpes zoster. Medscape from WebMD, 2020 https://emedicine.medscape.com/article/1132465-overview

Le P, Rothberg M. Herpes zoster infection. BMJ Best Practice, BMJ Publishing Group, 2020 https://bestpractice.bmj.com/topics/en-gb/23

Nair PA, Patel BC. Herpes Zoster (Shingles). Treasure Island (FL): StatPearls Publishing LLC; 2020 https://www.ncbi.nlm.nih.gov/books/NBK441824/

Saguil A, Kane S, Mercado M, Lauters R. Herpes Zoster and postherpetic neuralgia: prevention and management. *American Family Physician*. 2017 Nov 15;96(10):656–663.

9

Common Warts

Abstract Warts are fleshy lumps, usually on the hands and feet, which result from an infection with certain strains of the human papilloma virus. There are several types, depending on their shape, size and arrangement – common, plantar (verrucas), plane, filiform, periungal and mosaic warts. They can appear anywhere on the skin and are not usually painful although they may bleed or itch. However, warts can cause problems or be painful because of their location and, when on the sole of the foot, can impair walking. They aren't very contagious but can be transmitted by close skin-to-skin contact or indirectly from contaminated objects or surfaces such as the floors of communal changing areas. They can also spread to other parts of the body. Warts affect 7–12% of the global population but are most common among children and young adults. Because many warts disappear spontaneously, there's a strong case for not treating them, particularly in children who often can't tolerate the treatment. However, when the wart is painful or unsightly, it can be destroyed by applying salicylic acid or silver nitrate. Some warts may need to be removed professionally using cryotherapy or surgery. The risk of getting warts can be reduced by not touching, or sharing personal items with, someone who has a wart, keeping feet dry, changing socks every day, avoiding prolonged immersion in water and refraining from nail biting.

© Springer Nature Switzerland AG 2021
M. Wilson, P. J. K. Wilson, *Close Encounters of the Microbial Kind*,
https://doi.org/10.1007/978-3-030-56978-5_9

What Are Warts?

Warts (or "verrucae vulgaris") are small skin lumps that most often appear on the hands and feet. They are usually round and skin-coloured with a rough greyish-white or light brown surface. Although they may occur anywhere on the body, they tend to occur at sites prone to trauma, such as the knees and elbows. They are caused by infection of the skin with the human papilloma virus (HPV). Warts aren't usually painful on their own but can cause problems because of their location; they can occasionally itch or bleed. There are several different types of warts, all varying in size and shape (Fig. 9.1), and the main types are summarised in Table 9.1.

Who Gets Warts?

Factors that contribute to the risk of acquiring warts include: (i) prolonged water immersion due to swimming or regular dishwashing, (ii) regular handling of meat or fish, (iii) nail biting and (iv) being immunocompromised.

Warts aren't highly contagious, but they can be transmitted by close skin-to-skin contact. The virus is more likely to infect the skin if it's wet, soft, broken or has been in contact with a rough surface. The disease can also be transmitted indirectly from contaminated objects such as towels and shoes or from surfaces such as the area surrounding a swimming pool and the floors of communal changing areas. Warts can spread from one part of your body to another by scratching, knocking or biting a wart and by biting the nails or sucking fingers affected by warts.

Although common warts generally don't cause any great discomfort, they may cause cosmetic disfigurement, tenderness or problems depending on their location. Plantar warts can be painful if they rub on shoes or are on weight-bearing areas and, when on the sole of the foot, can even affect walking if they are extensive.

Approximately 23% of warts will resolve without treatment within 2 months, 30% within 3 months and 65–78% within 2 years. When this happens, no scarring is seen whereas treatment can sometimes leave a visible scar.

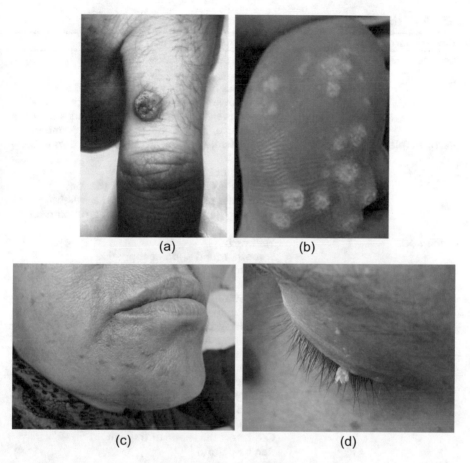

Fig. 9.1 Appearance of various types of warts. (**a**) Common wart on a finger. (Richard S. Hibbets, Centers for Disease Control and Prevention, USA). (**b**) A group of plantar warts on the toe. (CNX OpenStax / CC BY (https://creativecommons.org/licenses/by/4.0)). (**c**) Plane warts on the face. (Iffat Hassan, Taseer Bhat, Hinah Altaf, Farah Sameem, Qazi Masood / CC BY (https://creativecommons.org/licenses/by/4.0)). (**d**) A filiform wart on the eyelid. (Schweintechnik. I, the copyright holder of this work, release this work into the public domain. This applies worldwide. Via Wikimedia Commons)

How Common Are Warts?

Warts are widespread globally and are estimated to affect 7–12% of the population. Generally, they have a higher prevalence among children aged 5–10 years and young adults. In school-aged children, their prevalence has been estimated to be 5–30%.

Although they may affect any race, common warts are found approximately twice as frequently on Caucasian compared with black or Asian skin. Butchers

Table 9.1 The appearance and location of the main types of wart

Type of wart	Description and location
Common warts	Usually round or oval-shaped Firm and raised Has a rough, bumpy surface similar to a cauliflower Often develop on the knuckles, fingers and knees Vary in size, from <1 mm to >1 cm in diameter
Verrucas (plantar warts)	Usually white, often with small black dots in the wart Usually flat rather than raised Usually develop on the soles of the feet Sometimes painful if they are on a weight-bearing part of the foot or when rubbing on shoes
Plane warts	Yellowish Smooth, round and flat-topped 2–4 mm in diameter Several hundred warts may be present, sometimes develop in clusters Common in young children – mainly affecting the hands, face and legs Sometimes develop on the lower legs of women following shaving
Filiform warts (verruca filiformis)	Long and slender in appearance Often develop on the neck or face
Periungual warts	Develop under and around the fingernails or toenails Have a rough surface Can affect the shape of the nail Can be painful
Mosaic warts	Grow in clusters Form a tile-like pattern Often develop on the palms of the hands and soles of the feet

and meat-handlers have an increased risk of warts – approximately 50% are affected. They're also more prevalent in immunosuppressed individuals such as HIV patients, transplant recipients and patients receiving chemotherapy.

Box 9.1 "Warts and All"

The phrase "warts and all" is attributed to Oliver Cromwell who was about to have his portrait painted by Sir Peter Lely. Lely had painted the recently executed king of England, Charles I. His painting style was designed to flatter the person being portrayed – this was usual for that time. However, Cromwell, being a puritan, was opposed to all forms of personal vanity and said to him "Mr Lely, I desire you would use all your skill to paint my picture truly like me, and not flatter me at all; but remark all these roughnesses, pimples, warts and everything as you see me, otherwise I will never pay a farthing for it". The resulting portrait is not very flattering but appears to be a true likeness as it strongly resembles Cromwell's death mask.

Portrait of Oliver Cromwell by Sir Peter Lely (1654). (This work is in the public domain in its country of origin and other countries and areas where the copyright term is the author's life plus 100 years or fewer. Via Wikimedia Commons)

Are There Any Complications?

HPV skin infections in healthy people are usually only a nuisance and of minor concern; however in people with poor immunity, they can cause much more extensive symptoms that are sometimes resistant to treatment. Some strains can cause infections in places other than the skin, for instance, in the mouth, lungs or oesophagus. Certain HPV strains are carcinogenic and are a risk factor for developing squamous cell carcinoma (SCC). SCC is a type of cancer that makes squamous cells (the thin, flat cells of the epidermis) multiply uncontrollably. It can affect the skin, mouth, throat, larynx, nose, lungs, anus, cervix and vagina. HPV infection can also be passed from a mother to her baby while it's in the womb or from the birth canal during delivery. If this happens, it can result in warty growths in the baby's larynx that can obstruct its small airway.

What Causes Warts?

Warts are caused by certain strains of HPV which is a small, non-enveloped DNA-containing virus. Its DNA is contained within an icosahedral-shaped capsid made of two different proteins and this has a diameter of approximately 55 nm (Fig. 9.2).

There are more than 180 strains of HPV, but those most frequently responsible for warts are types 1, 2, 4, 27, 57 and 63. Different strains cause different problems (see Chap. 30) and they don't always produce symptoms. In butchers, HPV 7 is the most frequent cause although types 2, 4 and 27 may also be responsible. The main targets of HPV are basal keratinocytes, i.e.

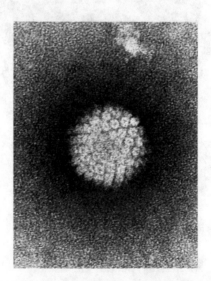

Fig. 9.2 Electron micrograph of the human papilloma virus. (Laboratory of Tumor Virus Biology, This image is a work of the National Institutes of Health, part of the US Department of Health and Human Services. As a work of the US federal government, the image is in the public domain. Via Wikimedia Commons)

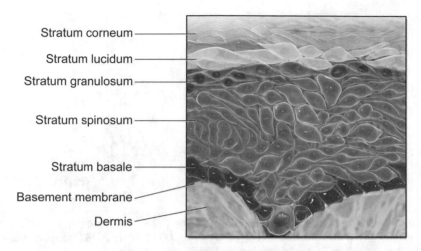

Fig. 9.3 The various layers of the human epidermis. (Blausen.com staff (2014). "Medical gallery of Blausen Medical 2014". *WikiJournal of Medicine* 1 (2). DOI: https://doi.org/10.15347/wjm/2014.010. ISSN 2002-4436. [CC BY 3.0 (https://creativecommons.org/licenses/by/3.0)])

keratinocytes that comprise the innermost layer of the epidermis (stratum basale) and are undergoing rapid cell division (Fig. 9.3).

The newly-formed keratinocytes are then pushed slowly towards the surface of the skin, and as they do so, they gradually die and produce keratin until eventually they form the dead, outer layer of the epidermis – the stratum corneum (Fig. 9.3). Because the cells of the stratum basale are protected by several layers of cells, the virus can only reach them following some form of damage to the epidermis. Once it has reached the basal layer, it invades a keratinocyte and then generally doesn't undergo replication for some time – between 3 weeks and 9 months. Invasion and replication occur in a manner similar to that described for other non-enveloped viruses (see Fig. 10.4 of Chap. 10). The virus then makes the keratinocyte undergo cell division, and, as the newly-formed keratinocyte (containing HPV) is propelled towards the outer layers of the skin, the virus is replicated. Cells infected with HPV have a different appearance (a larger nucleus and a vacuole) from normal keratinocytes and are known as koilocytes (Fig. 9.4).

HPV are released from koilocytes when they reach the skin surface. However, in some cases, the HPV-containing koilocyte is shed from the skin surface and can be transferred to another individual and initiate an infection once the viruses are released. The whole process from initial infection to HPV release from the stratum corneum takes about 3 weeks. The immune response initiated by HPV is not very effective at dealing with the virus, and it's

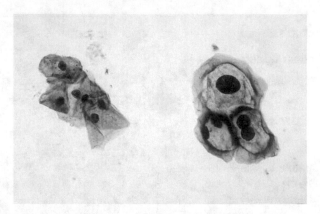

Fig. 9.4 A group of koilocytes are shown on the right as well as a group of normal cells on the left. (Photomicrograph by Ed Uthman, MD. 20 July 2006 Euthman 20:24, 29 November 2006 (UTC) / Public domain via Wikimedia Commons)

estimated that only 10–30% of infections with HPV resolve within 3 months of their onset without any treatment.

How Are Warts Diagnosed?

The infection is diagnosed on the basis of the appearance of the warts. Sometimes the fleshy lump might be confused with other causes of fleshy lumps such as a corn, if so the surface layers can be scraped off. A wart usually has characteristic brown or black dots within it. These are tiny blood vessels from the skin that have clotted; the dark colour is from the blood within them.

Box 9.2 What's in a Name?

Papilloma is a word derived from the Latin "papilla" which means a nipple in combination with the Greek "oma" which is a suffix that is used to form a noun to indicate a tumour (which may be malignant or benign). Papilloma is, therefore, a tumour that resembles a nipple.

The virus was first detected in rabbits in 1933 by Richard Shope and Weston Hurst. Working at the Rockefeller Institute for Medical Research in the USA, they showed that the virus was able to cause warts and carcinomas in rabbits. In 1983 the German virologist Harald zur Hausen showed that HPV was able to cause cervical cancer – he was awarded the Nobel Prize for this discovery in 2008.

There are more than 180 different strains of HPV, and these are referred to as "types" and denoted by a number – for example, human papilloma virus type 1 is abbreviated to HPV 1.

How Are Warts Treated?

Because many common warts resolve spontaneously, there's a strong case for not treating the disease. The option of watchful waiting is a reasonable approach to take and is often the best option in young children because they may not be able to tolerate treatment.

The average lifespan of a wart is 2 years, and up to 65% will resolve on their own within this timeframe. However, when the wart is painful (e.g. on the soles of the feet or near the nails) or is cosmetically unsightly, then treatment can be attempted. The main purpose of treatment is to remove the wart. Unfortunately the most commonly available treatments can take time and dedication; they are not always guaranteed to work and some can cause adverse effects. Even after the wart itself is removed, they can recur because the virus is still present. Treatment is more effective in recently occurring warts. Several specialist options are either in use or are under development.

The usual treatment for common warts is daily application of a chemical that destroys the wart, coupled with regular filing of the wart to reduce its size. There are many different options including salicylic acid-containing compounds which are keratolytic, which means they dissolve the keratin protein that sticks skin cells together. These can have a cure rate as high as 75%. When using them, it's important to protect the normal skin surrounding the wart.

Alternatively, you could apply silver nitrate which is available in different forms such as a solution or as an applicator stick. Silver nitrate acts as an antiseptic as well as destroying wart tissue. Another approach is to simply cover the wart with impermeable adhesive tape (such as duct tape) for at least 6 weeks. This simple measure is very effective – the success rate being 85% after 8 weeks. It's not clear why this works, but it may be because the wart is being deprived of oxygen and nutrients. Cryotherapy (see Chap. 29) has been found to be the most effective treatment although this has to be carried out by a medical practitioner, usually needs several applications, has some risks of adverse effects (such as scarring) and can be painful.

Some cases will require specialist input. These include facial warts, resistant cases, when large areas are affected and in those with reduced immunity. Some specialist approaches include using retinoids, formaldehyde or immunotherapy. The latter involves using imiquimod or diphenylcyclopropenone which activate the immune system to remove the abnormal, HPV-infected cells. Surgical approaches are available if medical treatments fail. These include **curettage** and different types of lasers as well as photodynamic therapy (see Chap. 3).

How Can I Avoid Getting Warts?

It's extremely difficult to avoid infection with HPV, but there are a number of precautions you can take to help reduce the risk, and these include (i) avoiding prolonged contact with affected skin; (ii) not sharing towels, flannels or other personal items with someone who has a wart; (iii) not sharing shoes or socks with someone who has a verruca; (iv) keeping your feet dry and changing your socks every day; (v) avoiding prolonged water immersion such as swimming and dishwashing; (vi) not handling meat or fish regularly; and (vii) not biting your nails.

If you have a wart or verruca, you should cover it up when taking part in communal activities as this will help to reduce passing on HPV to others.

Want to Know More?

American Academy of Dermatology https://www.aad.org/diseases/a-z/warts-overview
British Skin Foundation https://www.britishskinfoundation.org.uk/plantar-warts-verrucas
DermNet, New Zealand https://www.dermnetnz.org/topics/viral-wart/
Health Service Executive, Ireland https://www.hse.ie/eng/health/az/v/verrucas-and-warts/
Mayo Clinic, USA https://www.mayoclinic.org/diseases-conditions/common-warts/symptoms-causes/syc-20371125
National Health Service, UK https://www.nhs.uk/conditions/warts-and-verrucas/ https://www.nhsinform.scot/illnesses-and-conditions/skin-hair-and-nails/warts-and-verrucas
National Institute for Clinical Care and Excellence (NICE), UK. Warts and verrucae, 2020 https://cks.nice.org.uk/warts-and-verrucae#!topicSummary
Patient Info, UK https://patient.info/skin-conditions/warts-and-verrucas-leaflet
Primary Care Dermatology Society, UK http://www.pcds.org.uk/clinical-guidance/warts
Leung L. Recalcitrant nongenital warts. *Australian Family Physician*. 2011 Jan-Feb;40(1–2):40–2.
Lipke MM. An armamentarium of wart treatments. *Clinical Medicine and Research*. 2006 Dec;4(4):273–93.
Loo SK, Tang WY. Warts (non-genital). *BMJ Clinical Evidence*. 2014 Jun 12;2014. pii: 1710
Luria L, Cardoza-Favarato G. Human Papillomavirus. Treasure Island (FL): StatPearls Publishing LLC; 2020 https://www.ncbi.nlm.nih.gov/books/NBK448132/
Moore AY. Common warts. BMJ Best Practice, BMJ Publishing Group, 2019 https://bestpractice.bmj.com/topics/en-gb/615
Shenefelt PD. Nongenital warts. Medscape from WebMD, 2018 https://emedicine.medscape.com/article/1133317-overview

Part III

Infections of the Respiratory System

10

The Common Cold

Abstract The common cold is a highly contagious viral infection of the upper respiratory tract that occurs mainly during the winter in temperate climates. It causes sneezing, headache, sore throat and tiredness but usually lasts no longer than 10 days. Rhinoviruses are the main cause of the disease although a number of other viruses may be responsible including coronaviruses. A cold is sometimes followed by a bacterial infection of some region of the respiratory tract – this is particularly so in children. Currently there are few effective anti-viral agents against the viruses responsible, treatment is aimed at relief of the disease symptoms, but many widely used medications are of doubtful benefit. No vaccines are available for preventing colds, but hand washing as well as covering the mouth and nose when sneezing or coughing help to reduce the spread of the disease. Zinc supplements and exercise may help to protect against the infection.

A slight tickling at the back of your throat. Forget it, nothing to worry about. But then your head starts to ache and your nose starts to run – will it ever stop? You go through endless paper handkerchiefs, and your nose gets so sore it becomes red and even starts to throb. And the sneezing! Horrendous. I was at a barbecue last summer, and my host had an awful cold – he couldn't stop sneezing. I started off being very sensible and kept out of his way as much as possible, without making it too obvious. But, as the party warmed up and after I'd had a few drinks, I threw caution to the wind and sat next to him. What a mistake. I remember him taking a huge bite of hamburger, turning towards me and then sneezing! I was sprayed with partially chewed hamburger – and I'm a vegetarian! Needless to say, the next day I developed a cold.

The Common Cold: What Is It?

The common cold is a viral infection primarily of the nose although other regions of the upper respiratory tract (Fig. 10.1) such as the pharynx, larynx and sinuses may also be affected. It's short-lived and self-limiting and usually lasts no longer than 10 days. Other names for the condition include acute viral rhinopharyngitis, viral rhinitis and acute coryza.

The main symptoms of a cold are a runny nose (rhinorrhea), sneezing, headache and a general feeling of tiredness or malaise. Additional symptoms may include a sore throat (50% of cases) and a cough (40% of cases). Fever is another common symptom in children but is rare in adults. The main symptoms usually peak 1–3 days following the start of the infection and last around 7–10 days, although they occasionally linger for up to 3 weeks. You can pass on the viruses to other people 1–2 days before developing any signs of the illness. The severity and duration of the symptoms vary considerably due to differing susceptibility among people and which of the many possible viruses

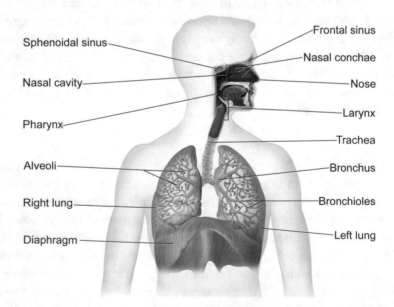

Fig. 10.1 Diagram showing the main regions of the respiratory system. (Blausen.com staff (2014). "Medical gallery of Blausen Medical 2014". *WikiJournal of Medicine* 1 (2). https://doi.org/10.15347/wjm/2014.010. ISSN 2002-4436)

is responsible. The incidence decreases with increasing age – children under 2 years experience between 6 and 8 infections a year, adults two to four infections while older people about one per year. It's the most common acute illness in the USA, being responsible for approximately 37 million (3%) ambulatory care visits each year. A study carried out in the USA in mid-October has shown that 23.6% of adults had experienced a cold in the previous 4 weeks. In temperate climates most infections occur during the winter. Neither sex nor ethnicity appear to affect susceptibility to the infection. Stress, smoking and lack of sleep increase the risk of contracting the infection in adults, whereas attendance at a day care centre increases the risk among preschool children.

The disease is highly contagious and can therefore spread rapidly where people gather together – this can result in considerable absenteeism from school and work. The viruses responsible for the infection can be spread by a variety of routes, the main ones being hand-to-hand contact, inhalation of droplets or aerosols (Fig. 10.2) and via contaminated surfaces from which the virus is transferred to the respiratory tract or mouth. The most important route is likely to involve hand-to-hand contact followed by transfer of the virus to the nostrils or eyes. There's little evidence to support the popular belief that colds arise because of exposure to low temperatures. However, low temperatures may reduce the ability of the nasal defence systems to protect us from viruses.

Fig. 10.2 Sneezing is an excellent way of dispersing virus-laden droplets and aerosols from the respiratory tract. (James Gathany and Brian Judd, Centers for Disease Control and Prevention, USA)

Which Viruses Are Responsible for the Infection?

A variety of viruses (more than 200 types) are responsible for colds, but because of the short duration and mild nature of the disease, the infecting agent hasn't been determined in most cases. When the infecting agent has been identified, rhinoviruses were found to be responsible for 52–76% of cases. Other known causes include coronavirus (10–15% of cases), influenza virus (5–15%), parainfluenza virus (5%), respiratory syncytial virus (5%), metapneumovirus, adenoviruses and enteroviruses.

Box 10.1 What's in a Name?

The word "cold" is derived from Saxon word "Kald". In the sixteenth century, it came to be used as the name for this common respiratory tract infection because the symptoms are similar to those experienced when we're exposed to low temperatures.

The rhinovirus is named after the body site where it causes problems – the nose. It's name is derived from the Greek word "rhinos" which means "of the nose" . The name of the coronavirus comes from the Latin word "corona" which means "crown or halo" and refers to its crown-like appearance when viewed through an electron microscope.

The respiratory syncytial virus gets its name because of the effect it has on human cells. It makes them fuse together to form a giant cell containing many nuclei – this is known as a syncytium.

Let's take a closer look at rhinoviruses, the most frequent cause of colds. They're very small, approximately 30 nm in diameter, and so are approximately 1/33 the diameter of a typical spherical bacterium such as *Staph. aureus*. They're therefore very small microbes. In comparison, human cells are gigantic. A cell from the respiratory epithelium (with which the rhinovirus interacts during an infection) is approximately 330 times wider than a rhinovirus particle. The structure of the virion is very simple and consists of a single molecule of RNA (carrying the genetic information of the virus) surrounded by a coat (known as a capsid) composed of three different proteins – a fourth protein anchors the RNA to the capsid. The overall shape of the particle is that of an icosahedron, i.e. it has 20 sides (Fig. 10.3).

The rhinovirus is described as being a "non-enveloped" virus because the virion is not enclosed within a membrane derived from the cell in which it was produced. There are more than 100 different types of rhinoviruses, and these are quite hardy – they can survive in the indoor environment for days and on the human skin for 2 hours. Once the virus has reached the respiratory

Fig. 10.3 Model of the outer coat of a rhinovirus particle which is composed of three different proteins. Inside this protein coat (known as a "capsid") is a single molecule of RNA anchored to the capsid by a fourth protein. (Image courtesy of David S. Goodsell, RCSB Protein Data Bank. CC BY 4.0)

tract, it invades the nasopharynx (the upper region of the throat behind the nose – Fig. 10.1), and the invaded cells are turned into virus factories (Fig. 10.4).

Newly produced virions are released with 10 hours of initial exposure to the virus, but peak release occurs after 24–48 hours. Viral invasion triggers a complicated immune response which results in the neighbouring blood vessels widening (known as vasodilation) and becoming more permeable – this brings more white blood cells and antibodies to the area to combat the viruses. However, this also results in the production of copious amounts of mucus from the cells that line the respiratory tract – this is the main symptom of a cold. For reasons that aren't clear, only approximately 75% of people infected with rhinoviruses actually develop cold symptoms, and so you can carry the virus without being obviously affected.

Coronaviruses are the second most common cause of colds. They're about 4 times larger than rhinoviruses with a diameter of approximately 125 nm and have a more complex structure. They're spherical with club-shaped spikes projecting from their surface (Fig. 10.5). Like the rhinoviruses, their genetic material is RNA rather than DNA.

Four different coronaviruses are responsible for colds, but three other types are associated with more serious respiratory infections – Middle East Respiratory Syndrome (MERS – see Box 10.2), severe acute respiratory syndrome (SARS) and coronavirus disease 2019 (see Chap. 12).

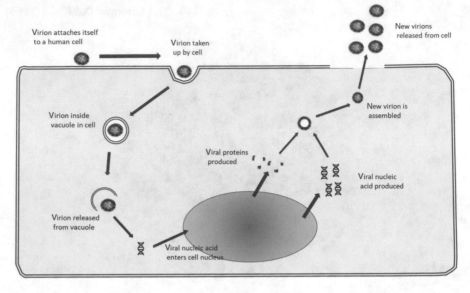

Fig. 10.4 The diagram gives a general outline of how many non-enveloped viruses replicate within human cells – the exact details vary from virus to virus. The rhinovirus induces the cell to take it inside where it becomes enclosed within a small sack-like structure (made from the cell's membrane) known as a vacuole. It's eventually released from the vacuole, and its nucleic acid enters the cell's nucleus where new viral proteins and nucleic acid are produced. The proteins and nucleic acid are then assembled into new virions which are then released from the cell. The release of virions often involves bursting (called "lysis") and death of the cell

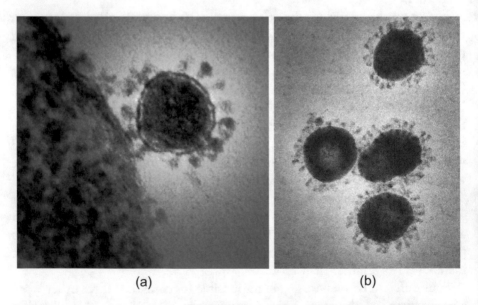

(a)　　　　　　　　　　　　　　(b)

Fig. 10.5 Images of coronaviruses. (**a**) Electron micrograph of a coronavirus attached to the surface of a human cell. The characteristic spikes projecting from the surface of the virus can be clearly seen. (Image courtesy of National Institute of Allergy and Infectious Diseases, USA). (**b**) Electron micrograph showing four coronavirus virions. (Courtesy of Dr. Fred Murphy and Sylvia Whitfield, Centers for Disease Control and Prevention, USA)

Box 10.2 Middle East Respiratory Syndrome (MERS)

MERS is a respiratory disease caused by a coronavirus known as Middle East Respiratory Syndrome Coronavirus (MERS-CoV) (Figure). It was first reported in Saudi Arabia in 2012, and cases have occurred mainly in countries in or near the Arabian Peninsula. The virus has been found in bats and camels which may be its natural reservoirs. Patients infected with the virus initially experience flu-like symptoms, but it can develop into severe pneumonia and respiratory failure. Its mortality rate is 30–40%. A total of 2458 laboratory-confirmed cases of infection with MERS-CoV have occurred since 2012, and these resulted in 848 deaths.

Very few cases have been reported in the UK or USA, and the main risk of acquiring the infection is by travelling to countries in the Arabian Peninsula. All travellers to these countries should practise good hygiene such as regular hand washing, especially after visiting farms, barns or market areas. Travellers should also:

- Avoid contact with camels.
- Avoid raw camel milk and/or camel products.
- Avoid eating or drinking any type of raw milk, raw milk products and any food that may be contaminated with animal secretions, unless it's been peeled and cleaned and/or thoroughly cooked.

The virus isn't readily transmissible between humans but can be spread via respiratory secretions.

MERS is considered to be an international threat to public health.

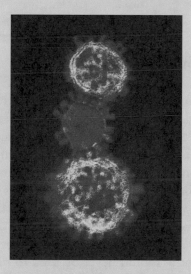

Electron micrograph showing three MERS-CoV virions. (National Institute of Allergy and Infectious Diseases, USA)

Influenza viruses are less frequent causes of the common cold, but they won't be discussed here because they'll be dealt with in Chap. 11.

In order to cause disease, these viruses must first of all overcome the antimicrobial defence mechanisms that operate in the nose. Large microbe-laden particles in air that's inhaled (approximately 10,000 litres per day) are removed by hairs in the nostrils, while any smaller particles and microbes become trapped in a layer of mucus that covers the epithelial lining of the nasal cavity. Sneezing is an effective means of expelling these mucus-entrapped microbes from the nose. However, mucus and the trapped microbes are continually being removed from the nose by an amazing system known as the "mucociliary escalator" (Fig. 10.6 – see also Fig. 1.19 in Chap. 1). In the deepest two thirds of the nasal cavity, the epithelial surface is covered by millions of fine, hair-like projections (known as cilia – Fig. 10.6b) which beat in unison to propel the mucus to the pharynx from where it's either swallowed or expelled by coughing or sneezing.

The mucus coating of the nasal cavity is completely removed and replaced every 15–20 minutes. However, in the front region of the nasal cavity (where the mucociliary escalator doesn't operate), entrapped microbes are killed mainly by antimicrobial compounds secreted into the mucus by cells that line the nose.

What About Complications?

Although colds are a nuisance they generally do no great harm to the otherwise healthy individual. However, they can sometimes be followed by more serious bacterial infections. This is because bacteria resident in the respiratory tract can occasionally take advantage of the damage inflicted by the virus and so cause further problems. In children, the most common bacterial complication is acute otitis media (see Chap. 16), which occurs in about 20% of those suffering from a viral upper respiratory infection. This usually occurs 3–4 days after the initial viral infection. The organisms involved are usually *Streptococcus pneumoniae, Haemophilus influenzae, Moraxella catarrhalis* and *Streptococcus pyogenes*. While antibiotics should never be used to treat the common cold, they're appropriate for some of these bacterial complications. Some, less frequent, bacterial infections arising from colds are sinusitis and pneumonia.

Colds can also result in short-term worsening (exacerbations) of lung conditions such as asthma, and these are particularly common when rhinoviruses are responsible for the cold. In elderly patients as many as two thirds have been found to develop diseases of the lower respiratory tract following colds,

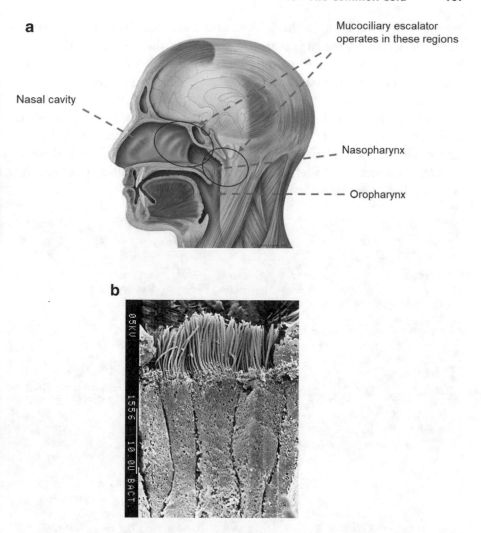

Fig. 10.6 The mucociliary escalator is an important antimicrobial defence system of the respiratory tract. (a) The mucociliary escalator operates in much of the nasal cavity as well as in the nasopharynx, sinuses and most of the lungs. (Modified from Patrick J. Lynch [CC BY 2.5 (https://creativecommons.org/licenses/by/2.5)]). (b) Epithelial cells covered in cilia. (Credit: David Gregory & Debbie Marshall. CC BY 4.0. Wellcome Image Library)

and younger children can develop wheezy responses to viral infections or inflammation in their smaller airways known as bronchiolitis.

Although the common cold isn't usually a serious illness, it has an enormous economic impact. In the UK in 2007, the annual number of visits to the doctor due to colds was 400,000, and the associated prescription costs amounted to £1.1 million. More data on the economic burden of colds are available from the USA. In the USA the common cold leads to 75–100 million visits to the doctor each year, and the cost of this has been estimated to be $7.7 billion per year. Americans spend $2.9 billion on over-the-counter drugs for colds and another $400 million on prescription medicines for symptom relief. It's been estimated that 22–189 million school days are missed annually because of colds. As a result, the parents of these children missed 126 million workdays to care for their children. When this is added to the 150 million workdays missed by employees suffering from a cold, the total economic impact of cold-related work loss exceeds $20 billion per year. This accounts for 40% of time lost from work in the USA.

Box 10.3 Antibiotics for the Common Cold: A Misuse of Vitally Important Drugs

Antibiotics, of course, are totally useless against viral infections and should never be used for treating colds. However, their administration to patients with colds is probably one of the greatest misuses of antibiotics and is a great worry because it contributes to the development of antibiotic resistance which is a huge problem. General practitioners (GPs) are under great pressure from their patients to prescribe antibiotics for colds and other respiratory tract infections due to viruses. A survey of 1000 GPs in the UK in 2014 reported that 55% felt under pressure, mainly from patients, to prescribe antibiotics, even if they weren't sure that they were necessary. Forty-four percent admitted that they had prescribed antibiotics to get a patient to leave the surgery. A similar proportion (45%) had prescribed antibiotics for a probable viral infection, suspecting that they wouldn't be effective.

In a study carried out in the USA in 1998, it was found that 55% (a total of 22.6 million) of the prescriptions issued for acute respiratory infections were for cases that were unlikely to have been caused by bacteria. The cost of these prescriptions amounted to $726 million. A similar study in 2014 concluded that 52% of antibiotic prescriptions given to children suffering from upper respiratory infections were unnecessary. Apart from the huge waste of money and resources this represents, the use of these large quantities of antibiotics will undoubtedly have contributed considerably to antibiotic resistance development in the indigenous microbiota of these individuals. Infections due to antibiotic-resistant bacteria are becoming an increasing problem, and in Europe approximately 25,000 patients die annually due to infections with antibiotic-resistant microorganisms.

A big part of this problem is distinguishing between viral and bacterial infections, which can be tricky in the family doctor's office without special tests being available, not least because symptoms overlap and infections can be caused by a mix of pathogens. Tools are being developed to assist in this, and hopefully these, alongside public awareness and education, will impact this global healthcare and economic problem.

In the UK a prize (the "Longitude Prize") of £10 million has been offered to anyone inventing an accurate and rapid near-patient test to identify bacterial infections that can be used in any health setting globally. The purpose of this test is to enable appropriate antibiotic prescribing and thus reduce antimicrobial resistance. See www.longitudeprize.org for more information.

(continued)

Box 10.3 (continued)

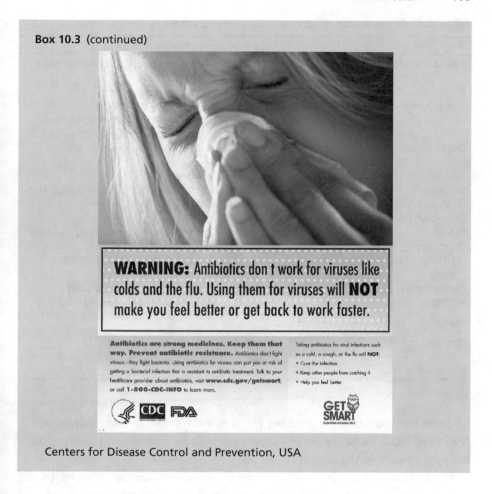

Centers for Disease Control and Prevention, USA

How Are Colds Treated?

No effective or licenced anti-viral treatment for colds has yet been developed, and this is mainly because of the large number of different viruses that may be involved. The development of anti-viral agents effective against the rhinoviruses, the predominant cause of the disease, is a promising approach. Examples of these include pleconaril and rupintrivir. Early studies with pleconaril have shown a 1–1.5 day reduction in the duration of the illness. Echinacea may well be of some benefit, but evidence is only limited at the moment – it may reduce symptom severity. It's thought it might do this by boosting the immune system. However, its long-term effects are not known.

Table 10.1   Effectiveness of various treatments for symptoms of the common cold

Treatment	Outcome and conclusion
Antihistamine	Overall symptoms and nasal obstruction not improved
	No clinically meaningful benefit
Antihistamine plus decongestant	Overall improvement in symptoms
	Likely to be beneficial in adults and older children but no effect in children ≤5 years
Decongestants	Small benefit but this is of uncertain clinical significance
	No data for children
Intranasal	Improved rhinorrhoea but not nasal congestion
Ipratropium	Probably beneficial
Over-the-counter	No benefit in children
Cough	Benefit unclear (but likely to be small) in adults
Treatments	
Non-steroidal anti-inflammatory drugs	Likely to be beneficial for pain and/or fever
	No benefit for other symptoms
Paracetamol	Likely to be effective for fever and pain
Honey	A single night-time dose of honey can have a small effect on cough and sleep in children over 12 months old
Oral zinc	Results in a reduced duration of cold
	Probably beneficial in adults but not in children
Vitamin C	No benefit

There may be a role for polymer gel nose sprays such as carrageenan which trap virus particles in the nose – this appears to be a promising treatment for early symptoms of the disease.

Most currently available treatments for colds are aimed at relief of the symptoms of the disease, and hundreds of such products are widely used. The effectiveness of the most commonly used treatments is summarised in Table 10.1.

How Can I Avoid Getting a Cold?

You can help to limit the spread of the infection to others by physical measures such as hand washing, covering your mouth and nose when sneezing or coughing (Fig. 10.7), properly disposing of tissues contaminated with nasal secretions and staying away from work or school. You should avoid contact with the nasal mucus or phlegm coughed up by people with the disease. You should wash your hands before eating as well as before touching your nose,

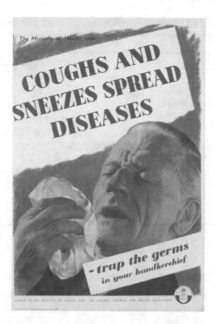

Fig. 10.7   Poster produced by the Ministry of Health and the Central Council for Health Education (1939–1945). (This work created by the UK Government is in the public domain. Via Wikimedia Commons)

eyes or mouth. Avoiding others with a cold, especially during the first few days of illness, reduces the chance of spread. Keeping your nose warm helps to maintain a healthy blood supply and mucus lining, and this will help to prevent you catching other peoples' viruses.

The wide variety of viruses responsible for the infection has hampered attempts at developing an effective vaccine. There's no immune protein (antigen) on the surface of the various rhinoviruses that is common to all of them so the prospects of a vaccine against the main viruses responsible for colds aren't good. However, effective vaccines are available against a less frequent cause of colds – the influenza virus. Furthermore, vaccines against other minor causes such as the respiratory syncytial virus and parainfluenza viruses are under development.

Table 10.2   Effectiveness of various medicaments used for the prevention of colds

Treatment	Outcome and conclusion
Zinc supplement	Significant reduction in frequency of colds
	Likely to be beneficial
Gargling with povidone-iodine	No reduction in frequency of colds
	No benefit
Ginseng	No significant reduction in colds
45 min of moderate intensity exercise 5 days per week	Significantly fewer self-reported colds
	Unclear benefit
Vitamin C	No benefit
Vitamin D	No benefit
Echinacea	No benefit

A wide variety of medicaments are available for the prevention of colds, and the effectiveness of the most popular of these is summarised in Table 10.2.

Want to Know More?

American Lung Association https://www.lung.org/lung-health-and-diseases/lung-disease-lookup/influenza/facts-about-the-common-cold.html

Centers for Disease Control and Prevention, USA https://www.cdc.gov/antibiotic-use/community/for-patients/common-illnesses/colds.html

Mayo Clinic, USA https://www.mayoclinic.org/diseases-conditions/common-cold/symptoms-causes/syc-20351605

MedicineNet https://www.medicinenet.com/common_cold/article.htm

MedlinePlus, U.S. National Library of Medicine https://medlineplus.gov/commoncold.html

National Health Service, UK https://www.nhs.uk/conditions/common-cold/

National Institute for Clinical Care and Excellence (NICE), UK. Common cold, 2016 https://cks.nice.org.uk/common-cold#!topicSummary

Patient Info, UK https://patient.info/doctor/common-cold-coryza

Allan GM, Arroll B. Prevention and treatment of the common cold: making sense of the evidence. *Canadian Medical Association Journal* 2014 Feb 18; 186(3): 190–199. doi: https://doi.org/10.1503/cmaj.121442

Arroll B, Kenealy T. Common cold. BMJ Best Practice, BMJ Publishing Group, 2020 https://bestpractice.bmj.com/topics/en-gb/252

Buensalido JAL. Rhinovirus (RV) Infection (Common Cold). Medscape from WebMD. 2019 https://emedicine.medscape.com/article/227820-overview

Casanova V, Sousa FH, Stevens C, Barlow PG. Antiviral therapeutic approaches for human rhinovirus infections. *Future Virology.* 2018 Jul;13(7):505–518. https://doi.org/10.2217/fvl-2018-0016 Epub 2018 Jun 12.

Fashner J, Ericson K, Werner S. Treatment of the common cold in children and adults. *American Family Physician.* 2012 Jul 15;86(2):153–9.

Thomas M, Koutsothanasis GA, Bomar PA. Upper Respiratory Tract Infection. Treasure Island (FL): StatPearls Publishing LLC; 2020 https://www.ncbi.nlm.nih.gov/books/NBK532961/

11

Flu

Abstract Influenza is a disease of the respiratory tract caused by influenza viruses. It usually occurs as epidemics during the winter and can affect a high proportion of the population. The main symptoms are fever, muscle pain, headache, coughing, sore throat and nasal congestion. It usually lasts for about a week but can sometimes be followed by a secondary bacterial infection such as pneumonia, sinusitis or otitis media. Treatment of the uncomplicated disease involves the use of a fever-reducing drug such as paracetamol, increased fluid intake and rest. Secondary infections may require antibiotics. A number of vaccines are available for preventing the disease.

What Is Influenza?

Influenza is a viral infection of the respiratory system. If you catch it, you'll suddenly come down with a fever and then experience muscle pain (myalgia), headache, coughing, sore throat and nasal congestion and generally feel pretty awful (Fig. 11.1). You might also suffer from nausea, eye symptoms, vomiting and diarrhoea – but these symptoms are less common.

Box 11.1 What's in a Name?

Influenza is an Italian word and was first used to describe an influenza epidemic in Florence, Italy, in 1357. At that time people believed the disease was due to unfavourable astrological "influences". The word was first used in English in 1703 by J. Hugger of the University of Edinburgh who wrote a thesis about the disease.

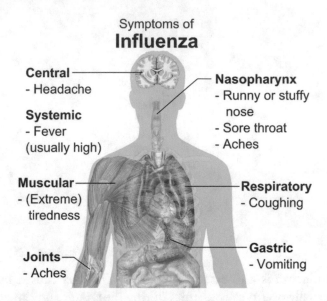

Fig. 11.1 The main symptoms of influenza. (Mikael Haggstrom. This file is made available under the Creative Commons CC0 1.0 Universal Public Domain Dedication, via Wikimedia Commons)

The incubation period for the disease is between 1 and 4 days, and it's usually self-limiting in healthy adults and lasts for 3–7 days. If you get it, you'll become a virus-emitter and a danger to others because you'll be shedding the virus over a long period – from about 1 day before you develop any symptoms until 5–7 days after you feel well again. Although it's an unpleasant experience for healthy adults, it's not generally a dangerous disease in these people. But that's not the case for everyone. For some people it can be serious and life-threatening such as the elderly, children under 6 months old, pregnant women and those with chronic conditions or who are immunosuppressed. Other complications include myocarditis, febrile convulsions, myositis, encephalitis and meningitis. Chronic conditions such as asthma, diabetes and heart disease can be aggravated.

The WHO estimates that each year as many as one billion people worldwide are infected with influenza and up to 500,000 people die from the disease, with most deaths occurring in young children and the elderly. The disease can occur as an epidemic or a pandemic. Epidemics usually occur annually in temperate regions – November to April in the northern hemisphere and May to October in the southern hemisphere. In the USA between 25 and 50 million people get the disease each year, while in the UK between 3 and 12 million are affected.

What Causes Influenza?

Influenza is a viral infection. There are four types of influenza virus – influenza A, B, C and D – but only types A, B and C cause human disease. Influenza A is found in humans, birds, pigs, horses, mink, dogs, seals and ferrets; influenza B in humans, seals and ferrets; and influenza C in humans, pigs and dogs. Most infections in humans are due to type A (Fig. 11.2a) which causes the most severe illness and is the one usually responsible for seasonal epidemics and pandemics. Influenza A affects people of all ages but causes more severe disease in older adults and people with underlying chronic health problems. Influenza B affects children the most. Influenza C affects all age groups but tends to cause only a mild illness.

The various influenza viruses have a similar structure (Fig. 11.2b) and are usually spherical or oval with a diameter of between 80 and 120 nanometres. The virion has an outer coating which, as we'll see later, is formed from the membrane (known as the "cytoplasmic membrane") of the human cell in which it was produced. This outer coating has a series of spikes sticking out of it. In the case of influenza A, most of these (80%) are made of a protein called haemagglutinin which is used by the virus to attach itself to a host cell that it's going to invade. The rest of the spikes consists of a different protein, neuraminidase, which is involved in the release of newly formed virions from the host cell. Inside the coat is the genetic material of the virus which consists of RNA molecules. The spikes on influenza B also consist of haemagglutinin and neuraminidase, whereas those of influenza C are made of only one protein – haemagglutinin-esterase-fusion glycoprotein (HEF).

Each of the three types of virus that cause disease in humans exists in a number of sub-types. In the case of influenza A, the various sub-types arise because of differences in the composition of the surface spikes. There are 18 different types of haemagglutinin and 11 types of neuraminidase. The various sub-types of influenza A are named on the basis of which of these molecules are present. For example, the virus responsible for the notorious Spanish flu in 1918 (see Box 11.2) has a type 1 haemagglutinin and a type 1 neuraminidase and so is known as influenza A(H1N1). Only two influenza A virus sub-types (H1N1, and H3N2) and two influenza B sub-types are currently circulating throughout the world.

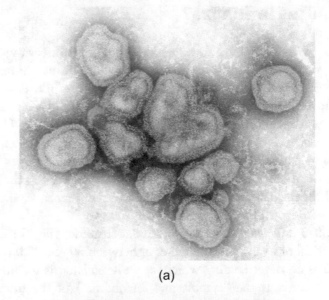

(a)

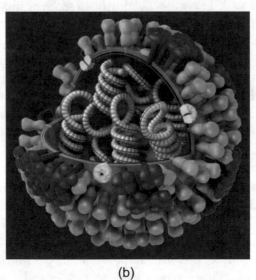

(b)

Fig. 11.2 The influenza virus. (a) Transmission electron micrograph of the influenza A virus. Each virion has a diameter of approximately 100 nm – approximately one tenth of the diameter of a typical bacterium such as *Staph. aureus*. (Image courtesy of F.A. Murphy, Centers for Disease Control and Prevention, USA). (b) Three-dimensional model of the influenza A virion showing details of its structure. The genetic material of the virus consists of molecules of RNA (coiled structures), and this is contained within an outer coat (orange). Spikes of haemagglutinin (pale blue) and neuraminidase (dark blue) protrude from the surface coating. (Image courtesy of Dan Higgins, Centers for Disease Control and Prevention, USA)

What Happens During an Infection?

So, what happens if you're unfortunate enough to be infected by influenza A? The haemagglutinin spikes on the surface of the virion bind to specific molecules (sialic acid) on the surface of the epithelial cells that line the respiratory tract – such molecules are called "receptors". Binding of the virion to its receptor on the epithelial cell triggers the ingestion of the virion by the cell – a process known as endocytosis (see Fig. 7.6 in Chap. 7). The virion is now within a small fluid-filled "balloon" inside the cell, and this is known as a vacuole (see Fig. 7.6). The virion then releases its genetic material (RNA) which enters the nucleus of the host cell (see Fig. 7.7). The nucleus produces more viral RNA and proteins which start to migrate to the surface of the host cell. When they get there, these molecules assemble to form a new virion which then causes the membrane to produce a bud that closes round it. The bud then separates from the host cell as a new virion. The neuraminidase on the surface of the virion is important in this process. The released virions can then infect more human cells or are expelled from the respiratory tract (by coughing and sneezing) and can go on to infect other individuals.

Our immune system, of course, senses that our epithelial cells have been invaded and responds to try to contain and destroy the invading viruses. The ways in which it does this result in many of the symptoms of the disease that we experience. The chemical messengers (known as cytokines) that are released to coordinate the immune response against the virus also, coincidentally, result in some of the characteristic symptoms of the infection such as fever, muscular pain, fatigue and headaches. The inflammation induced by the virus increases the diameter and permeability of the blood vessels of the respiratory tract so that more defensive white blood cells and antibodies can be brought to attack the virus. In the nasal cavity, this results in stuffiness and a runny nose (rhinorrhoea), while in the lower respiratory tract, it results in the production of lots of mucus which causes us to cough so that we can get rid of it.

Our defence system eventually gains the upper hand, and the virus is killed and/or expelled from the body. The antibodies that are produced during the course of the infection are effective against another attack by the same virus. However, over time the proteins on the surface of the virion gradually change as the virus naturally mutates until they're no longer recognised by these antibodies, and this leaves us vulnerable once again. This change is known as "antigenic drift" and explains why the virus is able to cause new epidemics each year. Sometimes there's a large and abrupt change in the structure of the viral surface proteins (known as antigenic shift) which gives rise to a new subtype of the virus which isn't recognised by our immune system – this can result in a pandemic.

Box 11.2 The Influenza Pandemic of 1918

The 1918 influenza pandemic, sometimes referred to as the "Spanish flu", affected one third of the world's population and killed an estimated 50 million people – approximately 5% of the world's population. In addition to its high mortality rate, another unusual feature of the virus responsible (which was an influenza A H1N1 type – see Figure a) was the high death rate it caused among young (15–34 years) healthy adults. It also killed its victims very rapidly – within a few hours or days.

This pandemic occurred in three waves starting in March 1918 and ending in the Spring of 1919. It had several names including Spanish flu, grippe, Spanish Lady, the 3-day fever and sandfly fever.

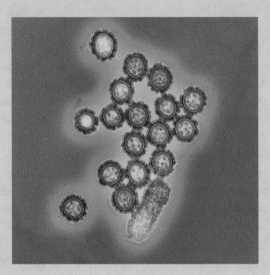

Figure (a) Electron micrograph of the influenza A(H1N1) virus. (National Institute of Allergy and Infectious Diseases, USA)

Its origins are uncertain. Two possibilities have been suggested – a military camp in Étaples, France, in 1917 and one in Kansas, USA, in 1918. The first well-documented case occurred on March 11, 1918, and involved Private Albert Gitchell, a company cook in Camp Funston in Kansas. This camp was where new recruits were trained before being sent to Europe to fight in World War I. He came down with symptoms of what appeared to be a bad cold, so he went to the infirmary and was isolated. But within an hour, several additional soldiers had the same symptoms and were also isolated. Despite these precautions, the disease spread rapidly, and after 5 weeks, 1127 soldiers at the fort had the disease and 46 died.

Soon other military camps around the USA reported cases, and it appeared in soldiers on transport ships to Europe. By mid-May French soldiers were also affected, and the flu quickly travelled across Europe, affecting nearly every country. Three quarters of French troops and more than half of British troops fell ill in the spring of 1918. When it arrived in Spain, the government announced the epidemic – the first country to do so publicly because there was no censorship of health reports in this neutral country. Since most people first heard about the flu affecting Spain, it became known as the Spanish flu. It then spread to Russia, India, China and Africa. By the end of July 1918, this first wave of the Spanish flu appeared to be dying out after having affected nearly every country in the world.

In August 1918, the second wave struck three port cities almost simultaneously – Boston in the USA, Brest in France and Freetown in Sierra Leone. This was a mutated form of the virus and was not only highly contagious but also more deadly. Early symptoms of the disease included extreme fatigue, fever and headache. Patients would start turning blue and would cough with such force that they tore their abdominal muscles. Foamy blood came from their mouths and noses and, in some cases, from their ears. It struck so suddenly and severely that many of its victims died within hours of coming down with their first symptom. Hospitals quickly became overwhelmed by the sheer numbers of patients (Figure b). The authorities couldn't cope with the huge number of bodies, and they had to resort to mass graves. By the end of September, it had spread to most of Europe. In New York the epidemic was declared to be over on November 5, but it persisted in Europe due to the deprivations of the war. By December most of the world was flu-free.

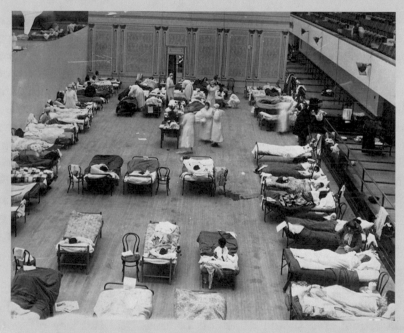

Figure (b) The Oakland Municipal Auditorium in use as a temporary hospital. The photograph shows volunteer nurses from the American Red Cross tending influenza sufferers in the Oakland Auditorium, Oakland, California, during the influenza pandemic of 1918. (Edward A. "Doc" Rogers, From the Joseph R. Knowland collection at the Oakland History Room, Oakland Public Library. Via Wikimedia Commons)

In the early months of 1919, the third wave of the pandemic affected most of Europe. A contributing factor may have been the ending of the war on November 11, 1918, and the resulting movement of large numbers of people as well as jubilation which would have resulted in close contact between non-combatants and returning soldiers. Although this wasn't as deadly as the second wave, it was worse than the first and also went around the world, killing many of its victims.

(continued)

Box 11.2 (continued)

This lasted until the spring of 1919 in the Northern hemisphere, although Japan experienced a later third wave which only ended in early 1920.

By the end of the pandemic, only one region in the entire world hadn't reported an outbreak: an isolated island called Marajo in Brazil's Amazon River Delta. This pandemic is regarded as having been the greatest human disaster in history. Between 10% and 20% of those infected died. The total numbers of dead were estimated to have been 12–17 million in India, 675,000 in the USA, 400,000 in Japan and 228,000 in the UK.

How Is Influenza Treated?

The surfaces of our body that are in contact with the external environment are covered in a tissue known as the epithelium. The skin is a dry epithelium on our external surface, but we also have moist epithelia (known as the mucosae) covering our internal surfaces – the respiratory, gastrointestinal and reproductive tracts. These epithelia constitute our first line of defence against microbial attack. Damage to the respiratory epithelium caused by the influenza virus makes it vulnerable to attack by some bacteria that live in the respiratory tract such as *Streptococcus pneumoniae, Haemophilus influenzae, Staphylococcus aureus, Streptococcus pyogenes* and *Moraxella catarrhalis*. Infections that arise as a result of the damage caused by some other disease are termed "secondary infections" or "superinfections". These infections, which include pneumonia, sinusitis and otitis media, usually arise within 6 days of the initial infection with influenza A. While antibiotics are, of course, of no value in the treatment of influenza, these secondary bacterial infections, which may be life-threatening, can require antibiotics.

If you're usually a healthy adult or child and don't get any superinfection, then all you need to do is take a fever-reducing drug (i.e. an antipyretic) such as paracetamol. You should drink lots of fluids, rest and stay off work or school so that you don't spread the disease. Anti-viral drugs such as neuraminidase inhibitors (e.g. oseltamivir and zanamivir) reduce viral release from infected cells and the rate of viral replication, but to be most effective, they have to be started within 48 hours of the onset of symptoms. However, they only have a moderate effect and reduce the duration of symptoms by only approximately 0.7 days. But they do reduce the risk of complications such as secondary infections, and they're recommended for anyone who may be at risk of these.

How Can I Avoid Getting Influenza?

If you get influenza, you're most contagious during the first 3–4 days after you develop symptoms. However, some people can infect others 1 day before their symptoms start and up to 5–7 days after becoming ill. Influenza viruses are transmitted mainly via the droplets we produce when we cough, sneeze or talk. These droplets can reach the mouths or noses of people nearby (up to about 6 feet away) and also can be inhaled into their lungs. The droplets can also land on surfaces from where they can then be transferred by touch to the mouth, nose or eyes of other people. It's been shown that the virus can remain alive for up to 7 days on non-porous surfaces such as steel and for 2 days on wooden surfaces. Important ways of reducing the spread of the infection include:

- Avoiding anyone with the disease
- Disinfecting any contaminated surfaces
- Regular hand washing
- Using disposable handkerchiefs during coughing and sneezing

The most effective way of preventing influenza is vaccination (see Chap. 1 for more about vaccination). However, the vaccines used have to be updated annually to take account of changes in the predominant strains that are circulating as well as antigenic drift in the virus. The most widely used vaccines are the inactivated trivalent (effective against the two influenza A sub-types and one of the B sub-types) and the inactivated quadrivalent (effective against both A sub-types and both B sub-types). Live, attenuated vaccines (LAIVs) which contain modified, but still alive, viruses are also available. Vaccination policies vary from country to country, but the WHO recommends that the following groups should be vaccinated: pregnant women, children aged 6–59 months, the elderly, individuals with specific chronic medical conditions and healthcare workers. Influenza vaccine side effects are generally mild and go away on their own within a few days. Some side effects that may occur from an influenza vaccination include soreness, redness and/or swelling where the shot was given, headache (low grade), fever, nausea, muscle aches and fatigue. The influenza vaccination, like other injections, can occasionally cause fainting. However, there's no truth in the idea that you can catch flu from the vaccine itself. Sometimes people will have been exposed to an influenza virus shortly before they're vaccinated or during the time that it takes the body to develop protection after being vaccinated. Such exposure could give rise to influenza before the vaccine has had time to promote a protective immune response – it's nothing to do with the vaccination itself.

Anti-viral agents such as the neuraminidase inhibitors are effective at preventing infection or rendering it less severe. They may, therefore, be useful in certain circumstances especially when the disease is circulating, for example, in at-risk groups and those who have had close contact with people with confirmed or suspected influenza, i.e. living in the same household or residential setting.

Want to Know More?

American Lung Association https://www.lung.org/lung-health-and-diseases/lung-disease-lookup/influenza/

Centers for Disease Control and Prevention, USA https://www.cdc.gov/flu/index.htm

Mayo Clinic, Mayo Foundation for Medical Education and Research, USA https://www.mayoclinic.org/diseases-conditions/flu/symptoms-causes/syc-20351719

MedlinePlus, U.S. National Library of Medicine https://medlineplus.gov/flu.html

Ministry of Health, New Zealand https://www.health.govt.nz/your-health/conditions-and-treatments/diseases-and-illnesses/influenza

National Health Service, UK https://www.nhs.uk/conditions/flu/

National Institute for Clinical Care and Excellence (NICE), UK. Influenza, 2019 https://cks.nice.org.uk/influenza-seasonal#!topicSummary

Virology Blog http://www.virology.ws/influenza-101/

World Health Organisation http://www.euro.who.int/en/health-topics/communicable-diseases/influenza

Boktor SW, Hafner JW. Influenza. Treasure Island (FL): StatPearls Publishing LLC; 2020 https://www.ncbi.nlm.nih.gov/books/NBK459363/

Committee on infectious diseases. Recommendations for prevention and control of influenza in children, 2018-2019. *Pediatrics*. 2018 Oct;142(4). pii: e20182367. doi: https://doi.org/10.1542/peds.2018-2367. Epub 2018 Sep 3.

Davidson S. Treating Influenza Infection, From Now and Into the Future. *Frontiers in Immunology* 2018 Sep 10;9:1946. doi: https://doi.org/10.3389/fimmu.2018.01946. eCollection 2018.

Ghebrehewet S, MacPherson P, Ho A Influenza.. *British Medical Journal*. 2016 Dec 7;355:i6258. doi: https://doi.org/10.1136/bmj.i6258.

Moghadami M. A narrative review of influenza: a seasonal and pandemic disease. *Iranian Journal of Medical Sciences*. 2017 Jan;42(1):2-13.

Nguyen HH. Influenza. Medscape from WebMD, 2020 https://emedicine.medscape.com/article/219557-overview

Radojicic C. Influenza infection. BMJ Best Practice, BMJ Publishing Group, 2019 https://bestpractice.bmj.com/topics/en-gb/6

12

Coronavirus Disease 2019 (COVID-19)

Abstract Coronavirus Disease 2019 (COVID-19) is a highly contagious viral infection of the respiratory tract that at the time of writing is responsible for a pandemic. The main symptoms include fever, coughing, being short of breath and loss or change in your sense of smell or taste. In most cases, it is a mild disease that lasts no longer than 2 weeks. However, it can cause severe illness, particularly in the elderly and those with a chronic medical condition. Complications include pneumonia and acute respiratory distress syndrome which is life-threatening. The disease has only been identified recently, since around December 2019, and so we know little about the virus responsible ("severe acute respiratory syndrome coronavirus 2" – SARS-CoV-2), how it is transmitted or how it causes illness, although we do understand more about some other coronaviruses. Currently there is no specific anti-viral agent against SARS-CoV-2, and no vaccine is available in Western countries – although research into these is being carried out and vaccination against the disease may soon be possible. Handwashing, social-distancing and covering the mouth and nose, particularly when sneezing or coughing, help to reduce the spread of the disease. Isolation and quarantine, especially of affected people, as well as environmental disinfection are also important protective measures. The medical, social and economic impacts of the disease, and governments' responses to it, have been enormous.

Warning *Because of the enormous, unprecedented research effort into COVID-19 that has been mounted across the world, it is likely that significant advances will rapidly be made in our knowledge of this disease and the virus responsible for it.*

© Springer Nature Switzerland AG 2021
M. Wilson, P. J. K. Wilson, *Close Encounters of the Microbial Kind*,
https://doi.org/10.1007/978-3-030-56978-5_12

As a result, some of the information in this chapter may be out of date by the time this book is published. It is very important, therefore, that you consult reliable sources (such as the websites of the Centers for Disease Control and Prevention, USA, and the National Health Service, UK) for more up-to-date information on this disease.

In around December 2019 a new disease emerged in China and was found to be due to a new coronavirus which was named "severe acute respiratory syndrome coronavirus 2" (SARS-CoV-2). In February 2020, the World Health Organisation named the disease caused by this novel virus "coronavirus disease 2019" (COVID-19). The medical, social and economic impacts of this disease, as well as the responses to it, have been enormous and continue to have a major impact worldwide.

Although the virus causes more significant mortality than seasonal influenza, it only results in high mortality rates in the older and at-risk population. It is thought that, because the illness is new and highly contagious, there is probably little or no background immunity to the infection, and so large numbers of people are falling ill at the same time. This overwhelms healthcare systems, meaning they are unable to offer treatment as efficiently as usual which results in even more mortality and morbidity. Added to this is an inadequate understanding of the disease and treatment options as well as limited and, at times, poor or patchy availability of equipment. There has been a global surge in demand for ventilators and personal protective equipment especially for health personnel for which most have been inadequately prepared. The isolation strategies used to try to slow the spread of the infection have had unintended negative consequences such as impeding flow of goods, panic buying and hoarding, increasing isolation of the vulnerable and marginalised and a universal impact on the workforce and businesses. The economic shock has been deeper and faster than the financial crisis of 2008 and the Great Depression of the 1920s. Reports abound that the final death toll may approach that of World War II.

What Are the Symptoms of the Disease?

COVID-19 is a viral disease of the respiratory tract. The main symptoms appear within 2–14 days (on average after 4–5 days) of becoming infected and include fever, coughing and shortness of breath. Other symptoms include fatigue and muscle or body aches. In addition, change in sense of taste or smell has been reported in up to 80% of those affected. In the first wave of the outbreak, in China, the following initial symptoms were reported:

- Fever (99% of cases)
- Fatigue (70%)
- Dry cough (59%)
- Anorexia (40%)
- Muscle pain (35%)
- Breathing difficulties (31%)

Other less common symptoms include headache, diarrhoea, sore throat and a runny nose. Some people may be entirely asymptomatic or have such mild symptoms that they do not recognise they have been infected.

If you are infected, you may experience only a mild illness, but severe symptoms and death are possibilities. People with the mild form of the disease usually recover within 1–2 weeks. The severe disease can involve either a viral or a bacterial pneumonia (due to a secondary infection) for which the symptoms include cough, difficulty breathing, persistent pain or pressure in the chest, fatigue, fainting or falling, confusion or inability to arouse and bluish lips or face. This severe form of the illness can last for 3–6 weeks.

Those who are elderly (over 65 years of age) or have severe underlying chronic medical conditions (such as hypertension, heart or lung disease, or diabetes) are at a greater risk of developing the more serious form of the disease. In some of these severely affected patients, the disease then progresses, usually in week 2, to what is known as "acute respiratory distress syndrome (ARDS)" where the lungs become so inflamed they can no longer sufficiently transfer gases such as oxygen between the air and the blood. Other possible complications include severe blood clotting, kidney injury, heart problems, secondary bacterial infections, pulmonary embolism, acute coronary syndrome, septic shock, strokes and delirium.

The WHO has classified the disease into four types depending on its severity:

(i) Mild illness. Patients with symptoms who do not develop hypoxia (shortness of oxygen in the tissues) or pneumonia.
(ii) Moderate disease. Clinical signs of pneumonia but no signs of severe pneumonia.
(iii) Severe disease. Clinical signs of pneumonia, severe respiratory distress and hypoxia.
(iv) Critical disease. Presence of ARDS, sepsis, or septic shock.

In the disease outbreak in China, it was found that 81% of those infected had the mild form of the disease, 14% had severe disease and 5% developed ARDS. The overall mortality rate was 2.3%. In a UK study of those who were hospitalised, there was a 17% likelihood of needing critical care. Most of the fatal cases occurred in elderly patients or those with underlying medical conditions such as cardiovascular disease, diabetes mellitus, chronic lung disease, hypertension and cancer.

How Common Is It?

This is a new disease that first appeared in Wuhan City, China, probably in late December 2019 and has since spread globally. The World Health Organisation has declared it to be a pandemic which means that it has spread to many countries and affected large numbers of people. As of September 14, 2020, over 29 million cases had been diagnosed in 213 countries with over 928 thousand fatalities (Table 12.1).

Although the disease affects people of all ages, middle-aged adults and the elderly are most likely to be affected. In the initial outbreak in China, only 2% of cases were in individuals younger than 20 years. In the USA, 80% of deaths due to the disease were in adults who are 65 years or older. The death rates of those with the disease increase with age. Compared to 18–29-year-olds, the death rates are 30X higher in 50–64-year-olds and 630X higher in those over 85 years of age. The risk of being hospitalised, or of death, appears to be higher in lower income groups and varies with ethnicity. This may be due to differences in work and housing circumstances as well as variation in the frequency of chronic health conditions in these communities.

Table 12.1   Number of confirmed cases of COVID-19 and deaths due to the disease (as of September 14, 2020) in various countries

Country	Number of cases	Number of deaths	Deaths per million population
USA	6,708,458	198,520	599
Brazil	4,330,455	131,663	619
Spain	576,697	29,747	636
UK	368,504	41,628	613
Italy	287,753	35,610	589
China	85,194	4,634	3

Which Microbe Causes the Disease?

The virus responsible is called "severe acute respiratory syndrome coronavirus 2" (SARS-CoV-2). This is a coronavirus (Fig. 12.1), but, because it is so new, we do not know much about it. It appears to be similar to the coronavirus responsible for severe acute respiratory syndrome (SARS) which caused a pandemic in 2003. It is currently thought to have originated in bats.

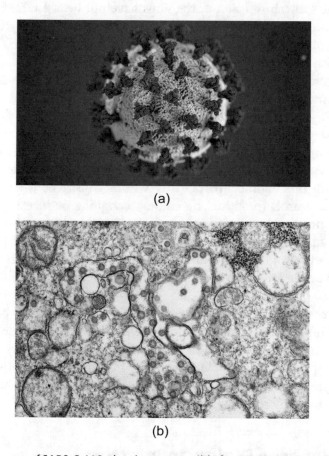

Fig. 12.1   Images of SARS-CoV-2, the virus responsible for COVID-19. Hannah A Bullock; Azaibi Tamin, Centers for Disease Control and Prevention, USA. (**a**) Model of SARS-CoV-2 showing the typical appearance of a coronavirus. The three different protein spikes on its surface are shown in red (S protein), yellow (E protein) and orange (M protein). Alissa Eckert, MS; Dan Higgins, MAMS, Centers for Disease Control and Prevention, USA. (**b**) Electron micrograph of SARS-CoV-2. The virions have been coloured blue and can be seen to have the typical appearance of coronaviruses (see Fig. 10.5 of Chap. 10)

How Is the Disease Transmitted?

Because the disease is so new, we have not had the chance to study all aspects of it and are still learning about it. However, we do know a lot about other coronaviruses, such as the ones that cause colds, severe acute respiratory syndrome (SARS) and Middle East respiratory syndrome (MERS). We are, therefore, assuming that SARS-CoV-2 is likely to behave in a similar manner.

The routes of transmission of the virus have not been established for certain. However, we do know that other coronaviruses are mainly transmitted by large respiratory droplets (with a diameter >5 μm) and by direct or indirect contact with infected secretions. Aerosols (i.e. droplets with a diameter of <5 μm) from the respiratory tract of infected individuals are likely to be another route of transmission. As well as being present in respiratory secretions, SARS-CoV-2 has been found in blood, urine, tears and faeces. These could, therefore, be other routes by which the SARS-CoV-2 is transmitted.

Current opinion is that person-to-person transmission is the main way in which the virus is spread. This occurs more readily when people are in close contact with one another (particularly within 6 feet) or when respiratory droplets or aerosols (produced by talking, coughing or sneezing) reach the mouths or noses or are inhaled by, nearby individuals. It may also be possible that you can become infected by touching a surface or object that has the virus on it and then touching your mouth, nose or eyes. Currently, this is not thought to be the primary way the virus spreads but is still a possible transmission route.

Whatever the routes of transmission, SARS-CoV-2 appears to be spreading readily within the population. There is some evidence that transmission is more likely to occur during the early stages of the disease, shortly after symptoms develop. It is also possible that people infected with the virus but who show no symptoms are still able to transmit the virus to others – this may be another reason why the virus is spreading so rapidly.

What Happens During an Infection?

It is thought that the epithelial cells lining the lung are the main target for the virus. The virus binds to a receptor on the surface of these cells, identified as being an enzyme called angiotensin-converting enzyme 2 (ACE2). This is followed by invasion of the cell which subsequently makes more virions that are released to infect other cells. The invasion is accompanied by an increase in the levels of pro-inflammatory cytokines (often called a "cytokine storm") and an inflammatory response that damages the lining of the lung and increases its leakiness. As a result, fluid and large numbers of white blood cells, particularly neutrophils, accumulate in the tiny air sacks (alveoli) of the lungs. This severely reduces the ability of the lungs to function and transfer oxygen into, and waste gases out of, the blood stream, resulting in respiratory failure.

In addition, it has been found that as many as one third of patients with the severe form of the disease develop blood clots that can damage many organs of the body. Other organs with high levels of ACE2 are the heart, oesophagus, kidneys, bladder and ileum which might explain why these can also be affected during the course of the disease. The lower levels of ACE2 in the respiratory epithelium of children younger than 10 years may explain why COVID-19 is less prevalent in this age group.

How Is the Disease Diagnosed?

Most patients with confirmed COVID-19 have fever, altered taste or smell, or symptoms of acute respiratory illness such as a new cough. Patients displaying these symptoms, therefore, should be regarded as potentially suffering from COVID-19 and should be tested for SARS-CoV-2. They should also be immediately isolated. Testing is usually carried out on a sample of fluid taken from the nasopharynx with a swab (see Fig. 10.6). The fluid is tested for the presence of RNA (the genetic material) from SARS-CoV-2 using PCR (see Chap. 1). If the test proves to be positive, the patient should remain isolated and his/her contacts traced for testing and possible isolation.

Antibody tests are being developed which will be a significant advantage over current swab-based testing in that they are faster and can also identify not only who has the infection but also who has had it previously and if they are likely to have developed immunity. Measuring the concentration of oxygen in your blood by pulse oximetry is a simple, rapid means of determining whether you are suffering from respiratory problems. Chest imaging by X-ray or computerised tomography (CT) is also useful as it helps in making the diagnosis and identifying complications.

How Is the Disease Treated?

Management of the disease currently depends mainly on its severity and the following description is based on guidelines published in *BMJ Best Practice* https://bestpractice.bmj.com/topics/en-gb/3000201

(a) Mild disease. Patients with mild disease, as well as those who are asymptomatic, should be isolated to contain virus transmission. In most cases, isolation will be at home but if this is not suitable or available, then other options can be considered such as a community or healthcare facility. For those isolated at home, patients and household members should follow appropriate infection prevention and control measures as outlined by the WHO or CDC:

https://www.who.int/publications/i/item/home-care-for-patients-with-suspected-novel-coronavirus-(ncov)-infection-presenting-with-mild-symptoms-and-management-of-contacts

https://www.cdc.gov/coronavirus/2019-ncov/hcp/guidance-home-care.html

Paracetamol is usually recommended for managing fever and pain.

Those over 1 years of age may consider taking a teaspoon of honey to ease coughing and you should avoid lying on your back as this makes coughing less effective. Loss of taste and smell often recovers spontaneously.

(b) Moderate disease. Patients with moderate disease should be isolated in a healthcare or community facility. Home isolation, with telemedicine or remote visits, may be possible in low-risk patients. Recommendations on when to stop isolation differ between countries. In the UK, the isolation period is 14 days from a positive test in hospitalised patients, and 10 days in patients with milder disease who are managed in the community. Infection prevention and control measures should be implemented as outlined by the WHO or CDC (see under "mild disease"). Symptoms should be managed as described above for "mild disease". In addition, antibiotics may be considered if a secondary bacterial pneumonia is suspected.

(c) Severe disease. Patients in this category are at serious risk of rapid clinical deterioration and should usually be isolated and managed in a healthcare facility.

(d) Critical disease. Such patients are usually admitted to an intensive/critical care unit.

Prognosis

Long-term outcomes of the disease are not yet clearly understood. Although most patients recover within 2 weeks, approximately 10% have symptoms after 3 weeks, and some may have symptoms for months – such patients are referred to as suffering from "long COVID". In approximately 90% of hospitalised patients, at least one symptom persisted 2 months after discharge, and 55% had three or more symptoms. Symptoms vary widely, may relapse and remit, and can occur in those who had only the mild disease. Long-term symptoms include cough, low-grade fever and fatigue. Dyspnoea, chest pain, myalgia, headaches, rashes, gastrointestinal symptoms, neurocognitive difficulties and mental health conditions have also been reported. Many patients recover spontaneously following a period of rest, symptomatic treatment and a gradual increase in normal activity.

How Can I Avoid Getting COVID-19?

Person-to-person spread of the disease is thought to occur mainly via respiratory droplets that are generated by coughing, sneezing and talking. However, evidence is accumulating to suggest that the virus may also be transmitted via aerosols produced by the respiratory tract – this is often referred to as "airborne transmission". Virions present in these droplets or aerosols can infect another person if they reach their mucous membranes. Respiratory droplets do not usually travel more than 6 feet (about 2 meters) and do not linger in the air; therefore, keeping away from people by more than this distance will reduce your risk of getting the disease. Virions present in aerosols, however, can remain suspended for long periods of time (>3 hours) and can travel further than 2 metres and remain viable for at least 3 hours. Crowded, indoor places would, therefore, likely contain a high concentration of virus-containing aerosols. It is important to remember that the virus can be spread by people who are not showing any symptoms of the disease. However, the infective dose of the virus is also currently not known so the risks of these situations is difficult to quantify.

Avoiding crowds, especially in confined spaces, will help to reduce the risk of meeting, and being infected by, someone with the disease. Respiratory droplets and aerosols from an infected individual will, of course, fall onto surfaces and if you touch such a contaminated surface and then touch your face, eyes, nose or mouth then you could become infected.

A study of a room in which a patient with COVID-19 was isolated showed that the virus was present on nearly all surfaces tested – handles, light switches, bed and handrails, interior doors and windows, toilet bowl and sink basin. Survival of the SARS-CoV-2 virus on surfaces depends on the nature of the surface – on plastic and steel it can survive for as long as 72 hours. Disinfection of surfaces, therefore, is an important protective measure as is washing your hands in order to remove any virions that you may have inadvertently picked up from contaminated surfaces. The CDCP recommends the following precautions in order to reduce the risk of you becoming infected:

- Wash your hands frequently with soap and water for at least 20 seconds especially after you have been in a public place, or after blowing your nose, coughing or sneezing.
- If soap and water are not readily available, use a hand sanitizer that contains at least 60% alcohol. Cover all surfaces of your hands, and rub them together until they feel dry.
- Avoid touching your eyes, nose and mouth with unwashed hands.
- Avoid close contact with people who are sick.
- Put distance between yourself and other people if COVID-19 is spreading in your community. This is especially important for people who are at higher risk of getting very sick.
- Stay at least 6 feet (about 2 arms' length) from other people.
- Do not gather in groups.
- Stay out of crowded places and avoid mass gatherings
- Stay home if you are sick, except to get medical care.
- Cover your mouth and nose with a cloth face cover when around others.
- Cloth face coverings should not be placed on young children under age 2 years, anyone who has trouble breathing, or is unconscious, incapacitated or otherwise unable to remove the mask without assistance.
- Do NOT use a facemask meant for a healthcare worker.
- Cover your mouth and nose with a tissue when you cough or sneeze, or use the inside of your elbow.
- Dispose of used tissues carefully and immediately wash your hands with soap and water for at least 20 seconds or clean your hands with a hand sanitizer that contains at least 60% alcohol.
- Clean and disinfect frequently touched surfaces daily. This includes tables, doorknobs, light switches, countertops, handles, desks, phones, keyboards, toilets, faucets and sinks.
- If surfaces are dirty, clean them using detergent or soap and water before you disinfect them.

Further information on personal protection is available from the following websites:

https://www.cdc.gov/coronavirus/2019-ncov/prevent-getting-sick/prevention.html
https://www.who.int/emergencies/diseases/novel-coronavirus-2019/advice-for-public

Currently, the governments of many countries have introduced measures to try to slow down the spread of the disease. These include closing schools, universities, restaurants, bars, places of entertainment, etc. Self-isolation and social distancing are being encouraged and, in some cases, enforced. Curbs on travel, both within and between, countries have also been introduced. Hopefully, this will help to inhibit the spread of COVID-19. However, these measures will also have profound social, health and economic consequences some unintended and disadvantageous. At the time of writing, we are probably not yet able to fully appreciate the balance of risks and benefits for some of these interventions.

Want to Know More?

Centers for Disease Control and Prevention, USA https://www.cdc.gov/coronavirus/2019-ncov/about/index.html
Government of Canada https://www.canada.ca/en/public-health/services/diseases/coronavirus-disease-covid-19.html
Mayo Clinic, USA https://www.mayoclinic.org/diseases-conditions/coronavirus/symptoms-causes/syc-20479963
National Health Service, UK https://www.nhs.uk/conditions/coronavirus-covid-19/
National Institute for Clinical Care and Excellence (NICE), UK. Coronavirus - COVID-19. 2020 https://cks.nice.org.uk/coronavirus-covid-19#!topicSummary
National Institutes of Health, USA https://www.nih.gov/health-information/coronavirus
Patient Info, UK https://patient.info/coronavirus-covid-19
World Health Organisation https://www.who.int/emergencies/diseases/novel-coronavirus-2019
Beeching NJ, Fletcher TE, Fowler R. Coronavirus disease 2019 (COVID-19). BMJ Best Practice, BMJ Publishing Group, 2020 https://bestpractice.bmj.com/topics/en-gb/3000201
Cascella M, Rajnik M, Cuomo A, Dulebohn SC, Di Napoli R. Features, Evaluation and Treatment Coronavirus (COVID-19). Treasure Island (FL): StatPearls Publishing LLC; 2020 https://www.ncbi.nlm.nih.gov/books/NBK554776/

Cennimo DJ. Coronavirus Disease 2019 (COVID-19). Medscape from WebMD, 2020 https://emedicine.medscape.com/article/2500114-overview

Chams N, Chams S, Badran R, Shams A, Araji A, Raad M, Mukhopadhyay S, Stroberg E, Duval EJ, Barton LM, Hajj Hussein I. COVID-19: A Multidisciplinary Review. *Frontiers in Public Health.* 2020 Jul 29;8:383. https://doi.org/10.3389/fpubh.2020.00383. eCollection 2020. PMID: 32850602

Chauhan S. Comprehensive review of coronavirus disease 2019 (COVID-19). *Biomedical Journal.* 2020 Jun 1:S2319-4170(20)30087-1. https://doi.org/10.1016/j.bj.2020.05.023.

Van Damme W et al. The COVID-19 pandemic: diverse contexts; different epidemics-how and why? *BMJ Global Health.* 2020 Jul;5(7):e003098. https://doi.org/10.1136/bmjgh-2020-003098.

Wolff D, Nee S, Hickey NS, Marschollek M. Risk factors for Covid-19 severity and fatality: a structured literature review. *Infection.* 2020 Aug 28. https://doi.org/10.1007/s15010-020-01509-1.

Mortenson LY, Malani NP, Ernst RD. Caring for Someone With COVID-19. *Journal of the American Medical Association* 2020;324(10):1016. https://doi.org/10.1001/jama.2020.15061

13

Sore Throat

Abstract A sore throat (pharyngitis) is a symptom of many diseases, but this chapter will focus on two of the main causes of the condition – *Streptococcus pyogenes* (a bacterium) and the Epstein-Barr virus. Infection with *Strep. pyogenes* is most frequent in 5–15-year-old children and is usually accompanied by fever, red and swollen tonsils and swollen glands in the neck. In some children a skin rash develops, and the condition is then known as scarlet fever. It usually occurs in winter or early spring and is highly contagious. The disease usually resolves without medication after about 7 days, but antibiotics may be needed in severe cases or for those at risk of complications. Complications are rare but can be serious and include rheumatic fever, rheumatic heart disease and glomerulonephritis. The Epstein-Barr virus causes infectious mononucleosis (glandular fever or the "kissing disease") mainly in adolescents and young adults. The main symptoms are sore throat, fever, malaise and enlarged lymph glands. It's usually a mild, self-limiting condition, and the symptoms can be relieved by antipyretics and analgesics. It's spread mainly in saliva.

What Are the Main Causes of a Sore Throat?

Infection of the throat (called the pharynx, which includes the tonsils located at the back of the mouth) invariably results in it becoming inflamed, and this leads to the characteristic symptoms – pain and scratchiness or irritation of the throat that often feels worse during swallowing. Inflammation of the throat is known as pharyngitis, and it can have non-infective as well as infective causes. A large variety of microbes are able to cause pharyngitis, and, depending on which of these is responsible, a number of other symptoms may be present. The common cold and influenza, for example, are invariably accompanied by a sore throat, and the other symptoms of these diseases are described in Chaps. 10 and 11, respectively. Other infective causes of pharyngitis include herpes, candida, measles and chlamydia, but, as these are relatively less common causes in Western countries and/or are discussed in other chapters, they won't be discussed further here. As the common cold and influenza are described in Chaps. 10 and 11, this chapter will focus on the other two main causes of infective pharyngitis in Western countries – *Streptococcus pyogenes* (a bacterium) and the Epstein-Barr virus (responsible for glandular fever).

Box 13.1 What's in a Name?

Streptococcus is derived from two Greek words – "streptos" and "kokkos" which mean "twisted or pliant" (i.e. like a chain) and "a grain", respectively. The name refers to the microscopic appearance of the organism which is often in the form of chains of cocci. The species name *pyogenes* comes from another two Greek words – "puon" and "gennao" which mean "pus" and "to produce", respectively. This refers to the fact that many infections due to the organism result in the accumulation of pus.

The organism was first described in 1874 by the Austrian surgeon, Theodor Billroth, who saw it in skin and wound infections. It was first isolated by Louis Pasteur in 1879 who grew the organism from samples obtained from women with puerperal fever. It was given its current name in 1884 by the German microbiologist Friedrich Julius Rosenbach.

What Are the Main Symptoms of Strep Throat (Streptococcal Pharyngitis)?

Between 20% and 30% of sore throats in children are caused by *Strep. pyogenes* whereas in adults it's between 5% and 15%. The main symptoms of pharyngitis (Fig. 13.1) due to this organism (often called "strep throat" or "streptococcal pharyngitis") are (i) a sore throat that starts very quickly; (ii) pain on swallowing; (iii) fever; (iv) red and swollen tonsils, sometimes with white patches or streaks of pus; (v) tiny, red spots (known as petechiae) on the

Fig. 13.1 (continued) (b) Showing inflammation and swelling (oedema) of the pharynx as well as petechiae on the soft palate. (Image courtesy of Dr. Heinz F. Eichenwald, Centers for Disease Control and Prevention). (c) Illustration of a child covered with the rash (scarlet fever or scarlatina) that sometimes accompanies pharyngitis due to *Strep. pyogenes*. (Image from page 291 of "Diseases of children for nurses" (1911). No known copyright restrictions)

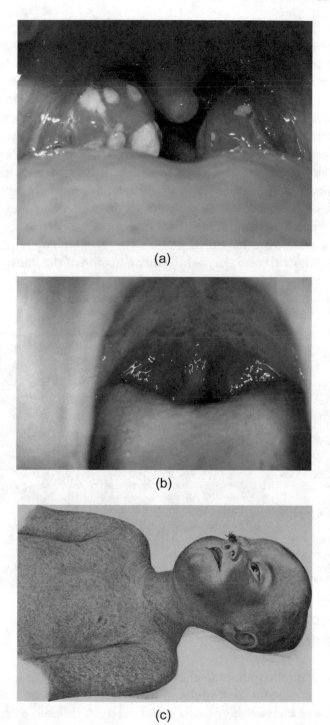

Fig. 13.1 Symptoms of pharyngitis due to *Strep. pyogenes* (i.e. strep throat). (a) Inflamed tonsils with patches of pus. (Michaelbladon at English Wikipedia/Public domain).

roof of the mouth; and (vi) swollen lymph glands at the front of the neck. Because the pharynx is located at the entrance to both the gut and the airways, pain and swelling here can cause serious problems, especially in children. Obstruction from swollen tissues can cause breathing problems, and failure to drink, because of pain on swallowing, can result in dehydration.

Additional, less frequent, symptoms include headache, stomach pain, nausea and vomiting (especially in children). Sometimes a child also develops a skin rash, and then the disease is known as scarlet fever (or scarlatina). Some people develop a red swollen bumpy tongue, perhaps with a whitish coating, which is described as "strawberry tongue". Symptoms usually become apparent 2–5 days following initial exposure to *Strep. pyogenes* and persist for 5–7 days. When the tonsils at the back of the mouth, as well as the throat, are affected, the disease is often called tonsillitis or tonsillopharyngitis.

Coughing, hoarse voice (laryngitis), inflammation of the lining of the nostrils and conjunctivitis aren't usual symptoms of streptococcal pharyngitis, and if these are present, they suggest that a virus is a more likely cause of the infection.

Box 13.2 Scarlet Fever

Scarlet fever (Fig. 13.1c) results from throat infections caused by particular strains of *Strep. pyogenes* that produce toxins known as erythrogenic (or pyrogenic) toxins that are responsible for the characteristic rash of scarlet fever. From 1840 until 1885, scarlet fever was one of the most common life-threatening infectious diseases of childhood with mortality rates as high as 30%. Two of Charles Darwin's children died of the disease, while another two recovered from it. John D. Rockefeller Sr.'s grandson died from scarlet fever in January 1901, and, because of this, he established the first biomedical research centre in the USA – the Rockefeller Institute. However, after 1885 the incidence of the disease, as well as the mortality associated with it, started to decline. Why this happened isn't clear but is likely to be due to a combination of factors including (i) the dominance of strains of the bacterium that did not produce erythrogenic toxins, (ii) an increased immunity to toxin-producing strains in the population and (iii) improved living conditions and hygiene standards.

Who Gets Strep Throat, and How Is It Transmitted?

Strep throat is most common in children aged 5–15 years, but it can affect all age groups. Certain groups of adults are at a greater risk of acquiring the infection, and these include the parents of school-aged children and those who have frequent contact with children such as teachers. The disease is highly contagious, and the bacterium is present in the droplets of aerosols produced when an infected individual coughs, sneezes or talks. Inhalation of these droplets can result in infection. Pathogen-containing droplets can also land on surfaces, and they can survive on these for several hours. Someone who touches these surfaces and then touches their mouth, eyes or nose can become infected. The pathogen can also be transmitted when someone drinks from the same cup or eats food from the same plate as an infected individual. The bacterium can also cause skin infections such as impetigo, and you may become infected after touching the skin of someone affected if you don't wash your hands. Crowded conditions such as those occurring in day care centres and schools increase the risk of becoming infected.

The disease occurs most frequently in winter and early spring, and in the USA, children typically experience strep throat once every 4 years. Acute pharyngitis accounts for 1.9% of outpatient visits to healthcare providers in the USA which in 2016 amounted to approximately 18 million patient visits. However, this represents only a small proportion of those actually infected as it's been shown that only approximately 15% of those afflicted seek medical care. The economic burden of streptococcal pharyngitis among children in the USA has been estimated to be $224 million to $539 million per year. In the UK, 4.1–5.7% of the population consult their doctors each year because of a sore throat, and it's been estimated that it's economic impact amounts to £400 million per year in consultations and lost productivity alone. However, studies have shown that in the UK only 6% of patients with a sore throat consult a medical practitioner, so, as in the USA, the disease burden is much greater than that implied by the above figures.

What About Complications?

Complications are rare but can be serious – they include abscesses in the affected areas (peritonsillar and retropharyngeal), acute rheumatic fever, rheumatic heart disease and glomerulonephritis. Abscesses can be treated by draining away the accumulated pus combined with antibiotic administration, whereas the three other conditions, known collectively as "non-suppurative complications", are more difficult to treat and have more serious, long-term consequences. Other complications include sinusitis, epiglottitis, mastoiditis, pneumonia, meningitis and otitis media.

Rheumatic fever occurs in up to 0.3% of patients with streptococcal pharyngitis and is the result of an immune response to the organism – it occurs

1–5 weeks after the throat infection. The accompanying inflammation can affect the heart, joints, skin or central nervous system and results in a variety of symptoms including:

- Fever
- Painful, tender joints – usually in the knees, ankles, elbows and wrists
- Chest pain, shortness of breath, fast heartbeat
- Fatigue
- Jerky, uncontrollable body movements (called "chorea")

Painless lumps (nodules) under the skin near joints and a rash are less frequent symptoms. Children between the ages of 5 and 14 years are most frequently affected, and worldwide there are over 500,000 new cases each year. Treatment of the condition focusses on treating the fever, pain and inflammation with drugs like paracetamol or ibuprofen together with antibiotics to combat *Strep. pyogenes*.

Between 30% and 50% of those with rheumatic fever develop rheumatic heart disease. This usually affects the mitral valve in the heart, and valve repair or replacement surgery may be necessary. More than 2.4 million children have rheumatic heart disease worldwide, and 94% of these are in developing countries. Approximately 223,000 deaths each year are attributable to rheumatic heart disease. In industrialised countries, however, it's no longer thought appropriate to routinely treat streptococcal pharyngitis in order to reduce the risk of rheumatic heart disease.

Between 10% and 15% of patients with streptococcal pharyngitis develop acute glomerulonephritis – usually 1–2 weeks after the initial infection. This is again an immune response to *Strep. pyogenes* and results in inflammation of the glomeruli, the blood-filtering units inside the kidney. As well as being triggered by some infections, glomerulonephritis can also be caused by a wide variety of other illnesses. Symptoms include blood in the urine (haematuria), protein in the urine (proteinuria), high blood pressure and swelling of the face, hands, feet and abdomen although, when it's mild, it may have no symptoms at all. It affects mainly school-age children, and about 470,000 cases occur annually resulting in 5000 deaths. Most cases occur in developing countries. Treatment is usually guided by kidney specialists (Nephrologists) and, depending on the cause and severity, can require controlling the body's fluid balance and the swelling symptoms, immune-suppressants and managing the high blood pressure.

What Happens During an Infection?

Strep. pyogenes, the main bacterial cause of pharyngitis, is a well-known pathogen of humans. It's also known as the "Group A streptococcus" (on the basis of a classification system that recognises 16 groups of bacteria belonging to the family *Streptococcaceae*) and the "flesh-eating bacterium" (because it's responsible for a tissue-destroying disease known as necrotising fasciitis). It's a Gram-positive bacterium that tends to form chains of cocci (Fig. 13.2).

This bacterium is a member of the indigenous microbiota of a relatively small proportion of healthy individuals and is found in various regions of the upper respiratory tract (Fig. 13.3).

The organism can live in the respiratory tract of healthy people without causing any problems. It survives on the nutrients that are present in the fluids that line the respiratory tract. However, it has a large number of virulence factors which shows that it does have the potential to cause disease if given the opportunity. These virulence factors include (i) a capsule that protects it against our phagocytic defence cells, (ii) enzymes that break down the antibodies that we produce to deal with it, (iii) toxins that can kill a range of different human cells, (iv) enzymes that can break down our tissues and (v) toxins that cause inflammation and fever. What triggers an infection by the organism which had previously not been problematic in an individual isn't fully understood. The first stage of an infection involves invasion of the epithelial cell to which it has attached itself. This induces an inflammatory response and the accompanying fever, redness and swelling that are characteristic of pharyngitis. Its array of tissue-destroying enzymes and toxins then cause further damage to respiratory tissues. A huge number of neutrophils are attracted to the site of the infection, and their eventual death results in pus accumulation. Some, but not all, strains of *Strep. pyogenes* produce a toxin called streptococcal pyrogenic exotoxin A which gives rise to a rash which is characteristic of scarlet fever (Fig. 13.1).

Because of its wide range of virulence factors, *Strep. pyogenes* is able to cause a variety of other diseases, some of which are life-threatening (Table 13.1 and Fig. 13.4). It can cause meningitis, bacteraemia, pneumonia, myositis, pericarditis and septic arthritis. Fortunately, all of these diseases (as well as those listed in Table 13.1) occur far less frequently than pharyngitis.

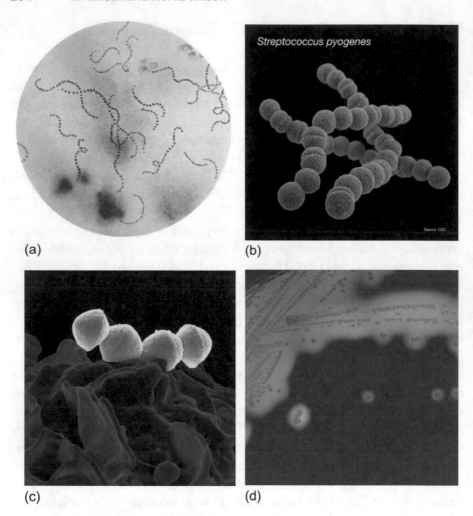

Fig. 13.2 Images of *Strep. pyogenes.* (a) Gram stain of *Strep. pyogenes* showing long chains of cocci. (Image courtesy of Centers for Disease Control and Prevention, USA). (b) A three-dimensional computer-generated image of a group of *Strep. pyogenes.* The artistic recreation is based upon scanning electron microscopic imagery. (Image courtesy of Jennifer Oosthuizen, Centers for Disease Control and Prevention, USA). (c) *Strep. pyogenes* on the surface of a human neutrophil. (Image courtesy of National Institute of Allergy and Infectious Diseases, USA). (d) Colonies of *Strep. pyogenes* growing on an agar plate containing blood. Note the clear zones around each colony. These clear zones are the result of the death of red blood cells around the colony due to the action of a toxin produced by the bacterium – it's described as being "haemolytic", i.e. it destroys red blood cells. (Nathan Reading from Halesowen, UK [CC BY 2.0 (https://creativecommons.org/licenses/by/2.0)])

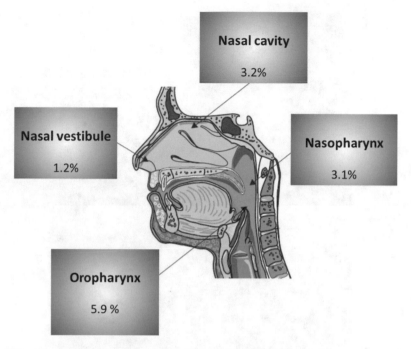

Fig. 13.3 Prevalence of *Strep. pyogenes* in various regions of the upper respiratory tract of healthy people. The figures show the proportion (%) of healthy individuals (adults and children) who carry the organism

Table 13.1 Some of the diseases, other than pharyngitis, that are caused by *Strep. pyogenes*

Disease	Features
Impetigo	A skin infection characterised by sores and blisters. A very common infection in children
Erysipelas	An infection of the upper layers of the skin – causes the skin to appear red, swollen and shiny
Necrotising fasciitis	A painful, life-threatening infection of the deeper layers of the skin that spreads rapidly and is accompanied by fever, vomiting and diarrhoea
Toxic shock syndrome	A painful, life-threatening condition caused by the massive toxin-induced release of cytokines

How Is Strep Throat Diagnosed?

A definitive diagnosis of streptococcal pharyngitis can be made by taking a throat swab and trying to grow *Strep. pyogenes* from this in the laboratory – a test that provides very reliable results in about 48 hours. Alternatively, an almost immediate result can be obtained using what's known as a "Rapid

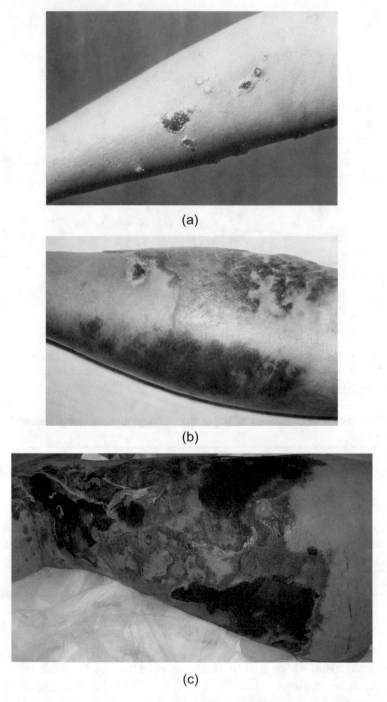

(a)

(b)

(c)

Fig. 13.4 Some of the diseases, other than pharyngitis, that are caused by *Strep. pyogenes*. (a) Impetigo due to *Strep. pyogenes*. (Dr. Herman Miranda, Univ. of Trujillo, Peru; A. Chambers, Centers for Disease Control and Prevention, USA).

Antigen Test". In this case, a sample of mucus from the throat is taken and tested for the presence of antigens of *Strep. pyogenes*. This is an example of "near patient" or "point-of-care" testing and is one of the tools that can be used to combat inappropriate antibiotic use. However, as mentioned earlier, some people have *Strep. pyogenes* as part of their normal microbiota, so a positive test only tells you that the bacterium is present and not that it's the cause of the current symptoms. Most authorities recommend antibiotic treatment if someone has a sore throat and a positive Rapid Antigen Test.

How Is Strep Throat Treated?

In the vast majority of cases, the infection resolves on its own – the sore throat lasts between 3 and 6 days and the fever around 7 days. Antibiotics only reduce the length of the symptoms by, on average, less than a day (16 hours) for 7 days of treatment. This small advantage has to be weighed against the adverse effects of antibiotics such as diarrhoea.

Antibiotics may be appropriate if someone is very unwell with severe symptoms or is at risk of, or develops, complications. Another advantage of antibiotic treatment is that it reduces the spread of the infection to others, which is a consideration in outbreak control.

Antibiotic administration has only a small to modest effect on the severity and duration of symptoms of the disease, and this is only in those most severely affected. They're generally used to reduce the likelihood of complications and secondary infections as well as the severity of symptoms in people who are very unwell. Fortunately, the organism hasn't yet developed resistance to penicillin, so this is the antibiotic generally used. In those who are allergic to penicillins, alternative antibiotics include clindamycin, cefalexin and erythromycin or clarithromycin. Sometimes, if the cause of the infection is unclear, a delayed prescribing strategy is advised for treating people with sore throat symptoms – antibiotics are used only if you're not improving as expected. Analgesics (such as paracetamol or ibuprofen) and local anaesthetics (such as lidocaine) can be taken to alleviate sore throat symptoms, headache and fever. You could also try salt water gargling because this can be helpful for pain relief.

Fig. 13.4 (continued) (**b**) Erysipelas due to *Strep. pyogenes*. (Dr. Thomas F. Sellers, Emory University, Centers for Disease Control and Prevention, USA). (**c**) Necrotising fasciitis due to *Strep. pyogenes*. (Piotr Smuszkiewicz, Iwona Trojanowska and Hanna Tomczak [CC BY 2.0 (https://creativecommons.org/licenses/by/2.0)])

How Can I Avoid Getting Strep Throat?

The main ways of avoiding being infected with *Strep. pyogenes* include proper handwashing and avoiding the sharing of items such as utensils, drinking glasses, water bottles and toothbrushes. If you do become infected, you should cover your mouth when coughing or sneezing and dispose of paper handkerchiefs properly, to help prevent transmitting it to others. Milk from cattle affected by mastitis due to *Strep. pyogenes* was once a source of the infection, but this route has been prevented by pasteurisation of milk.

What Is Glandular Fever?

Another frequent cause of pharyngitis is the Epstein-Barr virus (EBV) which causes glandular fever, also known as infectious mononucleosis (or "mono"). It's been shown that more than 95% of the world's population has been infected with EBV by the age of 40 years but often without having experienced any ill effects. However, in adolescents and young adults, it can cause glandular fever which is characterised by sore throat (84% of cases), fever (76% of cases), malaise (82% of cases) and enlarged lymph glands (94% of cases). Other symptoms can include a rash, nausea, joint aches and eyelid swelling (Fig. 13.5). Estimates suggest that it's responsible for 8% of sore throats in those between the ages of 16 and 20 years. Sore throat due to EBV, unlike that due to *Strep. pyogenes*, doesn't show any seasonal variation.

Symptoms of the disease develop 30–50 days following initial exposure and usually disappear after 2–4 weeks. However, feelings of malaise and fatigue can persist for several months. As well as a typical sore throat, the infection can provoke a wide variety of rarer physical, psychological and neurological symptoms such as "Alice in Wonderland" syndrome, an unusual set of symptoms of distorted perception. Infection with EBV in the past is associated with the risk of a number of other diseases developing in the future. These include Burkitt's lymphoma, some cancers of the pharynx and saliva glands and multiple sclerosis. It might also be a contributor to a variety of unusual and rare disorders mostly affecting the nervous system and blood production; however the evidence for much of this remains patchy, and the role of EBV is often not clearly understood. Other rare complications include chronic active EBV, inflammation of the pancreas, spleen rupture, anaemia, myocarditis, encephalitis, hearing loss, optic neuritis, and pneumonia.

The EBV, also known as human herpesvirus 4, is between 122 and 180 nm in diameter (Fig. 13.6a). It has a complex structure with DNA as its genetic

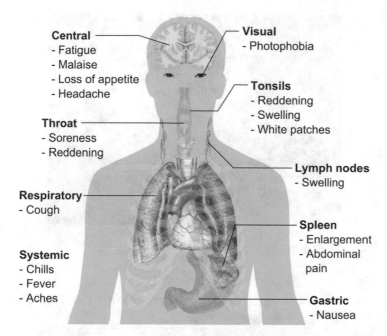

Fig. 13.5 The main symptoms of infectious mononucleosis (glandular fever). (Mikael Häggström. When using this image in external works, it may be cited as Häggström, Mikael (2014). Medical gallery of Mikael Häggström 2014; WikiJournal of Medicine 1(2). DOI:https://doi.org/10.15347/wjm/2014.008. ISSN 2002-4436. Public Domain.orBy Mikael Häggström, used with permission. / Public domain)

material, and this is contained within a protein coat (tegument) surrounded by a lipid envelope which has protein spikes (Fig. 13.6b).

The virus was discovered in 1964 by two researchers, Anthony Epstein and Yvonne Barr, working at the Middlesex Hospital Medical School in London. However, it wasn't until 1968 that it was found to be the cause of glandular fever.

What Happens if I Get Glandular Fever?

On arriving in the respiratory tract, the EBV invades white blood cells in the oropharynx, and these infected cells then carry it to the liver, spleen and lymph glands. The immune response to the virus gives rise to the symptoms of the disease. Glandular fever is diagnosed on the basis of the characteristic symptoms and the use of special blood tests to confirm the presence of antibodies to the EBV. A rapid test has been developed to identify DNA of the

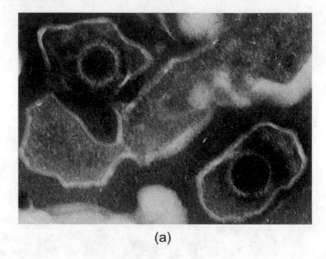

(a)

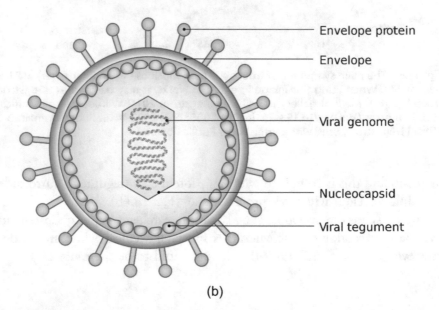

Envelope protein

Envelope

Viral genome

Nucleocapsid

Viral tegument

(b)

Fig. 13.6 The Epstein-Barr virus. (a) This electron micrograph shows two virions, each of which contains a round DNA-containing region that is surrounded by a loose membranous coating. (Liza Gross [CC BY 2.5 (https://creativecommons.org/licenses/by/2.5)]). (b) Diagram showing the structure of the Epstein-Barr virion. (Ben Taylor. I, the copyright holder of this work, release this work into the public domain. This applies worldwide. Via Wikimedia Commons)

EBV for high-risk cases who might benefit from prompt identification. In most cases, the blood of a patient with the disease has a characteristically large proportion of white cells called lymphocytes, many of which have an abnormal shape and resemble different white cells called monocytes – this has given rise to the alternative name (infectious mononucleosis) for the disease. 90% of cases of infective mononucleosis are caused by the EBV with most of the remainder being caused by other viruses including HIV and Cytomegalovirus.

How Is Glandular Fever Treated?

Glandular fever is in most cases a mild, self-limiting condition with no specific treatment other than the control of symptoms using antipyretics and analgesics such as paracetamol and non-steroidal anti-inflammatory drugs. Anti-viral agents and steroid treatment haven't, so far, been found to be helpful although steroids are sometimes used in rare complications or to reduce tonsil swelling. You should take plenty of rest and refrain from strenuous physical activity and contact sports during the first few weeks or months of illness – this is to avoid the possibility of a swollen spleen rupturing with trauma.

How Can I Avoid Getting Glandular Fever?

The disease isn't particularly contagious and is spread mainly via saliva and so is often known as the "kissing disease". People with no symptoms themselves can unknowingly infect others. You're most likely to catch the disease by kissing an infected person or using their toothbrush. There's some evidence that it can also be transmitted by sexual intercourse.

Want to Know More?

American Academy of Otolaryngology–Head and Neck Surgery Foundation. https://www.enthealth.org/conditions/sore-throats/
American Family Physician. https://www.aafp.org/afp/2004/0315/p1465.html, https://www.aafp.org/afp/2004/1001/p1279.html
Centers for Disease Control and Prevention, USA. https://www.cdc.gov/groupastrep/diseases-public/strep-throat.html?CDC_AA_refVal=https%3A%2

F%2Fwww.cdc.gov%2Ffeatures%2Fstrepthroat%2Findex.html, https://www.cdc. gov/epstein-barr/about-mono.html

Mayo Clinic, Mayo Foundation for Medical Education and Research, USA. https:// www.mayoclinic.org/diseases-conditions/sore-throat/symptoms-causes/ syc-20351635, https://www.mayoclinic.org/diseases-conditions/mononucleosis/ symptoms-causes/syc-20350328

MedlinePlus, U.S. National Library of Medicine. https://medlineplus.gov/sore-throat.html

National Health Service, UK. https://www.nhs.uk/conditions/sore-throat/, https:// www.nhs.uk/conditions/glandular-fever/

National Institute for Clinical Care and Excellence (NICE), UK. Sore throat, 2018. https://cks.nice.org.uk/sore-throat-acute#!topicSummary

Patient Info, UK. https://patient.info/doctor/infectious-mononucleosis, https:// patient.info/doctor/tonsillitis-pro

Acerra JR. Pharyngitis. Medscape from WebMD, 2020. https://emedicine.medscape. com/article/764304-overview#a1

Basetti S, Hodgson J, Rawson TM, Majeed A. Scarlet fever: a guide for general prac-titioners. *London Journal of Primary Care (Abingdon)*. 2017 Aug 11;9(5):77-79. doi: https://doi.org/10.1080/17571472.2017.1365677. eCollection 2017 Sep.

Donowitz JR. Acute pharyngitis. BMJ Best Practice, BMJ Publishing Group, 2020. https://bestpractice.bmj.com/topics/en-gb/5

Dunmire SK, Hogquist KA, Balfour HH. Infectious Mononucleosis. *Current Topics in Microbiology and Immunology.* 2015;390(Pt 1):211-40. doi: https://doi. org/10.1007/978-3-319-22822-8_9.

Kalra MG, Higgins KE, Perez ED. Common questions about streptococcal pharyn-gitis. *American Family Physician.* 2016 Jul 1;94(1):24-31.

Wessels MR. In: Ferretti JJ, Stevens DL, Fischetti VA, editors. Pharyngitis and scarlet fever. *Streptococcus pyogenes*: Basic Biology to Clinical Manifestations [Internet]. Oklahoma City (OK): University of Oklahoma Health Sciences Center; 2016-.2016 Feb 10 [updated 2016 Mar 25]

Wolford RW, Goyal A, Syed SYB; Schaefer TJ. Pharyngitis. Treasure Island (FL): StatPearls Publishing LLC; 2020. https://www.ncbi.nlm.nih.gov/books/ NBK519550/

Womack J, Jimenez M. Common questions about infectious mononucleosis. *American Family Physician.* 2015 Mar 15;91(6):372-6.

14

Acute Sinusitis

Abstract Acute sinusitis is an infection of the sinuses that lasts for no longer than 12 weeks. It usually follows a viral infection of the upper respiratory tract – the resulting inflammation impairs drainage of mucus from the sinus. However, poor drainage may be due to a variety of other factors. This blockage creates conditions suitable for the proliferation of viruses or bacteria within the sinuses and results in sinusitis. The main symptoms include nasal discharge, nasal blockage, facial pain, fever, toothache and a reduced sense of smell. It's very common, and up to 14% of adults experience one episode of the disease every year. The virus responsible is usually a rhinovirus. When bacteria are involved, the most frequently detected species are *Streptococcus pneumoniae*, *Haemophilus influenzae* and *Moraxella catarrhalis*. Treatment includes using paracetamol or ibuprofen to relieve fever and pain. Intranasal corticosteroids may unblock the nose and improve breathing problems. Antibiotics are usually not recommended unless the symptoms haven't started to improve after 10 days. Antibiotics should also be considered in patients who are immunocompromised, those with complications and those with severe illness.

What Is Sinusitis?

The paranasal sinuses (often referred to simply as "sinuses") are a group of four hollow air-filled cavities of the skull that are situated behind the forehead and cheekbones (Fig. 14.1). They're joined to the nasal cavity by small openings (only 1–3 mm in diameter) known as ostia.

© Springer Nature Switzerland AG 2021
M. Wilson, P. J. K. Wilson, *Close Encounters of the Microbial Kind*,
https://doi.org/10.1007/978-3-030-56978-5_14

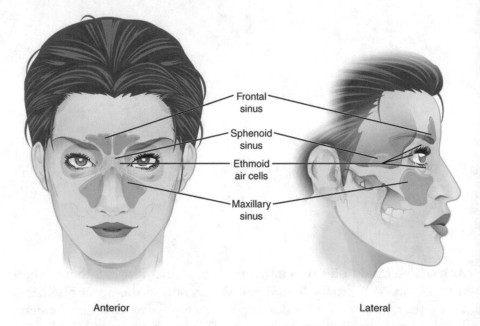

Fig. 14.1 Diagram showing the location of the four pairs of paranasal sinuses – frontal, sphenoid, ethmoid and maxillary. (Anatomy & Physiology, Connexions Web site. http://cnx.org/content/col11496/1.6/ OpenStax College [CC BY 3.0 (https://creativecommons.org/licenses/by/3.0)])

Sinusitis (often known colloquially as a "sinus infection") is an inflammation of the lining (i.e. the mucosa or epithelium) of one or more of the sinuses. Because the inflammation usually extends into the nasal cavity, the condition is often known as rhinosinusitis. Depending on which sinus is affected, four types of rhinosinusitis are recognised – maxillary, ethmoidal, frontal and sphenoidal. If the condition persists for no longer than 12 weeks, it's classified as acute rhinosinusitis; if it extends beyond 12 weeks, it's known as chronic rhinosinusitis (discussed in Chap. 15). The mucosa lining the sinuses continually produces mucus which is propelled by the mucociliary escalator (Fig. 1.18) through narrow gaps (1–3 mm in diameter) known as ostia into the nasal cavity. Inflammation results in swelling of the mucosa which blocks the ostia, and this prevents the mucus from draining out of the sinuses (Fig. 14.2). This results in the characteristic symptoms of rhinosinusitis which include:

- A green or yellow discharge from the nose
- A blocked nose
- Pain and tenderness around the cheeks, eyes or forehead
- Fever

maxillary sinus is completely blocked   maxillary sinus is no longer blocked

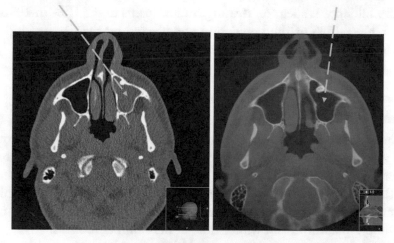

Fig. 14.2 CT scans of a patient with acute maxillary sinusitis before (left) and after (right) successful treatment. Note the complete blockage of the maxillary sinus before treatment. (Edvardsson, B. Cluster headache associated with acute maxillary sinusitis. SpringerPlus 2, 509 (2013). https://doi.org/10.1186/2193-1801-2-509. This article is distributed under the terms of the Creative Commons Attribution 2.0 International License (https://creativecommons.org/licenses/by/2.0), which permits unrestricted use, distribution, and reproduction in any medium, provided the original work is properly cited)

- Toothache in the upper teeth
- A reduced sense of smell
- Headache

The symptoms usually last for less than 4 weeks.

Complications of the disease are very rare and affect only 2.5–4.3 people per million per year. Children are more likely than adults to experience complications, and these include chronic rhinosinusitis (Chap. 15), orbital cellulitis (infection of the eye socket), orbital abscess, meningitis and osteomyelitis (infection of a bone).

How Common Is the Disease?

Acute rhinosinusitis affects an estimated 35 million people each year in the USA and accounts for around 16 million visits to medical practitioners per year. Approximately 14% of adults report having had one episode of the disease each year, and it's the fifth most common diagnosis for which antibiotics are prescribed. Women experience twice as many episodes of the disease than men. $3.39 billion was spent on treating rhinosinusitis in the USA in 1996.

In the UK, 6–15% of the population are affected by the disease, and it accounts for 1–2% of visits to general medical practitioners.

What Causes Rhinosinusitis?

Rhinosinusitis often follows a cold or other infection of the upper respiratory tract; it's therefore considered to be a secondary infection. It's estimated that 6–13% of children will have had one episode of the disease by the age of 3 years. The disease can affect any age group but may be less common in children because their sinuses aren't fully developed. In temperate climates, it's more common in autumn and winter than in the summer months.

Acute rhinosinusitis may be due to an infection with viruses or bacteria, with viral infection being the most frequent cause. The disease can also be precipitated by a number of non-infective factors including an allergy (such as allergic rhinitis or hay fever), asthma, smoking, dental procedures, mechanical obstruction (such as a foreign body or nasal polyps) and activities that can damage the lining of the sinuses such as diving. These can cause swelling of the mucosa of the sinuses (Fig. 14.2) and/or obstruction of the narrow ostia and therefore prevent drainage of mucus into the nasal cavity.

Acute rhinosinusitis is usually preceded by a viral infection of the upper respiratory tract – the virus then spreads into the sinuses. The virus responsible for the condition is usually a rhinovirus (approximately 50% of cases), influenza virus or parainfluenza virus (see Chaps. 10 and 11). However, it's important to emphasise that in the vast majority of cases, no attempt is made to identify the microbe responsible. If the disease symptoms last for less than 10 days, then this suggests that the infection is due to a virus.

Inflammation of the mucosa due to the viral infection results in blockage of the ostia causing mucus accumulation within the sinuses. This situation can also arise because of the other precipitating factors mentioned previously. This provides ideal conditions for the growth of bacteria that reside in the sinuses, some of which are pathobionts and have high pathogenic potential – this can result in a condition known as acute bacterial rhinosinusitis (ABRS). If the symptoms of the condition last for longer than 10 days with no improvement, it is more likely by then that a bacterium is responsible. In the USA it has been suggested that there are more than 20 million cases of ABRS each year, although estimates differ widely. The most frequently detected bacteria are those members of the respiratory microbiota that are often associated with infections: *Streptococcus pneumoniae* (20% to 43% of cases), *Haemophilus influenzae* (22% to 35% of cases) and *Moraxella catarrhalis* (2% to 10% of cases).

Strep. pneumoniae (Fig. 14.3) is a Gram-positive coccus that usually occurs in pairs and short chains.

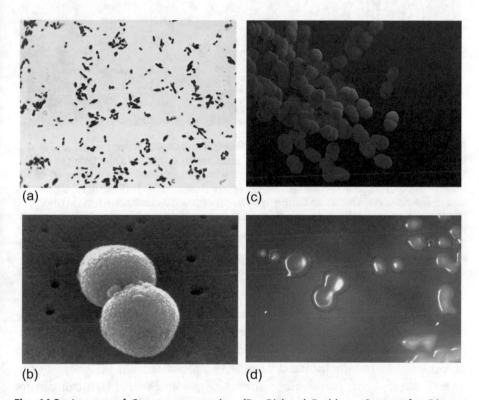

(a)   (c)

(b)   (d)

Fig. 14.3 Images of *Strep. pneumoniae*. (Dr. Richard Facklam, Centers for Disease Control and Prevention, USA). (a) Gram stain of *Strep. pneumoniae* showing typical Gram-positive cocci (purple-coloured) in pairs and short chains (×1700). (Image courtesy of Arnold Kaufman, Centers for Disease Control and Prevention, USA). (b) Electron micrograph of a pair of *Strep. pneumoniae* cells. (Dr. Richard Facklam, Centers for Disease Control and Prevention, USA). (c) A three-dimensional computer-generated image of a group of *Strep. pneumoniae* cells. The artistic recreation is based upon scanning electron microscopic imagery. (Antibiotic Resistance Coordination and Strategy Unit, Centers for Disease Control and Prevention, USA). (d). Colonies of *Strep. pneumoniae* on an agar plate (magnification ×10)

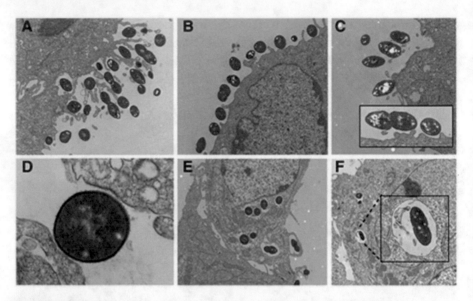

Fig. 14.4 Electron micrographs showing *Strep. pneumoniae* adhering to, and invading, human epithelial cells. The organism attaches to epithelial cells (images **a–d**). It then invades these cells (**e, f**). Original magnifications (**a**) 4400×; (**b**) 4400×; (**c**) 4400× (inset, 15000×); (**d**) 26000×; (**e**) 3200×; (**f**) 1650× (inset, 6500×). (Phenotypic, genomic and transcriptional characterization of *Streptococcus pneumoniae* interacting with human pharyngeal cells. Kimaro Mlacha SZ, Romero-Steiner S, Hotopp JC, Kumar N, Ishmael N, Riley DR, Farooq U, Creasy TH, Tallon LJ, Liu X, Goldsmith CS, Sampson J, Carlone GM, Hollingshead SK, Scott JA, Tettelin H - BMC Genomics 2013 14:383. This article is published under license to BioMed Central Ltd

It's found in several regions of the upper respiratory tract of healthy individuals, but its main habitat is the nasopharynx where it's present in approximately 25% of healthy people. It's a facultative anaerobe and has a wide range of virulence factors including a capsule to protect it from phagocytic cells, several enzymes that can damage human tissues and toxins that can disrupt the mucociliary escalator and kill human cells. It's capable of invading human epithelial cells (Fig. 14.4).

It's a very important pathogen of humans, being responsible for a range of serious, sometimes life-threatening, infections including meningitis, pneumonia and otitis media.

Box 14.1 What's in a Name?

The genus *Moraxella* was named in honour of Victor Morax, a Swiss ophthalmologist, who was the first to describe this genus. While working at the Pasteur Institute in Paris in the 1920s, he isolated a bacterium belonging to this genus from a case of conjunctivitis. The species name, *catarrhalis*, is derived from the Greek word "catarrh" which means "to flow down" which describes the watery discharge associated with colds. When it was first detected in 1896 (and originally named *Micrococcus catarrhalis*), it was thought to be responsible for the common cold.

The name *Haemophilus* is derived from the two Greek words "haema" and "philia" which mean "blood" and "fondness", respectively. This name reflects the fact that, in order to grow, bacteria belonging to this genus need certain compounds present in blood. In 1892 the German bacteriologist Richard Pfeiffer isolated a bacterium from the noses of patients with flu, and, because he thought it was responsible for the disease, he named it *Bacillus influenzae*. This bacterium was later recognised as belonging to the genus *Haemophilus* and so was re-named *Haemophilus influenzae* in 1917. However, we now know that it isn't responsible for influenza which is a viral infection.

H. influenzae is a small Gram-negative coccobacillus (i.e. it's intermediate between a round coccus and a long bacillus – Fig. 14.5) that can't grow in the absence of oxygen.

It can be found in several regions of the upper respiratory tract of healthy individuals, but its main habitat is the oropharynx where it's present in approximately 32% of healthy people. It has thin, hair-like structures (fimbriae) that enable it to attach to epithelial cells as well as other virulence factors including (a) a capsule that protects it from phagocytic cells (Fig. 14.6), (b) an enzyme that destroys antibodies, (c) components that disrupt the mucociliary escalator and (d) an endotoxin that induces an inflammatory response. It's capable of invading human epithelial cells (Fig. 14.7). It's a very important human pathogen and is responsible for a range of infections, some of which are life-threatening. These include meningitis, pneumonia, epiglottitis, otitis media, bronchitis and arthritis.

Mor. catarrhalis is a Gram-negative coccus (Fig. 14.8) that can't grow without oxygen. It's found in several regions of the upper respiratory tract of healthy people, but its main habitat is the nasopharynx where it's present in approximately 5% of healthy adults. It's found more frequently in healthy children – approximately 50% of 3–12 year olds. Its virulence factors include molecules that protect it against phagocytic cells and enzymes that can break down components of human tissues. As well as its involvement in sinusitis, it also causes conjunctivitis and otitis media in children and infections of the lower respiratory tract in adults.

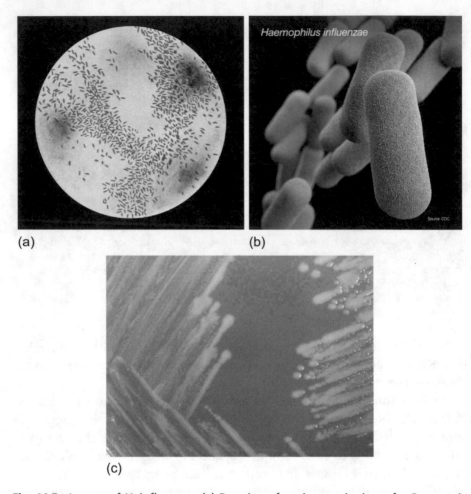

(a)   (b)

(c)

Fig. 14.5 Images of *H. influenzae*. (**a**) Drawing of a microscopic view of a Gram-stain of *H. influenzae* showing typical Gram-negative (coloured red) coccobacilli. (Centers for Disease Control and Prevention, USA). (**b**) Three-dimensional computer-generated image of a group of *H. influenzae* cells. The artistic recreation was based upon scanning electron microscopic imagery. (Jennifer Oosthuizen and Sarah Bailey Cutchin, Centers for Disease Control and Prevention, USA). (**c**) Colonies of *H. influenzae* growing on an agar plate. (Todd Parker, Ph.D., Assoc Director for Laboratory Science, Div. of Preparedness and Emerging Infections at CDC)

Fig. 14.7 (continued) invasion to bronchial epithelial cells. BMC Microbiology 2015; 15:26. This article is distributed under the terms of the Creative Commons Attribution 4.0 International License (http://creativecommons.org/licenses/by/4.0/), which permits unrestricted use, distribution and reproduction in any medium, provided you give appropriate credit to the original author(s) and the source, provide a link to the Creative Commons license and indicate if changes were made)

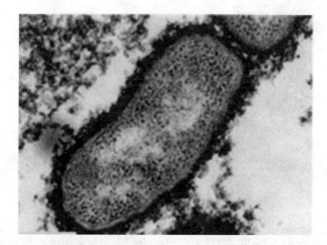

Fig. 14.6 Electron micrograph showing the capsule (stained black) surrounding a cell of *H. influenzae* (×67,000). This protects the bacterium from being engulfed by phagocytic cells. (Schouls L, van der Heide H, Witteveen S, Zomer B, van der Ende A, Burger M, Schot C. Two variants among Haemophilus influenzae serotype b strains with distinct bcs4, hcsA and hcsB genes display differences in expression of the polysaccharide capsule. BMC Microbiol. 2008; 8: 35

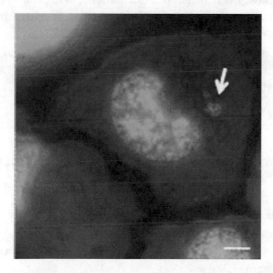

Fig. 14.7 Photomicrograph showing an epithelial cell (with a blue-stained nucleus) that has been invaded by *H. influenzae*. The white arrow indicates a clump of bacteria which are stained green (×2000). Bar = 5 μm. (Ikeda M, Enomoto N, Hashimoto D, Fujisawa T, Inui N, Nakamura Y, Suda T, Nagata T. Nontypeable Haemophilus influenzae exploits the interaction between protein-E and vitronectin for the adherence and

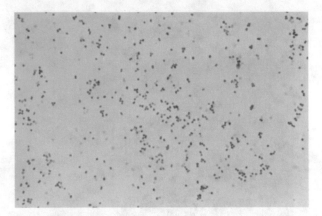

Fig. 14.8 Gram stain of *Mor. catarrhalis* showing typical Gram-negative cocci, often in pairs. (Image courtesy of Dr. W. A. Clark, Centers for Disease Control and Prevention, USA)

In two thirds of cases of ABRS, one of these three pathogens predominates, while in the remaining cases, a number of other bacterial species are responsible. *Streptococcus pyogenes*, *Staphylococcus aureus* and anaerobic bacteria are found in about 10% of patients with ABRS. Interestingly all three of the main pathogens that have been isolated from patients with the disease are able to disrupt the mucociliary escalator, and this contributes to mucus accumulation and so will exacerbate the condition.

How Is Acute Rhinosinusitis Diagnosed?

Diagnosis is based on patient history and symptoms. Important risk factors include a history of viral upper respiratory tract infection or allergic rhinitis. The following are indicators that help to distinguish between viral and bacterial sinusitis:

- Symptoms of viral infection tend to peak early and gradually resolve within 10 days.
- Symptoms present for more than 10 days without an improvement suggest a bacterial infection.
- Symptoms that worsen after an initial improvement (so-called double sickening) suggest secondary bacterial infection.
- A purulent nasal discharge, nasal obstruction, dental pain or facial pain/pressure/headache implies bacterial sinusitis.
- A clear nasal discharge, fever, sore throat and muscle pains usually suggest viral sinusitis.

If a bacterial cause is suspected and symptoms aren't responding well to antibiotics, then samples can be obtained by endoscopy and cultured in the laboratory to identify the pathogen responsible and to determine which antibiotics should be used for treating the infection. However, this is only required in unusual or persistent cases and requires specialist equipment and skills. Nasal swabs unfortunately don't supply sufficiently reliable results.

How Is Acute Rhinosinusitis Treated?

Most cases of the disease are caused by viruses, and the symptoms tend to improve after about 5 days and resolve within 2–3 weeks. Approximately 85% of patients have an improvement of symptoms within 7 to 15 days. Paracetamol or ibuprofen can be taken to relieve fever and pain. Unfortunately there is not much good evidence available to guide self-care treatment choices. Decongestant nasal sprays or drops are probably not helpful. Intranasal corticosteroids may help to unblock the nose and improve breathing problems but are not licensed for acute sinusitis. Warm packs applied to the face might help to relieve pain and to drain mucus from the sinuses. A salt water solution can be tried to clean the nostrils and hopefully ease congestion.

Antibiotics are not routinely recommended, especially during the first 10 days of the disease as there's a high likelihood that the microbe responsible is a virus. However, if the symptoms haven't started to improve by then, antibiotics might be considered, bearing in mind that they don't have much impact on the length of illness or complication rates and most bacterial causes are self-limiting and get better without antibiotics. Back up antibiotics might be offered to be taken if self-care measures don't improve the symptoms or if the symptoms worsen. Phenoxymethylpenicillin or amoxicillin/clavulanic acid are the antibiotics most frequently used unless a different antibiotic has been suggested by laboratory analysis – in the rare cases where this has been carried out. For those allergic to penicillins, other antibiotics such as doxycycline, clarithromycin or erythromycin might be used instead. Antibiotic therapy should be considered in the case of patients who are immunocompromised or those with, or at risk of, complications or with severe symptoms such as fever, moderate to severe facial or dental pain. Serious complications may require admission to hospital for the administration of intravenous antibiotics.

How Can I Avoid Getting Acute Rhinosinusitis?

The best way to protect yourself from acute rhinosinusitis is to avoid getting upper respiratory tract infections, and you can do this by frequent handwashing, staying away from people with colds or influenza and getting vaccinated against respiratory pathogens. You should try to maintain a healthy respiratory tract by avoiding tobacco smoke and other air pollutants. If you have allergies, do your best to stay clear of any allergy triggers.

Want to Know More?

American Academy of Otolaryngology – Head and Neck Surgery https://www.enthealth.org/conditions/sinusitis/
American Family Physician https://www.aafp.org/afp/2016/0715/p97.html
Centers for Disease Control and Prevention, USA https://www.cdc.gov/antibiotic-use/community/for-patients/common-illnesses/sinus-infection.html
Healthline https://www.healthline.com/health/acute-sinusitis
Mayo Clinic, USA https://www.mayoclinic.org/diseases-conditions/acute-sinusitis/symptoms-causes/syc-20351671
National Health Service, UK https://www.nhs.uk/conditions/sinusitis-sinus-infection/
National Institute for Clinical Care and Excellence (NICE), UK. Sinusitis, 2018 https://cks.nice.org.uk/sinusitis#!topicSummary
Patient Info, UK https://patient.info/ears-nose-throat-mouth/acute-sinusitis
Royal Australian College of General Practitioners https://www.racgp.org.au/afp/2016/june/sinusitis/
Royal College of Surgeons, UK https://www.entuk.org/sinus-infection-sinusitis
Aring AM, Chan MM. Current concepts in adult acute rhinosinusitis. *American Family Physician*. 2016 Jul 15;94(2):97-105.
Brook I. Acute Sinusitis. Medscape from WebMD, 2018 https://emedicine.medscape.com/article/232670-overview
DeBoer DL, Kwon E. Acute Sinusitis. Treasure Island (FL): StatPearls Publishing LLC; 2020 https://www.ncbi.nlm.nih.gov/books/NBK547701/
Sun GH. Acute sinusitis. BMJ Best Practice, BMJ Publishing Group, 2020 https://bestpractice.bmj.com/topics/en-gb/14

15

Chronic Sinusitis

Abstract Chronic Sinusitis is defined as an infection of the sinuses that lasts for more than 12 weeks. It's a very common disease and affects up to 5% of the population. The main symptoms include nasal blockage, facial pain and a reduced ability to smell. The condition results from the accumulation of mucus in the sinuses due to impaired drainage caused by a viral infection or a number of other factors. This enables biofilm formation within the sinuses by *Staphylococcus aureus* or any of a number of other bacteria which results in tissue damage due to a chronic inflammatory response. Treatment involves identifying and treating the underlying causes of the sinus blockage. Saline washouts and nasal steroids are recommended for reducing symptoms. Long-term use of antibiotics isn't appropriate, but short-term courses may be useful in severe cases. Sinus surgery may be necessary to relieve obstructions and restore drainage.

What Is Chronic Sinusitis?

As we described in Chap. 14, the paranasal sinuses are a group of four hollow air-filled cavities of the skull (Fig. 14.1). Inflammation of the mucosa of one or more of the sinuses is known as sinusitis, and, because the inflammation usually extends into the nasal cavity, the condition is often known as rhinosinusitis. If it lasts for longer than 12 weeks, it's classified as being chronic rhinosinusitis. The mucosa of the sinuses continually produces mucus which is propelled by the mucociliary escalator (described in Chap. 1) through narrow gaps (1–3 mm in diameter) known as ostia into the nasal cavity. Problems

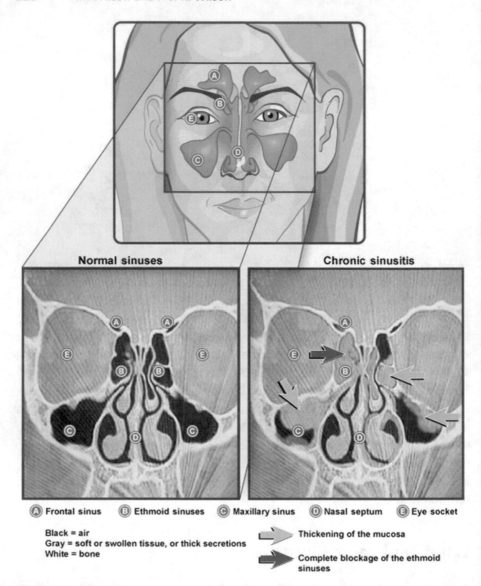

Fig. 15.1 Front view illustration and side-by-side CT scans of a healthy individual (left) and a patient suffering from chronic sinusitis (right). In the scan of the patient, thickening of the mucosa can be seen in her left maxillary, left ethmoid and right maxillary sinuses. Her right ethmoid sinus is completely blocked. (NIAID/CC BY (https://creative-commons.org/licenses/by/2.0))

with this drainage, usually due to some type of blockage of the ostia, is the underlying cause of the condition (Fig. 15.1).

In addition to an infection (see Chap. 14), there are a number of other ways in which such problems can arise: (i) factors that not only affect the sinuses such as cystic fibrosis, immunodeficiency, autoimmune disease and acid reflux; (ii) environmental factors such as smoking, environmental pollutants and allergies; and (iii) structural factors such as anatomical abnormalities or the presence of a foreign body. Any of these factors, or the inflammation accompanying a viral infection, enables mucus to accumulate, and this creates conditions suitable for the proliferation of one or more bacterial species that inhabit the sinuses. This can result in an ongoing chronic inflammatory response which damages the surrounding tissues and causes more thickening of the lining of the sinuses. Under these conditions, the bacteria in the sinuses can form structures known as biofilms (described in more detail in Chaps. 20 and 21) which consist of huge numbers of microbes enclosed within a protective polymeric matrix. Our antimicrobial defence systems find biofilms difficult to contend with, and so they persist for long periods of time. Antibiotics are also much less effective against biofilms, and this means the condition can be difficult to treat conventionally.

The main symptoms of chronic rhinosinusitis include:

- Facial pain/pressure
- Nasal blockage or obstruction
- Nasal discharge
- Postnasal drip (i.e. mucus dripping from the nasal cavity into the throat)
- Reduced ability to smell

Nasal obstruction is the most common symptom (81–95% of patients), followed by facial pain/pressure (70–85%) and a reduced ability to smell (61–69%). Other symptoms that may be present are headache, fever, halitosis, dental pain, fatigue, cough and ear pain. The condition may persist for several months and so can have a significant impact on quality of life due to (i) sleep problems, fatigue and depression; (ii) its impact on employment such as absenteeism, reduced effectiveness and productivity; (iii) reduction in social functioning; and (iv) high healthcare usage. Complications similar to those associated with acute rhinosinusitis may arise, and these are listed in Chap. 14.

How Common Is the Disease?

Chronic rhinosinusitis is a common disease worldwide, particularly in places with high levels of atmospheric pollution. Damp, temperate climates and high pollen concentrations are also associated with a higher prevalence of the condition. It's one of the more prevalent chronic diseases in the USA, affecting 1–5% of the population. This results in 18–22 million physician visits in the USA each year and an annual treatment cost of $3.4–5 billion. In the UK, one in ten adults are affected. Although it occurs in all age groups, its prevalence increases with age, and it's more likely to occur in women. Up to 64% of patients with AIDS develop chronic sinusitis.

Which Microbes Cause Chronic Sinusitis?

The bacteria associated with the disease differ from those involved in acute rhinosinusitis and include *Staphylococcus aureus* (50% of cases), various Gram-negative bacilli (20%), *Haemophilus influenzae* (4%), *Strep. pyogenes* (4%), *Strep. pneumoniae* (2%) and a variety of anaerobic bacteria. Often more than one microbe is involved. The main characteristics of these bacteria have been described in previous chapters.

How Is the Disease Diagnosed?

Diagnosis of chronic rhinosinusitis is based on the presence of nasal blockage or nasal discharge together with facial pain/pressure (or headache) and/or reduced (or loss) of the sense of smell that persist for at least 12 weeks without complete improvement. The nasal cavity can be examined using a nasal speculum (Fig. 15.2) to identify abnormalities such as inflammation or deformity. Other tools to aid diagnosis include imaging (CT or MRI), endoscopy to look inside the sinuses and allergy tests, if appropriate.

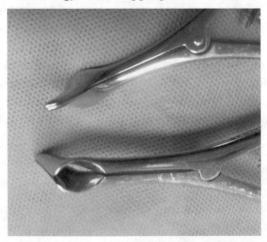

Fig. 15.2 Nasal speculum used to examine the nasal cavity. The end of the speculum is inserted into one of the patient's nostrils and viewed under bright illumination by means of a headlight. (SnowBink. This file is made available under the Creative Commons CC0 1.0 Universal Public Domain Dedication. Via Wikimedia Commons)

How Is the Disease Treated?

An important aspect of treatment is to identify and treat underlying causes such as allergies, asthma and structural abnormalities including foreign bodies and tumours. For a start, you should avoid known allergy triggers and smoking. Daily saline washouts and nasal steroid drug sprays or drops are helpful for reducing symptoms and improving your quality of life. Oral steroids are effective for short-term symptom control, but their use has to be balanced by possible side effects which include indigestion or heartburn, increased appetite, difficulty sleeping, mood changes and an increased risk of infections. There's no clear consensus regarding the use of antibiotics in treating the condition. Most authorities agree that their long-term use isn't appropriate and should be prescribed only on the advice of a specialist. Short-term courses (up to 3 weeks) may be useful for severe cases or when an infection seems to have been the cause of temporary worsening. However, it's not clear what the optimal drug or dosage is and which specific groups of people are most likely to benefit. It's difficult to obtain useful samples from the infected sinuses for analysis in the laboratory because they're so easily contaminated by fluids from the nasal cavity. The choice of drugs, therefore, can usually only be based on which microbes are most likely to be present. Frequently used agents include (i) a combination of a penicillin (such as amoxicillin) with clavulanic acid, (ii) a combination of metronidazole plus a macrolide (such as erythromycin) or a cephalosporin and (iii) a quinolone such as moxifloxacin.

Recently, an alternative approach has been developed based on the use of light-activated antimicrobial agents. This procedure, known as "antimicrobial photodynamic therapy", involves the application of a light-sensitive drug to the sinuses followed by illumination of the region with red light from a low-power laser for a few minutes (Fig. 15.3 – see also Box 3.4). Activation of the drug by the light results in the production of reactive oxygen molecules that kill the virus and/or bacteria that are in the sinuses.

The procedure has a number of important advantages over antibiotics: (i) it requires only a single application and is very rapid (a few minutes) in contrast to antibiotics which must be taken daily for several days or longer, (ii) it's effective against all types of microbes, and they can't develop resistance to the treatment, (iii) it's effective against microbial biofilms, and (iv) being an antimicrobial, it can also affect the immune response and can reduce inflammation. This approach has been shown to be effective in the treatment of periodontitis (Chap. 21) and in eliminating MRSA from the nostrils (Chap. 3). However, its use for the treatment of chronic sinusitis is still at an early stage.

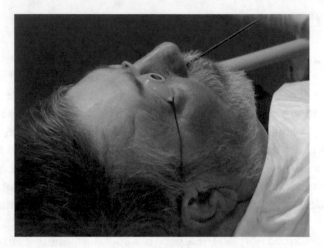

Fig. 15.3 A patient with chronic sinusitis being treated by antimicrobial photodynamic therapy. The sinuses were first treated with a photosensitiser and were then irradiated with red light from a laser via optical fibres. (Image courtesy of Ondine Biomedical Inc., Vancouver, Canada)

If medication doesn't result in improvement, then sinus surgery may be necessary in order to relieve obstructions and restore drainage and mucociliary clearance and thus improve symptoms.

How Can I Avoid Getting Chronic Rhinosinusitis?

Once any underlying contributory factor has been dealt with, you'll find that nasal irrigation with saline 3 or 4 times a day is helpful in preventing recurrences. If you have allergies, you should avoid the triggers for these. You should avoid tobacco smoke and other air pollutants.

Want to Know More?

American College of Physicians https://acpinternist.org/archives/2011/01/sinusitis.htm
American Family Physician https://www.aafp.org/afp/2017/1015/p500.html
Canadian Family Physician https://www.cfp.ca/content/59/12/1275
Mayo Clinic, USA https://www.mayoclinic.org/diseases-conditions/chronic-sinusitis/symptoms-causes/syc-20351661
Medical News Today https://www.medicalnewstoday.com/articles/320569.php

National Health Service, UK https://www.nhsinform.scot/illnesses-and-conditions/ears-nose-and-throat/sinusitis

National Institute for Clinical Care and Excellence (NICE), UK. Sinusitis, 2018 https://cks.nice.org.uk/sinusitis#!topicSummary

Patient Info, UK https://patient.info/ears-nose-throat-mouth/acute-sinusitis/chronic-sinusitis

Royal College of Surgeons, UK https://www.entuk.org/sinus-infection-sinusitis

World Allergy Organisation https://www.worldallergy.org/education-and-programs/education/allergic-disease-resource-center/professionals/rhinosinusitis-synopsis

Antisdel J, Sindwani R. Chronic sinusitis. BMJ Best Practice, BMJ Publishing Group, 2018 https://bestpractice.bmj.com/topics/en-gb/15

Brook I. Chronic Sinusitis. Medscape from WebMD, 2019 https://emedicine.medscape.com/article/232791-overview

Kwon E, O'Rourke MC. Chronic Sinusitis. Treasure Island (FL): StatPearls Publishing LLC; 2020 https://www.ncbi.nlm.nih.gov/books/NBK441934/

Marcus S, DelGaudio JM, Roland LT, Wise SK. Chronic Rhinosinusitis: Does Allergy Play a Role? *Medical Sciences (Basel).* 2019 Feb 18;7(2). pii: E30. doi: https://doi.org/10.3390/medsci7020030. PMID: 30781703

Morcom S, Phillips N, Pastuszek A, Timperley D. Sinusitis. *Australian Family Physician.* 2016 Jun;45(6):374-7.

Sedaghat AR. Chronic Rhinosinusitis. *American Family Physician.* 2017 Oct 15;96(8):500-506.

Sivasubramaniam R, Douglas R. The microbiome and chronic rhinosinusitis. *World Journal of Otorhinolaryngology - Head and Neck Surgery.* 2018 Oct 31;4(3):216-221. doi: https://doi.org/10.1016/j.wjorl.2018.08.004. eCollection 2018 Sep.

Vennik J, Eyles C, Thomas M, Hopkins C, Little P, Blackshaw H, Schilder A, Savage I, Philpott CM Chronic rhinosinusitis: a qualitative study of patient views and experiences of current management in primary and secondary care.. *British Medical Journal Open.* 2019; 9(4): e022644. Published online 2019 Apr 23. doi: https://doi.org/10.1136/bmjopen-2018-022644

16

Middle Ear Infections

Abstract Acute otitis media (middle ear infection) is very common in children with 80% experiencing at least one episode before they're 2 years old. The main symptoms include earache, fever, vomiting, lack of energy and slight hearing loss. The condition usually follows a viral infection of the upper respiratory tract, usually involving the respiratory syncytial virus, because this results in conditions suitable for the growth of certain pathogenic bacteria. The bacteria usually responsible are *Haemophilus influenzae*, *Streptococcus pneumoniae* and *Moraxella catarrhalis*. Treatment focuses on relieving the pain and fever that accompany the condition and paracetamol or ibuprofen are generally effective. Antibiotics aren't usually needed for otherwise healthy children but are appropriate in some situations including those who are immunocompromised and are generally unwell with a severe illness or when both ears are affected.

This can be a very painful and unpleasant condition, especially for children. I remember so many incidences when I was a child. Perhaps the worst was when I was about 10. It started off by being only slightly painful, but I felt very unwell. I was even accused of malingering by my father, who was about to go away for 2 days, which was very upsetting. I clearly remember feeling very triumphant when he returned 2 days later to find me confined to bed with a high temperature. The morning after he'd left, I'd woken up with my pillow covered in pus. My mother sent for the doctor (in those days they came to your house), and I was prescribed an antibiotic. I remember lying in bed feeling very sick for hours before it finally took effect. I had another episode

© Springer Nature Switzerland AG 2021
M. Wilson, P. J. K. Wilson, *Close Encounters of the Microbial Kind*,
https://doi.org/10.1007/978-3-030-56978-5_16

about 3 years later. This one was really painful, and I felt terrified because the pain seemed so close to my brain – did I have a brain tumour? I kept thinking "If only the pain could move somewhere else, even for a short time, I'd be able to tolerate it". Again, thanks to antibiotics, I soon recovered, and, mercifully, I didn't have a brain tumour.

What Is Acute Otitis Media?

Infection of the middle ear, also known as acute otitis media (AOM), is characterised by inflammation, swelling and the accumulation of fluid behind the ear drum (Fig. 16.1).

It's extremely common in young children with more than 80% experiencing at least one episode before they reach the age of 2 years; this is probably partly because the shape of their heads is different from adults. The peak incidence is between the ages of 6 and 18 months. Globally, it's been estimated that 709 million cases occur each year with 51% of these being in

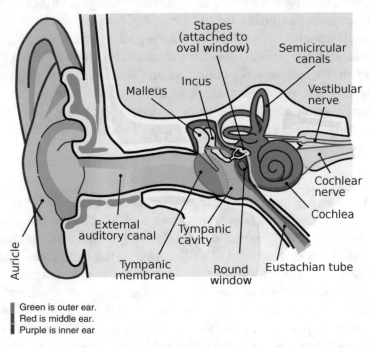

Green is outer ear.
Red is middle ear.
Purple is inner ear

Fig. 16.1 Diagram showing the various regions of the ear. When affected by otitis media, the middle ear becomes filled with fluid and is inflamed. (Lars Chittka; Axel Brockmann/CC BY (https://creativecommons.org/licenses/by/2.5))

under-fives. The overall disease burden is significant. The main symptoms of the infection in young children include earache, fever, vomiting, a lack of energy and slight hearing loss. However, in infants who can't talk, the symptoms include fever, vomiting, pulling, tugging or rubbing their ear, irritability, poor feeding, restlessness at night, a runny nose, unresponsiveness to quiet sounds and loss of balance. AOM isn't very common in adults with <1% being affected. Adult symptoms include ear pain, decreased hearing, and a discharge from the ear. Most of what we know about AOM comes from studying cases in children.

A bulging ear drum which is under pressure can rupture resulting in a hole – this can immediately reduce the pain but also impairs hearing. The majority of these heal on their own within a few weeks or months, but they can be repaired surgically if needed. This is helpful for restoring hearing and also replaces the protective barrier it provides against further middle ear infections. Longer-lasting infections following a perforated ear drum affect less than 1% of sufferers in the UK. Those at risk include children under 5 years of age, those with allergies and passive smokers. These infections may be caused by bacteria (often *Pseudomonas aeruginosa*) and fungi (often *Candida albicans*).

Some children suffer persistent infections which don't respond to treatment; this is usually due to antibiotic resistance in the infecting bacterium or because the infection is caused by a virus. Adults with persistent AOM should be assessed by a specialist for possible underlying conditions. Recurrent AOM is defined as three infections within 6 months or 4 within 12 months – such cases are best referred for specialist care.

Serious complications of AOM include meningitis, mastoiditis and brain abscesses, but these are rare. Globally, approximately 21,000 people die of these complications each year. Hearing loss due to AOM is also rare and affects 0.3% of the population worldwide. In developed countries, AOM is usually uncomplicated and self-limiting and rarely results in serious complications or hearing loss.

Which Microbes Cause Otitis Media?

AOM is usually preceded by a viral infection of the upper respiratory tract, and it's been shown that 37% of such infections are followed by AOM, usually 2–5 days later. The initial viral infection can be instigated by any of a wide range of viruses although those most frequently responsible include respiratory syncytial virus (RSV), rhinovirus, adenovirus, coronavirus, bocavirus, influenza virus, parainfluenza virus, enterovirus and human metapneumovirus. Most of these viruses have been described in other chapters of this book (see Chaps. 7–12); the remaining viruses are discussed briefly below.

RSV is the most common cause of acute respiratory tract infections in infants, and by the age of 2 years, more than 95% of all the world's children will have been infected by the virus. The virus is transmitted in droplets expelled from the respiratory tract and symptoms usually appear 4–5 days later – these can range from cold-like symptoms to severe diseases such as pneumonia and bronchitis. Approximately one third of children infected with RSV go on to develop AOM. The RSV is roughly spherical in shape, with a diameter of approximately 120–300 nm, but it can also exist in the form of long filaments (Fig. 16.2). The genetic material of the virus is RNA.

Bocaviruses are small, icosahedral, viruses with a diameter of 18–26 nm (Fig. 16.3) and contain DNA as their genetic material. They were discovered in 2005 and found to cause respiratory infections in all ages but particularly in those between 6 and 24 months.

The human metapneumovirus (HMPV) was discovered in 2001 and causes respiratory disease in people of all ages but particularly young children and older adults. The main symptoms of infections caused by this virus are cough, fever, nasal congestion and shortness of breath, and these appear 3–6 days following infection. The virus is spherical in shape with a diameter of about 150 nm and has an outer coat formed from the host cell in which it was produced, and its genetic material is RNA.

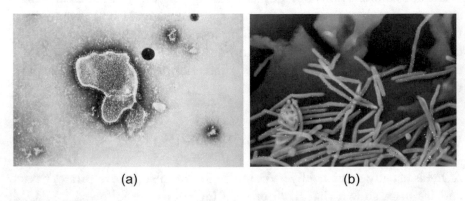

(a)   (b)

Fig. 16.2 Electron micrographs of the various forms of the RSV. (a) Spherical form of RSV. (Image courtesy of E. L. Palmer, Centers for Disease Control and Prevention, USA). (b) Filamentous form of RSV. (Image courtesy of National Institute of Allergy and Infectious Diseases [CC BY 2.0 (https://creativecommons.org/licenses/by/2.0)])

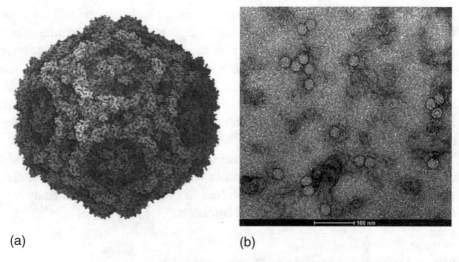

(a)   (b)

Fig. 16.3 Images of bocavirus. (**a**) Model showing the shape of a bocavirus. (NIH 3D print exchange, CC BY). (**b**) Electron micrograph of bocavirus virions. These have the typical icosahedral appearance of parvoviruses, each with a diameter of approximately 20 nm. (PLoS Pathog. 2012;8(8):e1002899. https://doi.org/10.1371/journal.ppat.1002899. Epub 2012 Aug 30. Establishment of a reverse genetics system for studying human bocavirus in human airway epithelia. Huang Q, Deng X, Yan Z, Cheng F, Luo Y, Shen W, Lei-Butters DC, Chen AY, Li Y, Tang L, Söderlund-Venermo M, Engelhardt JF, Qiu J

Box 16.1 Reducing the Spread of Viruses Responsible for Respiratory Diseases

Individuals can avoid being infected with respiratory viruses by following these steps that are generally referred to as "good respiratory hygiene measures":

- Washing their hands often with soap and water for at least 20 seconds
- Avoiding touching their eyes, nose or mouth with unwashed hands
- Avoiding close contact with people who are sick

Patients can avoid spreading the viruses by:

- Covering their mouth and nose when coughing and sneezing
- Washing their hands frequently with soap and water for at least 20 seconds
- Avoiding sharing their cups and eating utensils with others
- Refraining from kissing others
- Staying at home when they are sick

In addition, disinfection of possible contaminated surfaces (such as doorknobs, work surfaces and shared toys) helps to stop the spread of respiratory viruses. Suitable disinfectants include bleach and 70% alcohol.

The resulting inflammation of the nasopharynx and Eustachian tube arising from the initial viral infection creates an environment that encourages the growth of those pathogenic bacteria that are already present in the pharynx such as *H. influenzae*, *Strep. pneumoniae* and *Mor. catarrhalis* (these are described in Chap. 14). The Eustachian tube normally allows pressure inside the middle ear to be kept the same as outside. However, the damaged Eustachian tube results in negative pressure inside the middle ear which allows bacteria and viruses to gain access to this region which is usually microbe-free. The presence of large numbers of viruses and pathogenic bacteria in the middle ear causes inflammation and the other symptoms associated with the disease. Although bacteria are usually involved in the disease, approximately 5% of AOM cases are due solely to a viral infection.

How Is the Condition Diagnosed?

The infection is usually diagnosed by viewing the state of the ear drum (tympanic membrane) by means of an otoscope (Fig. 16.4). AOM is commonly associated with a bulging, red or opaque ear drum.

Any discharge from the ear can be collected and sent to the laboratory to be cultured to determine which bacteria are present and to which antibiotics they are sensitive. This is particularly important information to gain in persistent or recurrent infections.

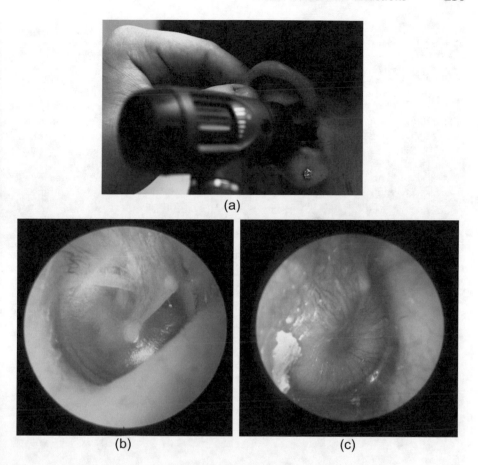

(a)

(b)   (c)

Fig. 16.4 Diagnosis of AOM by examination of the ear drum using an otoscope. (**a**) An otoscope being used to examine the ear drum of a patient. (US Air Force photo by Senior Airman Christopher Toon. This image or file is a work of a US Air Force Airman or employee, taken or made as part of that person's official duties. As a work of the US federal government, the image or file is in the public domain in the USA). (**b**) Appearance of the tympanic membrane (ear drum) of an individual without otitis media. Note the absence of pus and inflammation. (Michael Hawke MD/CC BY (https://creativecommons.org/licenses/by/4.0)). (**c**) Appearance of the ear drum of a patient with AOM. Cream-coloured pus is evident, and the dilated blood vessels of the ear drum can be seen as well as inflammation. (Michael Hawke MD [CC BY 4.0 (https://creativecommons.org/licenses/by/4.0)])

Box 16.2 How an Ear Infection Changed the History of Britain

In 1558 Mary Stuart, Queen of Scotland, married Francis who was heir to the French throne. She then became Queen consort of France when her husband ascended to the French throne in 1559 at the age of 15. Less than 18 months into his reign, King Francis II returned from a hunting trip in early December of 1560 complaining of a severe pain in his left ear. His physician, Ambroise Pare, thought that an abscess was forming and considered drilling into the king's skull to drain it, but this wasn't carried out. The King's condition gradually worsened, and, although Mary nursed her husband with great care, the infection spread to his brain, and the young King died on December 5, 1560. The French throne then passed to Francis's younger brother Charles IX, and Mary returned to Scotland. Four years later she married Lord Henry Darnley and gave birth to a son who became James VI of Scotland. James also later acceded to the throne of England and Ireland (as James I), and so the three countries became a single kingdom.

The death of Francis II was a major turning point in the history of Britain. If he hadn't died, Mary would have ruled Scotland jointly with her husband. She, or her descendant, would then also have ruled England and Ireland when Queen Elizabeth, who was childless, died in 1603. England, Scotland and Ireland would then have come under the rule of France.

Portrait of Francis II of France (1544–1560) by Francois Clouet in the Bibliotheque National de France. (This image comes from Gallica Digital Library and is available under the digital ID btv1b10544086n/f1. This work is in the public domain in its country of origin and other countries and areas where the copyright term is the author's life plus 100 years or fewer. Via Wikimedia Commons)

How Is the Disease Treated?

Treatment of AOM focusses on relieving the pain and fever that accompany the condition, and paracetamol or ibuprofen is generally effective. This is essential for both the benefit of pain control and also to reduce fever and the possible vomiting or fluid refusal that younger children can demonstrate. There's no strong evidence that topical drugs such as benzocaine and phenazone offer any great benefit. Decongestants, antihistamines and corticosteroids haven't shown any great benefit in relieving the symptoms of AOM and aren't recommended. In children who are otherwise healthy, AOM runs a favourable natural course without antibiotic treatment, with symptoms settling within a few days and complications being rare. Symptoms usually last around 3 days but can go on for as long as 1 week. Oral antibiotics reduce the duration of AOM symptoms, but generally by around only 1 day, and make little difference to the rates of the commonest complications. Taking antibiotics may, of course, have side effects such as diarrhoea and rashes. Another important concern is that the use of antibiotics for such a common disease will encourage the development of antibiotic-resistant bacteria. In view of the personal (adverse reactions) and community (antibiotic resistance development) risks in such cases, it's difficult to justify routine antibiotic administration. However, it's generally agreed that antibiotics should be offered to those children <2 years of age with AOM affecting both ears and in those of any age who have an ear discharge. Antibiotics are also recommended for those with AOM who are immunocompromised or have craniofacial malformations, as well as those who are generally unwell with a severe or lengthy illness. Amoxicillin or, in those allergic to penicillins, cefdinir, cefuroxime or clarithromycin are generally the most commonly used antibiotics. However, it's important to take into account the resistance patterns of AOM-associated bacteria in the geographical area of the patient. A useful strategy in mild cases involves withholding antibiotics for an observation period of 48 to 72 h to allow time for any spontaneous improvement to happen – this is termed "watchful waiting" or "deferred prescribing". It has been found that in those who aren't given antibiotics, AOM symptoms improve in 24 hours in 60% of children and within 3 days in 80% of children.

How Can the Disease Be Avoided?

Preventive measures include avoiding the risk factors associated with AOM, such as passive smoking, and observing good respiratory hygiene. Breastfeeding during the first 3 months is associated with a lower risk of AOM. There's evidence that the introduction of the PCV13 vaccine which is effective against 13 strains of *Strep. pneumoniae* has resulted in a decrease in AOM due to these 13 strains. However, there has been a parallel increase in AOM due to *H. influenzae*. There's evidence that the use of a vaccine against influenza has also reduced the incidence of AOM during the flu season.

Want to Know More?

American Family Physician https://www.aafp.org/afp/2013/1001/p435.html

Centers for Disease Control and Prevention, USA https://www.cdc.gov/antibiotic-use/community/downloads/Preventing-and-Treating-Ear-Infections-H.pdf

Healthline https://www.healthline.com/health/ear-infection-acute

Mayo Clinic, USA https://www.mayoclinic.org/diseases-conditions/ear-infections/symptoms-causes/syc-20351616

National Health Service, UK https://www.nhsinform.scot/illnesses-and-conditions/ears-nose-and-throat/middle-ear-infection-otitis-media

National Institute for Clinical Care and Excellence (NICE), UK. Otitis media – acute, 2018 https://cks.nice.org.uk/otitis-media-acute#!topicSummary

Patient Info, UK https://patient.info/doctor/acute-otitis-media-in-adults

Royal Children's Hospital, Melbourne, Australia https://www.rch.org.au/clinical-guide/guideline_index/Acute_otitis_media/

Stanford Childrens Health, USA https://www.stanfordchildrens.org/en/topic/default?id=otitis-media-middle-ear-infection-90-P02057

Armengol CE. Acute otitis media. BMJ Best Practice, BMJ Publishing Group, 2020 https://bestpractice.bmj.com/topics/en-gb/39

Danishyar A, Ashurst JV. Acute Otitis Media. Treasure Island (FL): StatPearls Publishing LLC; 2020 https://www.ncbi.nlm.nih.gov/books/NBK470332/

Donaldson JD. Acute Otitis Media. Medscape from WebMD, 2019 https://emedicine.medscape.com/article/859316-overview

Nokso-Koivisto J, Marom T, Chonmaitree T. Importance of viruses in acute otitis media. *Current Opinion in Pediatrics*. 2015 Feb;27(1):110-5. https://doi.org/10.1097/MOP.0000000000000184. PMID: 25514574

Rettig E, Tunkel DE. Contemporary concepts in management of acute otitis media in children. *Otolaryngology Clinics of North America*. 2014 Oct;47(5):651-72. https://doi.org/10.1016/j.otc.2014.06.006. Epub 2014 Aug 1.

Sakulchit T, Goldman RD.Antibiotic therapy for children with acute otitis media. *Canadian Family Physician*. 2017 Sep;63(9):685-687.

Venekamp RP, Damoiseaux RA, Schilder AG.Acute otitis media in children. *American Family Physician*. 2017 Jan 15;95(2):109-110.

17

Swimmer's Ear

Abstract Acute otitis externa, or swimmer's ear, is an infection of the ear canal that affects about 1% of the population overall but is most prevalent in older children and teenagers. The main symptoms of the disease include pain, itching, redness and swelling of the ear, hearing loss and an ear discharge. It's caused mainly by *Pseudomonas aeruginosa* or *Staphylococcus aureus* following some disruption of the ear's defence systems. Precipitating factors include obstruction of the ear canal, reduced ear wax due to repeated water exposure or overcleaning, damage to the lining of the ear as well as exposure to warm and damp environments. Treatment of the condition involves pain management, removal of debris from the ear canal, administration of topical medications to control swelling and infection and treatment/avoidance of contributing factors.

What Is Swimmer's Ear?

Otitis externa is inflammation of the external ear canal, which may also involve the auricle and the tympanic membrane (Fig. 17.1). It's commonly known as "swimmers ear" or "tropical ear". The disease occurs in two main forms: acute otitis externa (AOE) which has a rapid onset (usually less than 48 hours) and lasts for less than 3 weeks and chronic otitis externa (COE) which persists for more than 3 months. AOE is the most frequently encountered.

AOE usually affects only one ear and the main symptoms are:

© Springer Nature Switzerland AG 2021
M. Wilson, P. J. K. Wilson, *Close Encounters of the Microbial Kind*,
https://doi.org/10.1007/978-3-030-56978-5_17

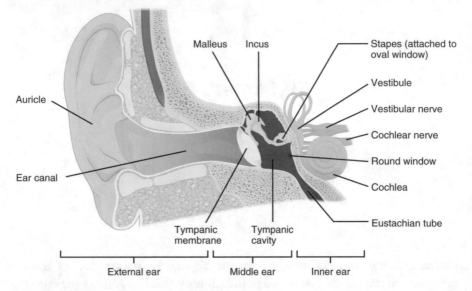

Fig. 17.1 Diagram showing the anatomy of the human ear. Otitis externa may affect the auricle, ear canal and the tympanic membrane. (OpenStax [CC BY 4.0 (https://creativecommons.org/licenses/by/4.0)])

- Ear pain
- Itching and irritation in and around the ear canal
- Redness and swelling of the auricle and ear canal
- A feeling of pressure and fullness inside the ear
- Scaly skin in and around the ear canal, which may peel off
- Discharge from the ear (Fig. 17.2)
- Tenderness on moving the ear or jaw
- Hearing loss

Complications are rare but may include:

- Abscess formation
- Progression to chronic otitis externa
- Spread of the infection to neighbouring tissues, including facial cellulitis
- Fibrosis (a build-up of scar tissue) leading to loss of hearing
- Myringitis (inflammation of the tympanic membrane)
- Perforation of the ear drum
- Malignant otitis externa, which is a very aggressive infection and can be fatal

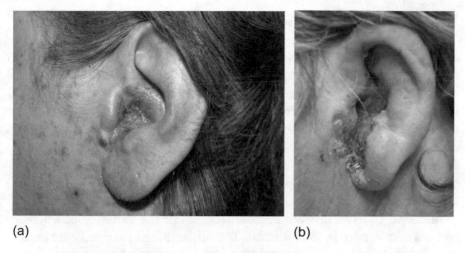

Fig. 17.2 Otitis externa. (a) An ear affected by otitis externa showing a discharge. (Klaus D. Peter, Gummersbach, Germany [CC BY 3.0 de (https://creativecommons.org/licenses/by/3.0/de/deed.en)]). (b) A severe case of otitis externa showing copious discharge and infection that has spread into the local area. (James Heilman, MD [CC BY 3.0 (https://creativecommons.org/licenses/by/3.0)])

There are a number of risk factors and triggers for the disease, and these include (i) obstruction of the ear canal or having a narrow ear canal, (ii) high environmental humidity (iii), warm environmental temperatures, (iv) contact with water, e.g. swimming, (v) local trauma, (vi) allergy, (vii) skin disease, (viii) diabetes and (ix) being in an immunocompromised state.

How Common Is the Disease?

Otitis externa affects people of all ages, but its incidence peaks at 7–12 years and declines in those aged over 50 years. It's slightly more common in women than in men up to the age of 65 years. In temperate climates, the prevalence increases at the end of the summer, especially in those aged 5–19 years. In the UK more than 1% of people are diagnosed with the condition each year, and 10% of the population will be affected at some time during their lives. In the USA it is one of the most common infections encountered by clinicians and affects 1 in 123 people. Each year there are about 2.4 million visits to medical practitioners for AOE (8.1 visits per 1000 population), and slightly less than half of all visits are for children 5 to 14 years of age. Direct costs of the disease are estimated to be about $0.5 billion annually.

Which Microbes Cause Acute Otitis Externa?

AOE is due to a bacterial infection in 98% of cases with either *Pseudomonas aeruginosa* (20–60% of cases) or *Staph. aureus* (20–60% of cases) being the most frequent cause. However, more than one bacterium is often involved. Occasionally fungi, particularly *Aspergillus* species or *Candida* species, may be responsible. Other, non-infectious causes, or contributors to the condition, include trauma (scratching, aggressive or cleaning), chemical irritants, allergy (commonly to antibiotic ear drops such as neomycin), high-humidity conditions, swimming, radiotherapy or skin disease (allergic dermatitis, atopic dermatitis, psoriasis).

Initiation of AOE involves some disruption of the antimicrobial defence systems that operate in the ear canal. The surface of the ear canal consists of a dry epidermis with the usual antimicrobial defence mechanisms of skin (see Chap. 1). In addition, ear wax is an important part of the defence system because it's acidic (which many microbes can't tolerate) and has a high content of antimicrobial compounds. The defence systems of the ear can be disrupted in a number of ways, and these are summarised in Table 17.1.

Table 17.1 Ways in which the defence systems of the ear can be disrupted

Disruptive mechanism	Consequences
Obstruction of the canal due to abnormal bone growth (surfer's ear) or the excessive accumulation of ear wax	Results in increased retention of water and debris after swimming, bathing or showering
Reduction in ear wax due to repeated water exposure or overcleaning of the ear canal	Allows bacteria to grow unchecked
Damage to the lining of the ear due to trauma	Enables invasion of microbes into the underlying tissues
Changes in the environment to either damper or warmer conditions	Promotes bacterial overgrowth.

Box 17.1 What's in a Name?

The name of the genus *Pseudomonas* was first used in 1894 by the German botanist Walter Migula. It's a combination of "pseudo-"a Greek word meaning false and *Monas* which is the name of a protozoan. The cells of *Pseudomonas* species resemble those of the protozoan *Monas* in both their size and how they move, and so the name "false-*Monas*" was used for the genus. The species name *aeruginosa* is derived from the Latin word "aerugo" which means verdigris (copper rust) which has a blue-green colour and refers to the fact that the organism produces a blue-green pigment (Figure). The organism was discovered in 1882 by the French bacteriologist Carle Gessard.

The blue-green pigment (pyocyanin) produced by *Pseudomonas aeruginosa*.
(Das T, Kutty SK, Kumar N, Manefield M (2013) Pyocyanin facilitates extracellular DNA binding to pseudomonas aeruginosa influencing cell surface properties and aggregation. PLoS ONE 8(3): e58299

P. aeruginosa is a Gram-negative, motile bacillus (Fig. 17.3) that prefers an oxygen-rich atmosphere. It's widely distributed in nature and is found in water and soil as well as in the intestinal tract of humans and other animals.

As well as causing AOE, the organism is responsible for a number of other diseases in humans including lung infections in patients with cystic fibrosis, burn and wound infections and catheter-associated infections. Unfortunately it has developed resistance to a range of antibiotics, and this makes infections due to this organism very difficult to treat. It has a wide range of virulence factors, and these are summarised in Table 17.2.

Staph. aureus is another well-known human pathogen and is found in the respiratory tract. Its virulence factors have been described in Chap. 3.

Infection by either organism results in an inflammatory response which causes inflammation and swelling of the ear tissues and the production of pus which exudes from the ear canal. Cuts or scratches in the ear canal can allow these microbes to invade neighbouring tissues, and this results in infection of the deeper layers under the skin, i.e. cellulitis.

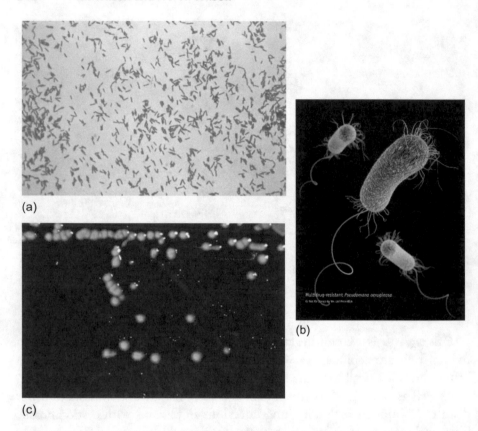

Fig. 17.3 Images of *P. aeruginosa*. (a) Gram stain of *P. aeruginosa* showing Gram-negative bacilli. (Dr. W.A. Clark, Centers for Disease Control and Prevention, USA). (b) A three-dimensional computer-generated image of three *P. aeruginosa* bacteria. The artistic recreation was based on scanning electron microscopic imagery. Note the presence of numbers of short, thin, fimbriae (involved in attachment to surfaces) as well as a single, long whip-like flagellum which enables the organism to move. (James Archer, Centers for Disease Control and Prevention, USA). (c) Colonies of *P. aeruginosa* growing on a blood agar plate. (Centers for Disease Control and Prevention, USA)

Table 17.2 Important virulence factors of *P. aeruginosa* and their function

Virulence factor	Function
Capsule	Protects it against phagocytic cells
Fimbriae	Enable it to adhere to ells and other surfaces
Flagella	Enable it to move
Toxins	Kill a range of human cells
Can form biofilms	Protects it against antimicrobial defence systems
Proteases	Break down human tissues and antibodies
Lipopolysaccharide	Induces the release of inflammatory cytokines

How Is the Condition Diagnosed?

Diagnosis of AOE is made on the basis of the symptoms, the presence of any of the risk factors described previously and examination findings. Culture of the discharge to identify the organism(s) responsible as well as to which antibiotics they are susceptible can be carried out, especially if the first attempt at treatment fails. The ear canal and ear drum should be examined using an otoscope (Fig. 17.4) to see if there's any damage to these structures. However, often the swelling and debris in the canal blocks a full view of the area affected.

How Is the Disease Treated?

Treatment of AOE involves pain management, removal of debris from the ear canal, administration of topical medications to control swelling and infection and treatment/avoidance of contributing factors. The condition can be painful and therefore, you may need to take paracetamol or a non-steroidal

Fig. 17.4 An otoscope being used to examine the ear of a patient. (Kate Whitley. Attribution 4.0 International (CC BY 4.0))

anti-inflammatory drug. Cleaning of the ear canal to remove any debris and wax is important in cases where these are preventing the medication from reaching the skin where it needs to work. This can be achieved by syringing or swabbing (both of which should be carried out by a health professional, not by yourself) or by a specialist using magnifying equipment and fine suction tubes, known as "microsuction". This also enables the ear drum to be examined to assess any potential damage. Antibiotic ear drops or sprays, often containing a corticosteroid to reduce inflammation and swelling, are then administered. There are many suitable agents for example a mixture of the two antibiotics neomycin and polymyxin B with hydrocortisone. More recently a mixture of the antibiotic ciprofloxacin and the corticosteroid dexamethasone has been shown to be effective. Topical 2% acetic acid is also effective especially if used early on. When applying the drops yourself, it's important that you do this while lying down with the affected ear upwards and then waiting for 5 to 10 minutes before getting up – this is to make sure that as much of the medication as possible reaches the affected area. If access to the ear canal is blocked by swelling, then an absorptive wick can be inserted by a clinician onto which drops can be applied. The choice of antibiotic should be guided by the results of sensitivity testing of the microbes found in samples taken from you. Oral antibiotics can be used but are rarely required and are usually reserved for severe cases and complications. During treatment and for 1–2 weeks after, you should keep the affected ear canal dry, and you should avoid swimming. You should wear a shower cap when showering or bathing. Treatment for any underlying skin conditions that may aggravate the infection, such as seborrhoeic dermatitis, psoriasis or eczema, should also be considered.

Occasionally, fungi such as *Aspergillus* species or *Candida* species may be responsible for AOE. In such cases treatment with acetic acid-containing ear drops with a corticosteroid is usually effective. If this fails, then ear drops containing an anti-fungal agent (such as clotrimazole) can be used.

How Can I Avoid Getting the Condition?

To prevent getting AOE, you should avoid the risk factors described previously, particularly the accumulation of water in your ear canal. You should avoid the frequent use of cotton-tipped applicators because this damages the lining of the ear canal and can compress ear wax into a plug that can block the canal. You can try applying acetic acid-containing ear drops before and after swimming as well as drying your ears with a cool hair-dryer after water exposure.

Want to Know More?

American Family Physician https://www.aafp.org/afp/2001/0301/p927.html

Canadian Paediatric Society https://www.cps.ca/en/documents/position/acute-otitis-externa

Centers for Disease Control and Prevention, USA https://www.cdc.gov/mmwr/preview/mmwrhtml/mm6019a2.htm?s_cid=mm6019a2_w

DermNet, New Zealand https://www.dermnetnz.org/topics/otitis-externa/

Health Service Executive, Ireland https://www.hse.ie/eng/health/az/e/ear-infection,-outer/treating-otitis-externa.html

KidsHealth, USA https://kidshealth.org/en/teens/swimmers-ear.html

Mayo Clinic, USA https://www.mayoclinic.org/diseases-conditions/swimmers-ear/symptoms-causes/syc-20351682

National Health Service, UK https://www.nhsinform.scot/illnesses-and-conditions/ears-nose-and-throat/otitis-externa,   https://www.guysandstthomas.nhs.uk/resources/patient-information/audiology/otitis-externa.pdf

National Institute for Clinical Care and Excellence (NICE), UK. Otitis externa, 2018 https://cks.nice.org.uk/otitis-externa#!topicSummary

Patient Info, UK https://patient.info/doctor/otitis-externa-and-painful-discharging-ears

Ghossaini S. Otitis externa. BMJ Best Practice, BMJ Publishing Group, 2019 https://bestpractice.bmj.com/topics/en-gb/40

Hajioff D, MacKeith S. Otitis externa. *BMJ Clinical Evidence.* 2015 Jun 15;2015. pii: 0510. PMID: 26074134

Medina-Blasini Y, Sharman T. Otitis Externa. Treasure Island (FL): StatPearls Publishing LLC; 2020 https://www.ncbi.nlm.nih.gov/books/NBK556055/

Schaefer P, Baugh RF. Acute otitis externa: an update. *American Family Physician.* 2012 Dec 1;86(11):1055-61.

Waitzman AA. Otitis Externa. Medscape from WebMD, 2020 https://emedicine.medscape.com/article/994550-overview

18

Acute Bronchitis

Abstract Acute bronchitis is an infection of the bronchi in the lungs that results in coughing and the production of sputum. It's usually caused by viruses – particularly influenza A and B, rhinoviruses and enteroviruses. Occasionally, bacteria are responsible – mainly *Streptococcus pneumoniae* and *Haemophilus influenzae*. Infection results in inflammation and narrowing of the bronchi and mucus accumulation. The disease is self-limiting and usually lasts no longer than 6 weeks. Antibiotics aren't usually necessary, and the symptoms can be relieved by increasing fluid intake and taking painkillers.

What Is Acute Bronchitis?

Acute bronchitis is an infection of the bronchi, the main airways of the respiratory tract (Fig. 18.1), which results in them becoming inflamed. In some patients, the trachea may also be involved and this results in tracheitis.

Acute bronchitis (often referred to as a "chest cold") is a very common disease and is one of the main reasons why patients seek medical care. "Acute" means it started recently and hasn't lasted for long (less than 21 days). Longer (over 3 months), or ongoing episodes of bronchitis, are called "chronic" and have different causes – mainly smoking.

The main symptoms of acute bronchitis are a cough (the most common) and the production of sputum (phlegm) from the airways. In addition to these major symptoms, a number of others may occur which overlap with those we experience with the common cold or sinusitis. This can make it tricky for patients to tell these problems apart. They include:

© Springer Nature Switzerland AG 2021
M. Wilson, P. J. K. Wilson, *Close Encounters of the Microbial Kind*,
https://doi.org/10.1007/978-3-030-56978-5_18

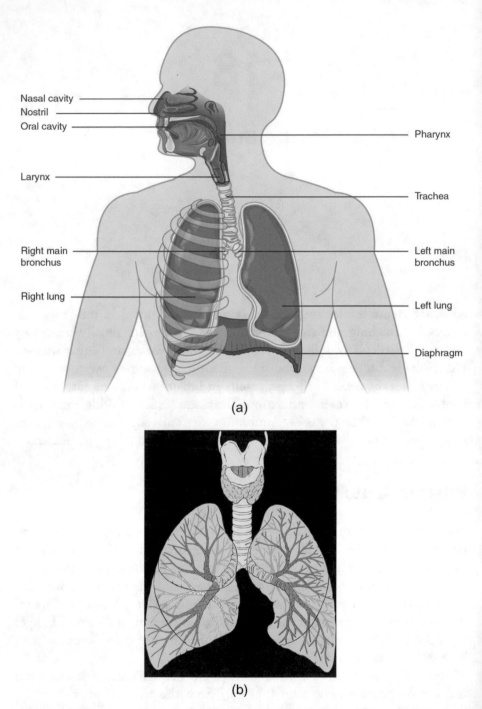

(a)

(b)

Fig. 18.1 The human respiratory system. (**a**) Diagram showing the main components of the respiratory tract. (Anatomy & Physiology, Connexions Web site. http://cnx.org/content/col11496/1.6/ OpenStax College / CC BY (https://creativecommons.org/licenses/by/3.0)). (**b**) Diagram showing the highly branched nature of the bronchi which become progressively narrower. The narrowest tubes are known as bronchioles. (Patrick J. Lynch, medical illustrator / CC BY (https://creativecommons.org/licenses/by/2.5))

- General malaise
- Fever
- Chest pain (in severe cases)
- Difficulty in breathing (known as "dyspnoea")
- Runny or stuffy nose
- Headache
- Wheezing

Coughing is usually worse at night or during exercise and lasts for more than 2 weeks in 50% of cases but can go on for as long as 4 weeks in 25% of patients. The continual coughing can make the throat or chest and stomach muscles sore. The cough can sometimes continue for several weeks after the other symptoms have settled. In temperate climates, most episodes (>80%) occur during autumn or winter. Complications aren't common but include an ongoing cough that may persist for several months. Pneumonia can follow in around 1 in 20 people, especially in older or frail patients.

Box 18.1 What's in a Name?

The term bronchitis was first used by a London physician, Charles Badham (1780–1845). He introduced the term in a publication in 1808 titled "Observations on the Inflammatory Affections of the Mucous Membrane of the Bronchiæ". In this paper he described the disease as an "inflammatory affection of that part of the mucous membrane which lines the bronchial tubes", and he made a distinction between the acute and chronic forms of the disease. As well as being a physician, Badham was also a classical scholar and in 1818 produced a translation of the Satires of Juvenal.

How Common Is It?

Acute bronchitis is common throughout the world and is one of the top 5 reasons why people seek medical care. In the UK the annual incidence is 44 per 1000 adult population. It's seen 2–4 times more often in people aged >60 years than in those aged <50 years. In the USA it's one of the most common adult outpatient diagnoses. At least one episode of acute bronchitis is reported in 5% of the USA population each year. The majority of these patients seek medical attention, and this accounts for >10 million office visits per year.

What Causes the Disease?

Some cases of acute bronchitis are due to breathing in irritant substances such as chemicals and dust which inflame the airways. However, it's usually due to an infection. There's some uncertainty about which microbes are responsible, and this is because only a small proportion of cases are investigated microbiologically, for example, by sending a sputum sample to the laboratory. Even in those that are, it's often difficult to identify the microbe involved because the microbiota of the upper respiratory tract is very complex and contains many potential pathogens that may, or may not, be the cause. The pathogen responsible for an episode of the disease has been identified in fewer than half of those cases that have had tests carried out. Nevertheless, it's generally agreed by scientists that approximately 90% of episodes are caused by viruses.

The most commonly identified viruses (in decreasing order) are influenza A and B, rhinoviruses, enteroviruses, human metapneumovirus, coronaviruses, parainfluenza viruses and respiratory syncytial virus. These have been described in previous chapters (Chaps. 10, 11, and 12). Of these, approximately 50% of cases are due to influenza A and B, rhinoviruses and enteroviruses.

When bacteria are responsible for the infection, the most frequently identified species are *Streptococcus pneumoniae* (66% of cases), *Haemophilus influenzae* (17%), *Moraxella catarrhalis* (4%), *Streptococcus pyogenes* (2%), *Haemophilus parainfluenzae* (2%) and *Staphylococcus aureus* (2%). These bacteria are found in the throats of many healthy individuals but are also responsible for a number of infections of the respiratory tract and other body sites – they are all well-recognised as being opportunistic pathogens (see Chap. 1). All of these species have been described in previous chapters (Chaps. 3, 13 and 14).

What Happens During an Infection?

The immune system reacts to the infecting virus or bacterium by mounting an inflammatory response, and this results in swelling of the walls of the bronchi and an increase in mucus production (Fig. 18.2). This narrowing of the airways can cause a wheezing sound as the air passes through them and can lead to breathlessness.

In addition, damage to the cilia (Fig. 18.3) on the cells that line the bronchi prevents the mucociliary escalator (see Fig. 1.18) from functioning properly. This means that mucus and debris aren't easily expelled, although the coughing reflex helps to remove some by forcing it out of the airways. For two of the

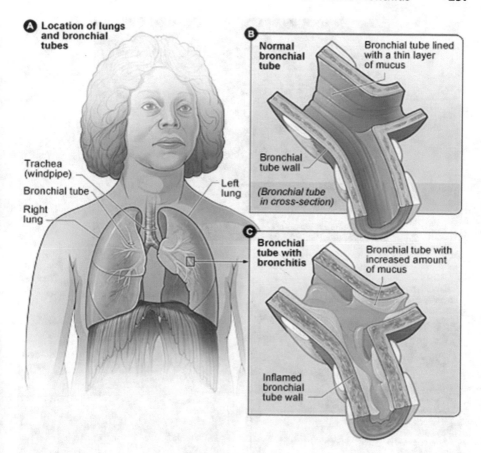

Fig. 18.2 Inflammation of the bronchi results in swelling of the walls of the bronchi, narrowing of the bronchi and increased mucus production. (Image courtesy of the National Heart Lung and Blood Institute, Public Domain via Wikimedia Commons)

main bacteria that cause acute bronchitis, we know how this damage to cilia occurs. *Strep. pneumoniae* produces a toxin known as pneumolysin. This toxin reduces the beating of cilia and can kill ciliated bronchial lining cells. Certain molecules in the cell wall of *H. influenzae* are also able to reduce ciliary action and kill bronchial epithelial cells.

Acute bronchitis is diagnosed by assessing the symptoms and examining the patient to rule out other possible causes. Microbiological tests aren't usually carried out because, as mentioned earlier, it can be difficult to identify which of the many microbes present in the respiratory tract is responsible.

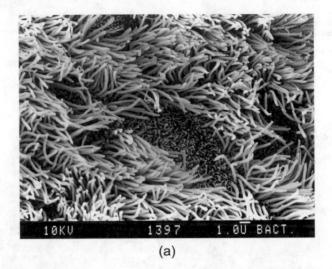

(a)

Cilia move together in wave-like motions to transport mucus towards the mouth. The mucus contains debris and microbes

Cilia are damaged and so do not produce waves. This results in a build-up of mucus containing debris and microbes

Normal mucus layer

normal cilia

Cells covered in cilia line the bronchi

damaged cilia

mucus build-up with debris and microbes

(b)

Fig. 18.3 Ciliated cell that line the bronchi are vital to the proper functioning of the mucociliary escalator. Disruption of this results in mucus accumulation. (a) Cilia on the bronchial mucosa. (Credit: David Gregory & Debbie Marshall. CC BY 4.0. Wellcome Image Library). (b) Pathogenic microbes damage the cilia and sometimes kill the ciliated cells – this disrupts the mucociliary escalator. (Modified image from National Heart, Lung and Blood Institute, Public Domain via Wikimedia Commons)

How Is Bronchitis Treated?

In healthy adults the illness is self-limiting and requires no specific treatment – it usually lasts for less than 6 weeks. The aim of medication is to reduce the symptoms and keep you comfortable until the illness resolves. Good fluid intake helps to prevent you from becoming dehydrated. Breathing in steam may loosen the mucus making it easier to cough up. Headaches and aches and pains can be treated with painkillers.

Mucolytics or expectorants have limited effect and aren't usually recommended. If you have more severe symptoms, treatment options include cough suppressants or inhalers to open up narrow airways ("bronchodilators") if wheezing is present.

Most of the major regulatory bodies don't recommend the routine use of antibiotics in treating acute bronchitis in otherwise healthy people. This is because (i) the disease is usually caused by a virus, (ii) inappropriate antibiotic use can have adverse side effects and (iii) inappropriate use contributes to the development of antibiotic resistance. Acute bronchitis leads to more inappropriate antibiotic prescribing than any other acute respiratory tract infection. Disappointingly, 65–80% of patients with acute bronchitis receive an antibiotic despite evidence indicating that, with few exceptions, they aren't effective. However, antibiotics may be prescribed to either treat, or reduce the risk of, serious complications, especially in frail, elderly, patients or those who are immune-impaired or those with underlying chest or other conditions that make them more vulnerable.

How Can I Avoid Getting Bronchitis?

To reduce the risk of getting the disease, and of passing on the infection if you have it, it's important to practice good hand and respiratory hygiene. This is because the microbes are carried in the drops of sputum that are produced when we cough or sneeze. Cold and flu viruses are thought to survive outside of the body for possibly upto 72 hours, while bacteria can survive for much longer. We can lower the risk of becoming infected by avoiding those with the disease and washing our hands often. If we do get the disease, then we can reduce the chances of spreading it to others by (i) staying away from school or work until we've recovered, (ii) coughing or sneezing into a tissue and discarding this responsibly and (iii) washing our hands frequently.

Passive smoking is a risk factor for acute bronchitis in non-smokers, because of the damage caused by smoke to the airways, and so care should be taken to avoid exposure to this. The most effective way for smokers to avoid the disease is to discontinue the habit.

Want to Know More?

American Academy of Family Physicians https://familydoctor.org/condition/acute-bronchitis/

American Lung Association https://www.lung.org/lung-health-and-diseases/lung-disease-lookup/acute-bronchitis/learn-about-acute-bronchitis.html

Centers for Disease Control and Prevention, USA https://www.cdc.gov/antibiotic-use/community/for-patients/common-illnesses/bronchitis.html

Chest Foundation, American College of Chest Physicians https://foundation.chest-net.org/patient-education-resources/acute-bronchitis/

Johns Hopkins Medicine, USA https://www.hopkinsmedicine.org/health/conditions-and-diseases/acute-bronchitis

Lung Foundation, Australia https://lungfoundation.com.au/wp-content/uploads/2018/09/Factsheet-Acute-Bronchitis-Aug2018.pdf

Mayo Clinic, Mayo Foundation for Medical Education and Research, USA https://www.mayoclinic.org/diseases-conditions/bronchitis/symptoms-causes/syc-20355566

MedlinePlus, U.S. National Library of Medicine https://medlineplus.gov/acutebronchitis.html

National Health Service, UK https://www.nhs.uk/conditions/bronchitis/

National Institute for Clinical Care and Excellence (NICE), UK. Chest infections – adult, 2020 https://cks.nice.org.uk/chest-infections-adult#!topicSummary

Physiopedia - physiotherapy knowledge resource https://www.physio-pedia.com/Bronchitis

Fayyaz J. Bronchitis. Medscape from WebMD, 2019 https://emedicine.medscape.com/article/297108-overview

Hueston WJ. Acute bronchitis. BMJ Best Practice, BMJ Publishing Group, 2020 https://bestpractice.bmj.com/topics/en-gb/135

Kinkade S, Long NA. Acute Bronchitis. *American Family Physician*. 2016 Oct 1;94(7):560-565.

Singh A, Avula A, Zahn E. Acute Bronchitis. Treasure Island (FL): StatPearls Publishing LLC; 2019 https://www.ncbi.nlm.nih.gov/books/NBK448067/

Smith SM, Fahey T, Smucny J, Becker LA. Antibiotics for acute bronchitis. *Cochrane Database of Systematic Reviews*. 2017 Jun 19;6:CD000245. doi: https://doi.org/10.1002/14651858.CD000245.pub4.

Wark P. Bronchitis (acute). *BMJ Clinical Evidence*. 2015 Jul 17;2015. pii: 1508. PMID: 26186368

Part IV

Infections of the Eye

19

Conjunctivitis

Abstract Conjunctivitis is inflammation of the conjunctiva which can be caused by a variety of viruses or bacteria and affects approximately 2% of the global population. Viruses, particularly adenoviruses, are the main cause of the disease in adults, while in children, bacteria are largely responsible, mainly *Streptococcus pneumoniae*, *Staphylococcus aureus*, *Moraxella catarrhalis* and *Haemophilus influenzae*. The main symptoms of viral conjunctivitis include a watery discharge, adherent eyelids, fever, itching and enlarged lymph glands. Red eye, pus formation and swelling are characteristic of the condition when bacteria are responsible. Topical antibiotics can shorten the duration of bacterial conjunctivitis but aren't usually necessary for treating mild or uncomplicated cases. Symptoms can be eased by cleaning the discharge from the eyes with sterile saline or boiled then cooled water and using cool compresses. Alleviation of some of the symptoms of viral conjunctivitis can be achieved by using artificial tears, topical antihistamines and cold compresses. Good hygiene measures are important in preventing the transmission of both bacterial and viral conjunctivitis.

What Is Conjunctivitis?

The conjunctiva is an epithelial layer that covers the cornea (the transparent, light-focussing front of the eye) and also lines the surface of the underneath of the eyelid (Fig. 19.1). Inflammation of this epithelial layer is known as conjunctivitis, and this can arise for a variety of reasons including infection, physical irritation, an allergic response (such as hay fever) or following the

© Springer Nature Switzerland AG 2021
M. Wilson, P. J. K. Wilson, *Close Encounters of the Microbial Kind*,
https://doi.org/10.1007/978-3-030-56978-5_19

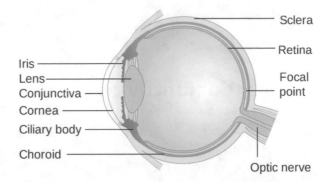

Fig. 19.1 The main structures of the eye showing the conjunctiva which covers the cornea and extends to line the inside of the eyelids. (Cancer Research UK, This image has been released as part of an open knowledge project by Cancer Research UK. If reused, attribute to Cancer Research UK / Wikimedia Commons)

administration of some medicines. In this chapter we'll discuss only the infectious causes of conjunctivitis. In order to protect it from infections, the eye has a number of defence mechanisms.

First of all, the eyelids protect the eyes from microbes in three important ways: (i) blinking (which occurs every 5 seconds) protects against microbe-laden foreign objects, (ii) they remove any debris as they move over the cornea, and (iii) they distribute tears (which contain many antimicrobial compounds) over the whole eye surface as a thin film which is known as the "tear film". The tear film has a thickness of 3–46 um and protects the conjunctiva from dehydration, microbes and particulate matter. It contains a network of large molecules of mucin that bind to any microbes present and prevents them from sticking to the conjunctiva itself. Blinking rolls the mucin network into small balls (containing trapped microbes and particles) which are pushed towards the inner corner of the eye. It then becomes compacted into a small clump and is pushed onto the skin where it dries and eventually either falls off or is removed by rubbing. In this way, microbes, foreign particles and debris are regularly removed from the conjunctival surface. The tear film also contains all of the components of the immune system (cells and antibodies) as well as antimicrobial compounds that can kill any microbes present.

The main symptoms of bacterial conjunctivitis include red eye, pus formation and swelling. In addition, the eyelids often become temporarily stuck together. Generally both eyes are affected, or it spreads quickly from one to both, and there's usually some itching. The condition usually lasts between 7 and 10 days, and the majority of cases get better without antibiotic treatment. When viruses are responsible, the symptoms usually include a watery discharge (rather than pus), adherent eyelids, fever, itching and enlarged lymph glands in front of the ears. Viral conjunctivitis usually lasts for 2 to 3 weeks and usually resolves on its own. Identifying whether a bacterium or a virus is the cause of the infection using symptoms alone is unreliable as there's a lot of overlap. Figure 19.2 shows examples of conjunctivitis due to bacteria and viruses.

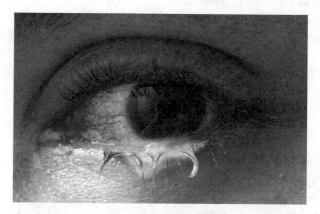

Tanalai at English Wikipedia [CC BY 3.0 (https://creativecommons.org/licenses/by/3.0)]

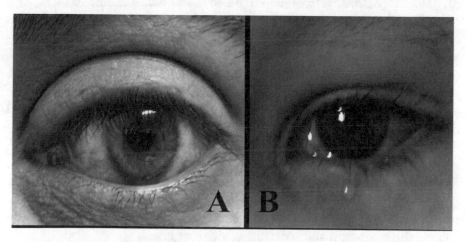

Fig. 19.2 The appearance of the eye affected by bacterial (top image) and viral conjunctivitis (bottom two images) – showing two different examples. (Clinical and public health management of conjunctivitis (bottom two images) in the Israel Defense Forces. Efros, O, Zahavi A, Levine H, Hartal M. Disaster Mil Med. 2015; 1: 12

How Common Is Conjunctivitis?

Conjunctivitis affects approximately 2% of the population worldwide. In the USA approximately six million people suffer from the disease annually, and it accounts for approximately 1% of all primary care visits. In the UK, conjunctivitis is the reason for 1.2–3% of visits to general practitioners and is estimated to affect 13 of every 1000 people each year.

With regard to infectious causes of conjunctivitis, viruses are responsible for approximately 70% of cases, with bacteria being the next most frequent cause. However, the type of microbe causing the infection varies according to the person's age, with bacteria being more commonly responsible in children (between 50 and 75% of cases), while in adults viruses are the predominant cause.

Which Bacteria Are Responsible for Conjunctivitis?

A wide range of bacteria are able to cause conjunctivitis, and the commonest of these are *Streptococcus pneumoniae* (see Chap. 10), *Staph. aureus* (see Chap. 3), *Moraxella catarrhalis* (see Chap. 10) and *Haemophilus influenzae* (see Chap. 10). All of these organisms are present in the upper respiratory tract of many healthy people. Rarer causes are *Neisseria gonorrhoeae* and *Chlamydia trachomatis*, and when these are responsible, there can be serious complications – often oral antibiotics are necessary for treatment. Newborn babies can catch unusual types of infections during the birthing process, and these can cause significant problems if not treated promptly. In the rest of this chapter, we'll focus on the most common causes of conjunctivitis.

The bacterium responsible for the disease can be identified by trying to grow it in the laboratory from a swab taken from the infected conjunctiva. Swabs aren't routinely taken but are reserved for those with severe symptoms, if a less common infection is suspected, if there are complications or if treatment fails to improve symptoms. Urgent swabs should be sent for newborn babies with symptoms and specialist advice sought.

The symptoms of bacterial conjunctivitis are a consequence of the immune response to the invasion of conjunctival cells by the bacteria. Inflammation causes the blood vessels on the surface of the eye to dilate, the conjunctiva to swell and a discharge to be produced.

Bacterial conjunctivitis can be picked up from infected individuals or animals.

Contaminated fingers and objects are common routes of transmission, and this can be prevented by appropriate hygiene measures such as adequate hand-washing and avoiding shared towels and bed linen.

How Is Bacterial Conjunctivitis Treated?

You can ease the symptoms by using a cold compress and by cleaning the discharge from your eyes with sterile saline or with water that has been boiled (to kill any microbes present) and then fully cooled. Topical administration of antibiotic drops or ointments can shorten the illness, but the routine use of antibiotics in infections that usually settle without treatment should be discouraged. However, as well as shortening the duration of the symptoms, using antibiotics also reduces the chance of infecting others and should be considered in those at risk of complications. Infections in those who have recently had eye surgery, contact lens wearers and those with impaired immunity may be treated with antibiotics if they're at risk of complications. People suffering severe or prolonged symptoms should have swabs sent for analysis and be offered antibiotic treatment. Most broad-spectrum antibiotics seem to be effective and are used for 7 days. The most widely used antibiotics are chloramphenicol, fusidic acid, gentamycin, ciprofloxacin, bacitracin and polymyxin/bacitracin.

Complications of bacterial conjunctivitis are rare but include scarring or perforation of the cornea. Urgent specialist advice should be sought if your vision is affected or if you fail to improve with treatment. Sometimes the deeper layers of the eye can become affected – inflammation of the cornea (keratitis) can develop especially in contact lens wearers. Immediate removal of the contact lenses is advisable, and they shouldn't be used again until symptoms have completely gone.

Which Viruses Cause Conjunctivitis?

Viral conjunctivitis can be caused by adenovirus, herpes simplex, Epstein-Barr virus, varicella zoster, molluscum contagiosum, Coxsackie and enteroviruses. However, adenoviruses are responsible for the majority (between 65% and 90%) of cases. The disease is highly contagious and has an incubation period of 5 to 12 days. Infected individuals can shed the virus for up to

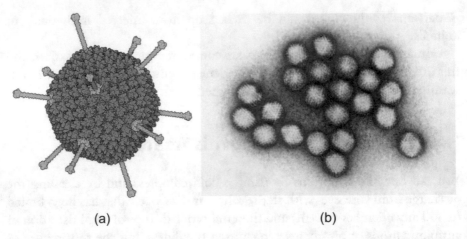

(a)   (b)

Fig. 19.3 Images of adenoviruses. (a) A model of an adenovirus showing its icosahedral shape and the protein spikes that protrude from its proteinaceous coat. (David S. Goodsell, RCSB Protein Data Bank. Attribution 4.0 International (CC BY 4.0)). (b) Digitally colourised electron micrograph of a group of adenoviruses. (Image courtesy of Dr. G. William Gary, Jr, Centers for Disease Control and prevention)

14 days following the appearance of symptoms. The adenovirus is between 70 and 100 nm in diameter and is icosahedral-shaped with a number of proteinaceous spikes (Fig. 19.3). Inside the outer protein coat is a molecule of DNA. It's an unenveloped virus which means that the outer layer doesn't consist of the cytoplasmic membrane of the host cell in which it was produced.

Box 19.1 What's in a Name?

Adenoviruses were first isolated from human adenoids by Wallace Rowe and his colleagues in 1953 at the National Institutes of Health, USA. They surgically removed the adenoids from 53 children and grew the tissues in the laboratory. After about 4 weeks, many of the cells appeared damaged, and a virus was isolated from these and called "adenoid-degenerating agent". A year later Maurice Hilleman and colleagues at the Walter Reed Army Institute of Research isolated the same agent from patients with influenza. It was later re-named the adenovirus because of its initial isolation form adenoids. It's now known that seven species of adenovirus (denoted A to G) are responsible for infections in humans.

Once it reached the surface of the eye, the virus is taken up by the epithelial cells of the conjunctiva which produce more virions (see Fig. 10.4). This starts an inflammatory response which results in the characteristic symptoms of the disease. Identification of an adenovirus as the cause of the conjunctivitis can be carried out by taking a sample of tear fluid and testing it for the presence of antigens from the virus – this is a type of test known as an "immunoassay". Results can be obtained in about 10 minutes. The test has been evaluated in the UK by the National Institute for Health and Care Excellence (NICE), but its role is still being researched and it's not in routine use in the UK.

As well as causing conjunctivitis, adenoviruses are responsible for a broad range of infections affecting many parts of the body. These include the common cold, sore throat, bronchitis, pneumonia, diarrhoea, inflammation of the stomach and intestines, urethritis and encephalitis.

How Is Viral Conjunctivitis Treated?

Viral conjunctivitis is highly contagious, and the virus spreads through direct contact via contaminated fingers, medical instruments, swimming pool water and personal items. Adenoviruses are very stable to chemical or physical factors, and this means that they can survive outside the body for long periods of time. Because of the high rates of transmission, effective handwashing is very important in preventing its spread, as well as avoiding shared towels and bed linen. Anti-viral medications can be useful in some cases but are less helpful when adenoviruses are involved. Steroid drops may help the symptoms by reducing inflammation but have to be used with care because of the risk of adverse effects. Referral for specialist input is advisable if the symptoms last for more than 7 days. You can alleviate some of the symptoms by using artificial tears, topical antihistamines and cold compresses. Obviously, antibiotics shouldn't be administered in infections that are known to be viral. If you normally wear contact lenses, then you should stop using them for at least 2 weeks.

Public Health England advises that children with conjunctivitis shouldn't routinely stay off school unless there's an outbreak of the infection. In order to reduce the spread of the infection, you should practice good hand hygiene and avoid close contact with others until the discharge has dried up.

Want to Know More?

American Academy of Ophthalmology https://www.aao.org/eye-health/diseases/pink-eye-quick-home-remedies

American Optometric Association https://www.aoa.org/patients-and-public/eye-and-vision-problems/glossary-of-eye-and-vision-conditions/conjunctivitis

Centers for Disease Control and Prevention, USA https://www.cdc.gov/conjunctivitis/index.html

College of Optometrists, London, UK https://www.college-optometrists.org/guidance/clinical-management-guidelines/conjunctivitis-bacterial-.html

Derm Net, New Zealand https://www.dermnetnz.org/topics/conjunctivitis/

KidsHealth, USA https://kidshealth.org/en/parents/conjunctivitis.html

Mayo Clinic, USA https://www.mayoclinic.org/diseases-conditions/pink-eye/symptoms-causes/syc-20376355

National Health Service, UK https://www.nhsinform.scot/illnesses-and-conditions/eyes/conjunctivitis

National Institute for Clinical Care and Excellence (NICE), UK. Conjunctivitis, 2018 https://cks.nice.org.uk/conjunctivitis-infective#!topicSummary

Patient Info, UK https://patient.info/doctor/conjunctivitis

Azari AA, Barney NP. Conjunctivitis: a systematic review of diagnosis and treatment. *Journal of the American Medical Association*. 2013 Oct 23;310(16):1721-9. doi: https://doi.org/10.1001/jama.2013.280318.

Epling J. Bacterial conjunctivitis. *BMJ Clinical Evidence* 2012 Feb 20;2012. pii: 0704. PMID: 22348418

Ryder EC, Benson S. Conjunctivitis. Treasure Island (FL): StatPearls Publishing LLC; 2020 https://www.ncbi.nlm.nih.gov/books/NBK541034/

Sambursky R. Acute conjunctivitis. BMJ Best Practice, BMJ Publishing Group, 2019 https://bestpractice.bmj.com/topics/en-gb/68

Scott IU. Viral Conjunctivitis (Pink Eye). Medscape from WebMD, 2020 https://emedicine.medscape.com/article/1191370-overview

Solano D, Virgile J, Czyz CN. Viral Conjunctivitis. Treasure Island (FL): StatPearls Publishing LLC; 2020 https://www.ncbi.nlm.nih.gov/books/NBK470271/

Watson S, Cabrera-Aguas M, Khoo P. Common eye infections. *Australian Prescriber*. 2018 Jun;41(3):67-72. doi: https://doi.org/10.18773/austprescr.2018.016. Epub 2018 Jun 1.

Yeung KK. Bacterial Conjunctivitis (Pink Eye). Medscape from WebMD, 2019 https://emedicine.medscape.com/article/1191730-overview

Part V

Infections of the Oral Cavity

20

Tooth Decay

Abstract Dental caries (tooth decay) involves the destruction of tooth enamel and dentine by the acids produced by certain bacteria when they are supplied with high levels of sugars from our diet. The bacteria responsible are mainly species belonging to the genera *Streptococcus, Lactobacillus* and *Actinomyces* that live in biofilms (dental plaque) on the tooth surface. It's a very common disease, particularly in developed countries, and affects 35% of the global population. It's treated by mechanical removal of the decayed regions of the tooth and replacing these with any of a variety of restorative materials. It can be prevented by regular removal of dental plaque using a toothbrush and dental floss. Fluoride-containing toothpastes and mouthwashes are also very useful as they make tooth enamel more resistant to damage by acids.

Is there any sound more terrifying than the dentist's drill? You sit in the dental chair, tilted backwards so you feel completely helpless and vulnerable. And the dentist, no matter how kindly and reassuring, seems to loom menacingly over you holding that dreadful device. Then it's in your mouth and the pitch increases as it touches your tooth – the horror. You've been anaesthetised so you know it won't hurt, but still you tremble. Your knuckles whiten as you clutch tightly at the chair handles trying not to show your fear. How long will this take? Will the drill ever stop? Surely your whole tooth will have been drilled away by now? And then it's over. Immense relief as the assistant gives you the mouth rinse. You almost leap out of the chair and stampede out of the door. Then you feel embarrassed at not having thanked the dentist enough. You swear that you'll cut down on sugar. You'll brush your teeth after every

M. Wilson, P. J. K. Wilson, *Close Encounters of the Microbial Kind*,
https://doi.org/10.1007/978-3-030-56978-5_20

meal; you'll floss three times a day. Your resolutions last maybe 2 days, and then it's back to the old caries-inducing lifestyle.

What Is Dental Caries?

Tooth decay (dental caries) is a disease which, if untreated, can result in tooth loss and complications such as dental abscesses.

A tooth (Fig. 20.1) consists mainly of a bone-like material known as dentine within which is a pulp cavity containing blood vessels, lymphatics and nerves. The dentine is covered by a layer of enamel on the region that protrudes into the mouth (known as the crown). Dentine consists mainly of a mineral, hydroxyapatite (the same mineral that's present in bone), and a protein called collagen. Enamel is also made up mainly of hydroxyapatite and is the hardest substance present in the body. The tooth is held in the jaw bone by one or more roots. The tooth gets its nutrients via blood vessels in the pulp.

Caries is a Latin word meaning "decay or rottenness" and was first used in English in the 1630s. Symptoms of caries include toothache; tooth sensitivity (tenderness or pain when eating or drinking something hot, cold or sweet); the appearance of white, grey, brown or black spots on the teeth; bad breath; and an unpleasant taste in the mouth.

Caries is multifactorial and results from the increased growth of particular tooth-damaging bacterial species (known as cariogenic species) because of the presence of sugars such as sucrose that these bacteria prefer. These cariogenic bacteria are usually present in low numbers in the mouth of healthy,

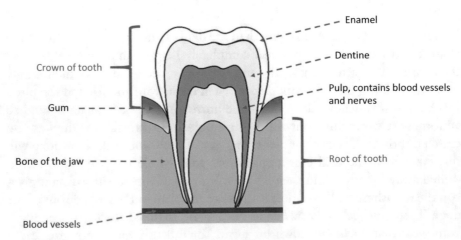

Fig. 20.1 Diagram showing the main features of a tooth

caries-free people, but excessive dietary sucrose (or other sugars) results in an increase in their numbers. Cariogenic bacteria produce acids from the sucrose, and these can dissolve the minerals present in teeth and produce a cavity (Fig. 20.2) that can increase in size until so much of the tooth has dissolved that it becomes loose in the jaw and eventually falls out.

In order to survive in the mouth, these cariogenic bacteria have to be able to withstand the antimicrobial defence systems that operate there (Box 20.1).

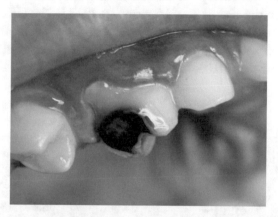

Fig. 20.2 A cavity due to dental caries in a tooth of a young boy. (Suyash.dwivedi [CC BY-SA 4.0 (https://creativecommons.org/licenses/by-sa/4.0)])

Box 20.1 Antimicrobial Defence Mechanisms in the Oral Cavity

Substantial physical forces are produced inside the mouth as a result of biting, chewing, tongue movements and the flow of saliva. These forces dislodge microbes from all the surfaces in the mouth (the teeth and the mucosa) and so discourage colonisation. Furthermore, mucins and various proteins in saliva bind to microbes and/or encourage them to stick together (known as co-aggregation) so that they stay in the saliva and are eventually swallowed. Swallowing transfers the dislodged microbes to the stomach, where most are killed by the very acidic conditions there. As many as 10^{11} microbes are removed each day from the oral cavity in this way.

Any microbes that become attached to the soft mucosal surfaces of the mouth are disposed of because the outer layers of these are regularly shed. Replacement of the outermost cells of the mucosa occurs every 24–48 h.

Saliva also contains a large variety of antimicrobial compounds that are effective against many microbes. A fluid, gingival crevicular fluid continually enters the mouth from the gingival crevice, and this contains important components of the immune system such as white blood cells and antibodies. Together these are effective at killing microbes and also preventing them from sticking to oral surfaces.

How Common Is Dental Caries?

Dental caries is a major disease in industrialised countries and affects 60–90% of children and almost all adults to some extent. In 2010 untreated caries in permanent teeth was the most prevalent disease condition, affecting 35% of the global population, or 2·4 billion people worldwide. It's less prevalent in developing countries. In industrialised countries, it's the most prevalent chronic disease of children, and surveys in the USA have shown that it affects 23% of children between the ages of 2 and 5 years. In 2017 in the UK, 23% of 5-year-old children were found to suffer from dental caries. With regard to older individuals, specifically those old enough to have permanent teeth, 35% of these had untreated caries in 2010 – the total number affected was 2.4 billion globally.

Box 20.2 Willoughby D. Miller and the Chemo-parasitic Theory of Caries

Willoughby Miller (1853–1907) was born in Ohio and studied maths, physics and dentistry at university. He travelled to Edinburgh and Berlin and was appointed Professor of Operative Dentistry at the University of Berlin. He then carried out research in microbiology in the Berlin laboratory of Robert Koch. In 1890 he formulated what he called the "Chemo-parasitic Theory of Caries" in which he proposed that the disease was due to the production of acid from sugars by oral bacteria. He also developed what became known as the "Focal Infection Theory". In this he suggested that oral microbes, or their products, can cause diseases at sites other than the oral cavity. Examples of such diseases include brain abscesses, pulmonary diseases and gastric problems. In the late twentieth century, evidence started accumulating in support of the Focal Infection Theory, and oral bacteria are now known to be responsible for (or contributors to) a number of conditions including coronary artery diseases, endocarditis, brain abscesses, osteomyelitis and pneumonia.

Photograph of Willoughby Miller in 1900. (This image was originally posted to Flickr by Internet Archive Book Images at https://flickr.com/photos/126377022@N07/14801496613. It was reviewed on September 9, 2015 by FlickreviewR and was confirmed to be licensed under the terms of the No known copyright restrictions)

What Microbes Are Responsible for Caries?

Like several other infectious diseases, such as acne, caries isn't due simply to the presence of a particular microbe in an individual. Caries is a disease that's certainly due to the activities of particular microbes in the mouth, but it only happens because of a change in the environment of the mouth that we ourselves bring about. The driver of this change is sugar. The main organisms responsible for dental caries are certain bacteria belonging to the genera *Streptococcus*, *Lactobacillus* and *Actinomyces*. These microbes are present in the mouths of many people but only in low proportions. They live on the surface of the tooth within a coating known as "dental plaque" (Fig. 20.3).

Plaque consists of bacteria embedded in a jelly-like substance (known as the plaque matrix) which is a mixture of large molecules (polysaccharides and proteins) produced by microbes and by our salivary glands (Fig. 20.4). It's an example of what is known as a "biofilm" which is defined as "a community of microbes embedded in a matrix and attached to a surface".

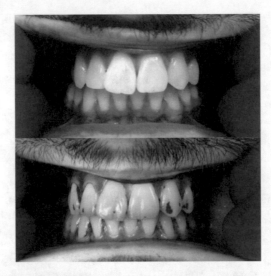

Fig. 20.3 Dental plaque revealed by using a coloured "disclosing" agent. Plaque is white or yellowish and so is difficult to see on the surface of white teeth. However, it's possible to show its presence by using a mouthwash containing a dye that sticks to the plaque. The top and bottom images show the appearance of the teeth before and after the person has used the disclosing agent. (Ajeverett [CC BY-SA 4.0 (https://creativecommons.org/licenses/by-sa/4.0)])

Plaque is a soft, white substance that covers the exposed surface of the tooth in a thin coating – tongue movements, chewing and talking constantly remove the outer layers of the plaque, and this prevents it from becoming too thick. However, in regions that are protected from these forces, such as between the teeth, between the tooth and the gum and in the crevices (fissures) on the surfaces of molars and premolars, plaque can accumulate as thicker layers (Fig. 20.5).

How Do Bacteria Cause Tooth Decay?

The microbes that live in plaque form stable communities and obtain nutrients from saliva, from the food we consume and from one another. However, as outlined earlier, the regular intake of sucrose and other sugars in our diet can unbalance these communities and encourages the growth of those species that can use these sugars as nutrients. These particular microbes produce acids from the dietary sugars, and this creates an environment within the biofilm that many other species can't tolerate. These sugar-utilising microbes can withstand the acid conditions they create and therefore eventually come to dominate the biofilm. The acid produced by the microbes dissolves the outer

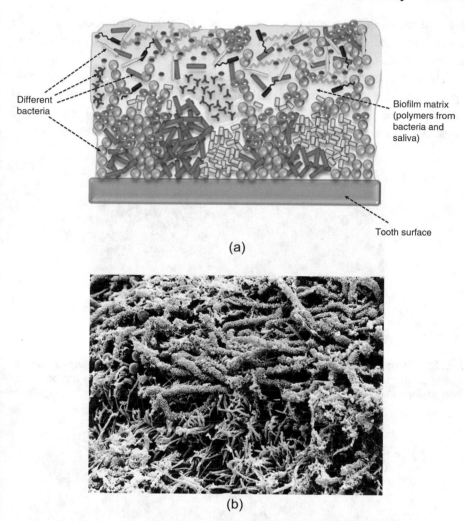

Different bacteria

Biofilm matrix (polymers from bacteria and saliva)

Tooth surface

(a)

(b)

Fig. 20.4 Dental plaque. (**a**) Diagram of the biofilm (known as plaque) on the surface of the tooth. The microbial community is enclosed within a gel-like matrix which is attached to the tooth surface. (Advancement of the 10-species subgingival Zurich Biofilm model by examining different nutritional conditions and defining the structure of the in vitro biofilms. Ammann et al. *BMC Microbiology* 2012, 12:227. (**b**) Dental plaque as seen through an electron microscope. The various types of bacteria (different sized cocci and bacilli) are packed tightly together. (Credit: David Gregory & Debbie Marshall. CC BY 4.0. Wellcome Image Library)

layer (enamel) of the tooth (a process known as "demineralisation"), and this eventually results in the formation of a cavity that can penetrate deep into the tooth (Fig. 20.6).

The microbes responsible for this process are referred to as being cariogenic species, and the most important of these are *Streptococcus mutans, Streptococcus sobrinus*, various *Lactobacillus* species and various *Actinomyces* species. All of these bacteria are members of our indigenous microbiota and are members of

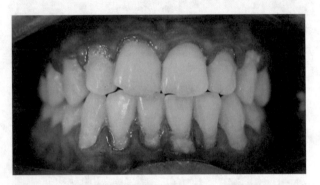

Fig. 20.5 Note the accumulation of plaque where the teeth meet the gums as well as between the teeth. (Shaimaa Abdellatif [CC BY-SA 4.0 (https://creativecommons.org/licenses/by-sa/4.0)])

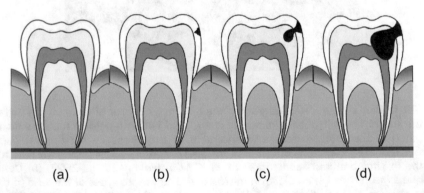

(a)   (b)   (c)   (d)

Fig. 20.6 Stages in the development of caries (black regions). The various regions of the tooth are described in Fig. 20.1. (**a**) A healthy tooth. (**b**) Early caries affecting only the enamel. (**c**) Later stages of caries showing involvement of the dentine. (**d**) The caries lesion has spread into the pulp region. (Images courtesy of Free Art Licence via Wikimedia Commons. This work of art is free; you can redistribute it and/or modify it according to terms of the Free Art License

the normal microbial communities that inhabit the mouth and, in some cases, other body sites.

Strep. mutans (Fig. 20.7) was one of the first species to be implicated in caries and has been the subject of intense investigation in this respect since 1924.

Strep. mutans is a bacterium that's present in the mouth, pharynx and colon of many healthy people including those who are free of caries. It can grow in the presence or absence of oxygen, and its most significant characteristics from the point of view of caries are that it can rapidly convert sucrose into acids and can survive in a very acidic environment. It can also use sucrose to construct large polysaccharide molecules (glucan and fructan) which help the bacterium to stick to the tooth surface and also form part of the plaque matrix. It's involved mainly in cavity formation in enamel. *Strep. sobrinus* is very closely related to *Strep. mutans*.

Box 20.3 What's in a Name?

The derivation of the genus name, *Streptococcus*, has already been described in Chap. 13. The species name, *mutans*, was coined by the British microbiologist J. Kilian Clarke who was the first to isolate the bacterium (in 1924) and considered it to be a mutant form of streptococcus. Since then another six species of streptococci have been found to be capable of causing caries and these are collectively known as "mutans streptococci".

The genus name *Lactobacillus* comes from the ability of these rod-shaped bacteria (i.e. bacilli) to convert sugars to lactic acid.

The name *Actinomyces* is derived from the Greek "actino" (meaning ray) and "mykes" (meaning a fungus). They are so-called because their colonies on an agar plate have a ray-like appearance and some species appear as filaments under the microscope and so were originally thought to be fungi rather than bacteria.

Lactobacillus species are Gram-positive bacilli (Fig. 20.8) and were first reported to be involved in caries in 1915. Since then, a number of different species have been implicated in caries, and these include *L. casei*, *L. paracasei*, *L. fermentum*, *L. rhamnosus*, *L. gasseri*, *L. salivarius*, *L. plantarum*, *L. oris* and *L. vaginalis*. They all prefer an oxygen-free environment although they can tolerate oxygen. Like the cariogenic streptococci, these *Lactobacillus* species can all metabolise sugars and produce lactic acid as well as survive in an acidic environment. However, unlike the streptococci, they don't produce polysaccharides and so don't contribute molecules to the plaque matrix.

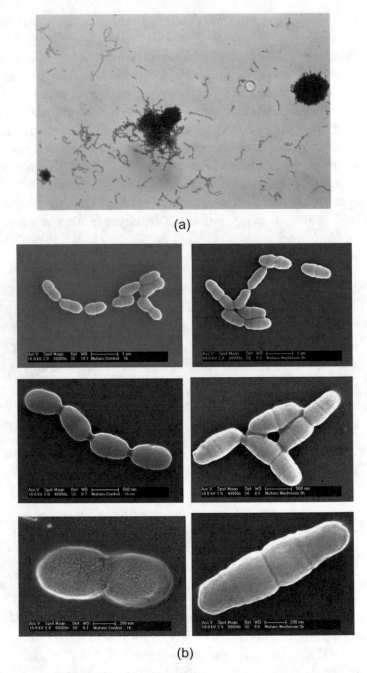

(a)

(b)

Fig. 20.7 *Strep. mutans*, an important cariogenic organism. (a) Gram stain of *Strep. mutans* showing characteristic Gram-positive cocci in chains. (Image courtesy of Dr. Richard Facklam, Centers for Disease Control and Prevention, USA). (b) Scanning electron micrographs of *Strep. mutans* at different magnifications. (*BMC Complement Altern Med.* 2013; 13: 117. Effects of mushroom and chicory extracts on the shape,

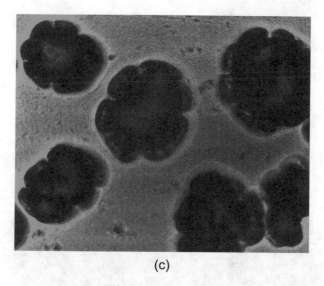

(c)

Fig. 20.7 (continued) physiology and proteome of the cariogenic bacterium *Streptococcus mutans*. Caterina Signoretto et al. (c) Colonies of *Strep. mutans* growing on an agar plate (×400). (Ronit611 [CC BY-SA 3.0 (https://creativecommons.org/licenses/by-sa/3.0)])

Actinomyces species are Gram-positive bacilli (Fig. 20.9) that can live in the presence or absence of oxygen. When using sugars as nutrients, they produce acids and can tolerate the resulting acidic conditions. The main species involved in caries are *A. naeslundii* and *A. viscosus*.

Actinomyces species are mainly involved in a particular type of caries that affects the roots of the teeth – this is known as "root caries" (RC). RC affects mainly elderly people because the gums have often receded leaving the tooth roots exposed within the mouth. Biofilms can then grow on the exposed surfaces, and, if sugar intake is high, then this can result in caries. Cavity formation tends to occur at a faster rate in root caries than in enamel caries because of the lower mineral content of the root (which contains a higher proportion of collagen) compared to the crown of the tooth. Recent studies have shown that, worldwide, 42% of adults suffer from RC. The prevalence increases with age, and a study in the UK found that the proportions of individuals with

(a)

(b)

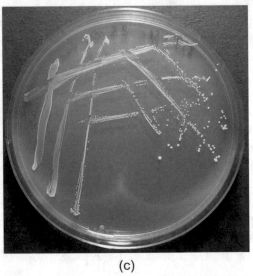

(c)

Fig. 20.8 Images of lactobacilli. (**a**) Gram stain showing Gram-positive lactobacilli as well as an epithelial cell. (Image courtesy of Dr. Mike Miller, Centers for Disease Control and Prevention, USA). (**b**) *L. paracasei* as seen through a scanning electron microscope. (Image reproduced from *AMB Express*. 2015; 5: 78. Complete nucleotide sequence of the 16S rRNA from *Lactobacillus paracasei* HS-05 isolated from women's hands). Woon Yong Choi and Hyeon Yong Lee. (**c**) Colonies of a *Lactobacillus* species growing on an agar plate. (A doubt [CC BY-SA 3.0 (https://creativecommons.org/licenses/by-sa/3.0)])

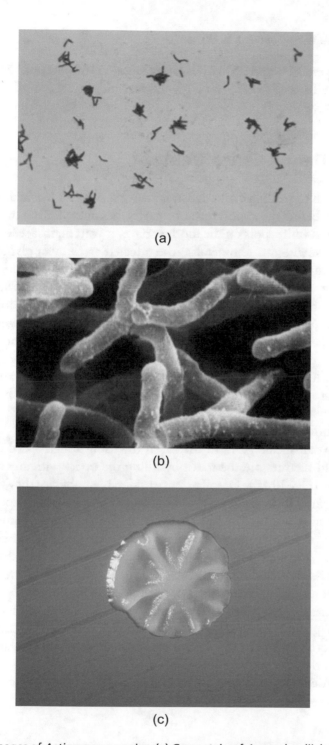

Fig. 20.9 Images of *Actinomyces* species. (**a**) Gram stain of *A. naeslundii* showing characteristic Gram-positive bacilli (magnification ×900). (Image courtesy of Dr. Lucille K. Georg, Centers for Disease Control and Prevention, USA). (**b**) Scanning electron micrograph of *Actinomyces israelii*. (GrahamColm at English Wikipedia [CC BY 3.0 (https://creativecommons.org/licenses/by/3.0)]). (**c**) Colony of an *Actinomyces* species growing on an agar plate. (Dr. Lucille K. Georg, Centers for Disease Control and Prevention, USA)

active RC were 11% and 20% in the age groups 55–64 years and 75–84 years, respectively. In the USA 15% of 35–49-year-olds and 22% of 50–64-year-olds had active RC.

How Is Dental Caries Treated?

The approach to treating caries depends on what stage it's reached. In the very early stages, before the appearance of a cavity, demineralisation of the enamel can be seen as white spots on the tooth surface. If your dentist sees these white spots, then she/he can remineralise the affected areas by applying formulations containing calcium phosphate. You can then prevent further demineralisation by applying fluoride to the surfaces of your teeth. Fluoride ions become incorporated into enamel to form fluorapatite which is more resistant to bacterial acids than hydroxyapatite. Furthermore, it interferes with the metabolism of plaque bacteria and reduces acid production by them. You can get fluoride onto your teeth by using a fluoride-containing mouthwash or toothpaste. Also, your dentist can apply a fluoride-containing varnish to the tooth surface. You should, of course, also practice good oral hygiene and reduce your sugar consumption.

If you suffer from the later stages of caries, then your dentist will have to remove the damaged regions of the tooth (by using a high-speed dental drill or laser) and then restore the tooth by filling the cavity with any of a variety of materials (Fig. 20.10). Despite the considerable publicity they've received, there's currently little evidence that lasers are better than high-speed drilling as a means of removing diseased tooth tissue. Suitable tooth-filling materials (known as "restorative materials") include mercury amalgam, composites, glass ionomers and resin ionomers. The choice of material used depends on a number of issues including the location of the tooth being restored, aesthetic considerations, cost and the durability of the restoration.

Sometimes the decay is so extensive that what's left of the tooth wouldn't be strong enough after being filled with a restorative material. In such cases, the restored tooth needs to be capped with a crown made of gold, ceramic, porcelain or porcelain/metal (Fig. 20.11).

Root caries is treated mainly by the application of fluoride to the diseased area.

How Can I Avoid Getting Caries?

Preventing caries is very straightforward – give up sugar and clean your teeth properly at least twice a day. What could be simpler? Given the prevalence of the disease in children and adults, the above preventive measures appear to be extremely difficult to carry out. Mechanically removing plaque from the surfaces of the teeth by proper toothbrushing and flossing removes the plaque biofilm and so helps to keep the numbers of cariogenic bacteria at low levels. Reducing the amount of sucrose in the diet and, more importantly, reducing

---

Box 20.4 George Washington's False Teeth

George Washington suffered from poor dental health from his early twenties and consequently spent a lot of his life in pain. He had his first tooth extracted in 1756 when he was 24, and by 1781 he was wearing partial dentures. By the time he was inaugurated President, in 1789, he had only one tooth left. Dr. John Greenwood, a New York dentist, made him a set of dentures out of hippopotamus ivory as well as human teeth (Figure a).

Figure (a) Photograph of George Washington's dentures. (Rights Info: No known restrictions on publication. Repository: Library of Congress, Prints and Photographs Division, Washington, D.C. 20540 USA, hdl.loc.gov/loc.pnp/pp.print)

(continued)

Box 20.4 (continued)

The first false teeth were made by the Etruscans in 700 BCE (Figure b). Teeth from another person or an animal were fixed into a band of gold and fitted on to the remaining teeth.

Figure (b) A copy of an original denture with two teeth which was found in a tomb in Etruria, Italy. (Credit: Science Museum, London. CC BY 4.0)

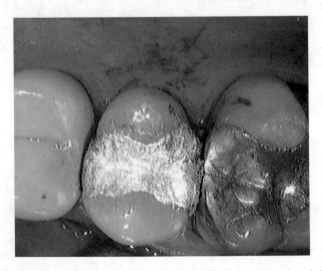

Fig. 20.10   Teeth restored using mercury amalgam. (Michael Ottenbruch [CC BY-SA 3.0 (http://creativecommons.org/licenses/by-sa/3.0/)])

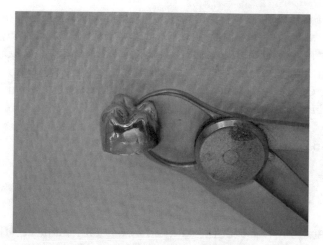

Fig. 20.11 Crown made of gold prior to fitting in the patient's mouth. (Sterilgutassistentin [GPL (http://www.gnu.org/licenses/gpl.html)])

the frequency of its consumption restrict the availability of this important nutrient to any cariogenic bacteria that are present in the mouth. Consequently, less acid will be produced, and this will ensure that enamel demineralisation doesn't occur or is greatly reduced. It's also important to be aware of the presence of "hidden" sugar in processed foods that we don't usually think of as being sweet – these include tomato ketchup, sauces, yoghurts and salad dressing. The World Health Organization recommends that sugars should comprise less than 10% of the total energy intake in our diet so as to reduce the risk of both caries and obesity. However, our craving for sweetness is difficult to resist. One approach that's been shown to address this need while helping to reduce caries is to replace sucrose with artificial sweeteners such as xylitol. Xylitol has a sweet taste but can't be converted to acids by cariogenic bacteria and is available in a wide range of products including chewing gum, candies, sweets, fruit drinks, peanut butter, desserts, etc.

Another important preventive measure is the use of fluoride. As mentioned previously this reduces the risk of caries by strengthening tooth enamel and by reducing acid production by cariogenic bacteria. Many water suppliers add fluoride to domestic water as a preventive measure. In addition, most toothpaste manufacturers offer fluoride-containing toothpastes, and fluoride-supplemented milk and salt are widely available. Fluoridation of the water supply is a controversial issue, and doubts have been expressed as to its safety and efficacy. However, the American Association for Dentistry currently carries this statement on its website: "More than 70 years of scientific research has consistently shown that an optimal level of fluoride in community water is safe and effective in preventing tooth decay by at least 25% in both children and adults. Simply by drinking water, Americans can benefit from fluoride's cavity protection whether they are at home, work or school. The Centers for Disease Control and Prevention named community water fluoridation one of 10 great public health achievements of the twentieth century" (https://www.ada.org/en/public-programs/advocating-for-the-public/fluoride-and-fluoridation).

Want to Know More?

American Academy of Pediatrics https://www.healthychildren.org/English/ages-stages/baby/teething-tooth-care/Pages/How-to-Prevent-Tooth-Decay-in-Your-Baby.aspx

American Dental Association https://www.mouthhealthy.org/en/az-topics/d/decay

British Dental Association https://bda.org/sugar

Dental Health Foundation, Ireland https://www.dentalhealth.ie/dentalhealth/causes/dentalcaries.html

Mayo Clinic, USA https://www.mayoclinic.org/diseases-conditions/cavities/symptoms-causes/syc-20352892

National Health Service, UK https://www.nhs.uk/conditions/tooth-decay/

National Institute of Dental and Craniofacial Research, USA https://www.nidcr.nih.gov/health-info/tooth-decay/more-info

World Health Organization https://apps.who.int/iris/bitstream/handle/10665/259413/WHO-NMH-NHD-17.12-eng.pdf;jsessionid=161361AF759ACED1E18B2D4C83C06505?sequence=1

Abranches J, Zeng L, Kajfasz JK, Palmer SR, Chakraborty B, Wen ZT, Richards VP, Brady LJ, Lemos JA. Biology of oral streptococci. *Microbiology Spectrum* 2018 Oct;6(5). doi: https://doi.org/10.1128/microbiolspec.GPP3-0042-2018

Aoun A, Darwiche F, Al Hayek S, Doumit J. The fluoride debate: the pros and cons of fluoridation. *Preventive Nutrition and Food Science*. 2018 Sep;23(3):171-180. https://doi.org/10.3746/pnf.2018.23.3.171. Epub 2018 Sep 30.

Arifa MK, Ephraim R, Rajamani T. Recent advances in dental hard tissue remineralization: a review of literature.. *International Journal of Clinical Pediatric Dentistry* 2019 Mar-Apr;12(2):139-144. https://doi.org/10.5005/jp-journals-10005-1603.

Burgess J. Diet and Oral Health. Medscape from WebMD, 2015 https://emedicine.medscape.com/article/2066208-overview

Moynihan P. Sugars and dental caries: evidence for setting a recommended threshold for intake. *Advances in Nutrition*. 2016 Jan 15;7(1):149-56. https://doi.org/10.3945/an.115.009365. Print 2016 Jan.

Rathee M, Sapra A. Dental Caries. Treasure Island (FL): StatPearls Publishing LLC; 2019 https://www.ncbi.nlm.nih.gov/books/NBK551699/

Sharma G, Puranik MP, KRS. Approaches to arresting dental caries: an update. *Journal of Clinical and Diagnostic Research*. 2015 May;9(5):ZE08-11. https://doi.org/10.7860/JCDR/2015/12774.5943. Epub 2015 May 1.

Sicca C, Bobbio E, Quartuccio N, Nicolò G, Cistaro A. Prevention of dental caries: A review of effective treatments.. *Journal of Clinical and Experimental Dentistry* 2016 Dec 1;8(5):e604-e610. eCollection 2016 Dec.

21

Gum Diseases

Abstract The main diseases of the gums are gingivitis and chronic periodontitis (CP), and both are due to the accumulation of bacterial biofilms (i.e. dental plaque) at the gingival margin where the teeth meet the gingivae (gums). Gingivitis results in redness and swelling of the gums which bleed when the teeth are brushed. It's very common and affects up to 90% of adults. In CP, dental plaque accumulates on the root surface of the tooth below the gingival margin (known as subgingival plaque), and the teeth become loose and may be lost because of the destruction of the bone that holds them in place. Up to 45% of adults in developed countries suffer from CP, and it's the main cause of tooth loss in adults. In both diseases the dental plaques responsible have very complex microbial communities that are dominated by Gram-negative anaerobic bacteria. CP is associated with some other long-term illnesses (cardiovascular disease and diabetes) and with adverse pregnancy outcomes.

Both conditions can be prevented by good oral hygiene involving the meticulous removal of dental plaque, and this may be supplemented with antimicrobial and/or anti-plaque mouthwashes. Gingivitis is treated by dental professionals removing plaque from the gum margins. CP is treated similarly, but this also requires the removal of subgingival plaque and sometimes involves the placement of antibiotics or antiseptics into the gap between the gums and teeth.

You brush your teeth and spit into the sink. What's that red colour? Surely it can't be blood? You look in the mirror. Yes, there's blood around your gums. You think about it for a while but it doesn't hurt so it can't be anything to worry about, can it? After a few weeks, the bleeding gets more pronounced

© Springer Nature Switzerland AG 2021
M. Wilson, P. J. K. Wilson, *Close Encounters of the Microbial Kind*,
https://doi.org/10.1007/978-3-030-56978-5_21

but it still doesn't hurt so why worry? Then your partner tells you that your breath smells. You're embarrassed so you start using a breath freshener but the smell and the bleeding persist. Eventually you go to the dentist and are told that you have gingivitis. Sounds bad, but so what? It's only my gums – who cares about gums? But then your dentist explains that if you don't start to take care of your gums, the disease will get worse and could lead to your teeth dropping out. My teeth dropping out! So now you're really worried. This can't be happening to me – surely only old folk lose their teeth.

What Is Gingivitis?

The gums (gingivae) are the soft tissues that surround the teeth (Figs. 21.1 and 21.2).

In healthy individuals the gums have a coral-pink colour (Fig. 21.2).

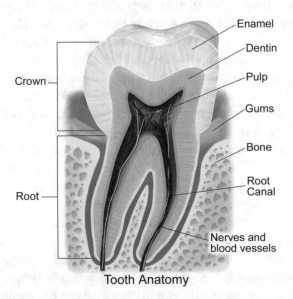

Tooth Anatomy

Fig. 21.1 Diagram of the tooth and its supporting structures. The crown is that portion of the tooth that protrudes from the gums, while the root is the region that is anchored into the jaw bone. The enamel is the outer mineralised surface of the tooth, while dentine is the inner, mineralised region of the tooth. The pulp chamber contains blood vessels and nerves. The bone (known as alveolar bone) is a ridge of bone that contains the tooth sockets. (BruceBlaus. When using this image in external sources, it can be cited as Blausen.com staff (2014). Medical gallery of Blausen Medical 2014. WikiJournal of Medicine 1 (2). DOI:https://doi.org/10.15347/wjm/2014.010. ISSN 2002-4436./CC BY (https://creativecommons.org/licenses/by/3.0))

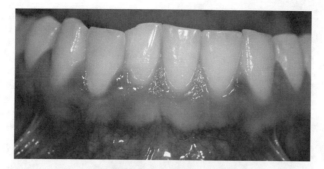

Fig. 21.2 The appearance of healthy gingivae. (Mohamed Hamze/Public domain. I, the copyright holder of this work, release this work into the public domain. This applies worldwide. In some countries this may not be legally possible; if so, I grant anyone the right to use this work for any purpose, without any conditions, unless such conditions are required by law. Via Wikimedia Commons)

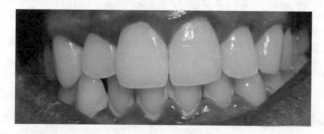

Fig. 21.3 Gingivae of a patient with gingivitis. Note the redness of the gums as well as the presence of plaque (pale yellow-coloured) at the junction between the teeth and gums (i.e. the gingival margin). (Onetimeuseaccount. This file is made available under the Creative Commons CC0 1.0 Universal Public Domain Dedication. Via Wikimedia Commons)

Inflammation of the gingivae is known as gingivitis and results in redness, swelling and bleeding when the teeth are brushed (Fig. 21.3). Other possible symptoms include bad breath and tenderness of the gums.

Gingivitis is a very common condition, the prevalence of which varies considerably from country to country – between 50% and 90% of adults are affected worldwide. It's due to plaque (Fig. 21.3) accumulating where the tooth is in contact with the gum – a region known as the "gingival margin", "gingival sulcus" or "gingival crevice" (Fig. 21.4).

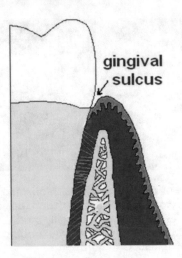

Fig. 21.4 The gingival sulcus (or crevice) is the shallow depression that is present at the junction of the tooth and gum. (Mikael Haggstrom. I, the copyright holder of this work, release this work into the public domain. This applies worldwide. In some countries this may not be legally possible; if so, I grant anyone the right to use this work for any purpose, without any conditions, unless such conditions are required by law. Via Wikimedia Commons)

What Causes Gingivitis?

If you don't regularly remove plaque from the gingival crevice (by toothbrushing and flossing), it will accumulate there, and you can end up with 10–20 times more than that found in a healthy crevice. Because it's been left undisturbed, changes in the microbial community within the plaque occur, and Gram-negative anaerobic bacteria take over from Gram-positive species as the dominant group. This causes inflammation and an increased flow of a liquid known as gingival crevicular fluid (GCF) into the gingival crevice. GCF has a composition similar to that of serum and so is very rich in nutrients and stimulates the growth of the Gram-negative anaerobes. The fluid also contains huge numbers of white blood cells including phagocytes. Unfortunately, the phagocytes can't dispose of the bacteria in the plaque because these are protected by the polymers present in the matrix of the biofilm. The phagocytes then discharge their antimicrobial compounds into the crevice (a process known as degranulation), and this exacerbates the inflammation. Gingival inflammation may also be a consequence of pregnancy. The high levels of progesterone that occur during pregnancy increase the response of the gingivae to plaque bacteria, and this results in them becoming more inflamed – this is known as "pregnancy gingivitis". More than half of pregnant women are affected by pregnancy gingivitis.

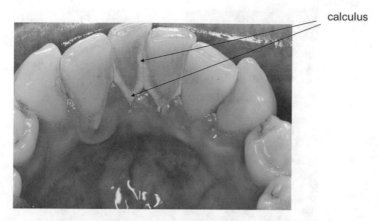

calculus

Fig. 21.5   Calculus formed in the gingival sulcus and between the teeth. (Bugmore. I, the copyright holder of this work, release this work into the public domain. This applies worldwide. In some countries this may not be legally possible; if so, I grant anyone the right to use this work for any purpose, without any conditions, unless such conditions are required by law. Via Wikimedia Commons)

The microbial communities associated with gingivitis are very complex and can consist of between 600 and 800 different bacterial species. Many of these belong to the following genera – *Campylobacter, Fusobacterium, Lautropia, Leptotrichia, Porphyromonas, Prevotella, Selenomonas, Actinomyces* and *Tannerella*. All of these genera are Gram-negative except for *Actinomyces* which is Gram-positive. Although the complexity of these communities makes them extremely interesting, it's beyond the scope of this book to describe the many different bacterial species involved in the disease.

Minerals in saliva and GCF may be deposited in the plaque that has accumulated in the gingival sulcus (as well as at other sites), and this results in the formation of hard deposits known as "calculus" or "tartar" (Fig. 21.5). 14–47% of adults in developed countries have calculus deposits compared with 36–63% of adults in developing nations.

The main risk factors associated with gingivitis are poor oral hygiene, pregnancy, diabetes, smoking and malnutrition.

How Is Gingivitis Treated?

Treatment of gingivitis involves removing the plaque and calculus that has accumulated in the gingival sulcus. This usually has to be carried out by a dentist or dental hygienist, especially if calculus is present, because this can't be removed by toothbrushing. This cleansing process is carried out using

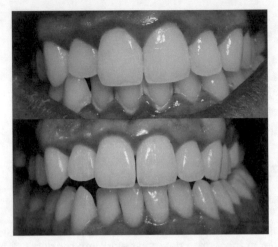

Fig. 21.6 Professional tooth-cleaning (debridement) showing the appearance of the teeth before (above) and after (below) the procedure. The lower photograph shows the absence of plaque and calculus. (Onetimeuseaccount. This file is made available under the Creative Commons CC0 1.0 Universal Public Domain Dedication. Via Wikimedia Commons)

various hand- and power-driven tools (such as ultrasonic devices) and is known as "debridement" or "scaling" (Fig. 21.6).

How Can I Avoid Getting Gingivitis?

You can stop getting gingivitis very easily – all you have to do is remove dental plaque from your teeth on a regular basis. For easily accessible regions of the teeth, you can do this using a toothbrush, but you'll have to use floss and inter-dental brushes for less-accessible areas such as between the teeth. You can supplement these mechanical methods by using mouth rinses containing antimicrobial and/or anti-plaque agents. The antibacterial agents kill, or prevent the growth of, many of the organisms present in plaque, while the anti-plaque agents disrupt, or prevent the formation of, biofilms without necessarily killing the bacteria present in them. A wide range of antimicrobial or anti-plaque agents have been incorporated into mouthwashes, and the most popular of these include chlorhexidine, triclosan, phenols, cetylpyridinium chloride, amine fluorides, stannous fluoride, zinc citrate, sanguinarine and delmopinol. Some of these agents, such as chlorhexidine, have both antibacterial and anti-plaque properties. Some have also been incorporated into toothpastes. Smoking is also a very significant risk factor, so this is best avoided.

What Is Chronic Periodontitis?

If left untreated, gingivitis can, in some individuals, progress to a disease known as chronic periodontitis (CP). Periodontitis means "inflammation of the periodontium" – the periodontium being the collection of tissues that surround and support the teeth, i.e. the gums, periodontal ligament, cementum and alveolar bone (Fig. 21.7). The word periodontium is derived from the Greek "peri" meaning "around" and "odont" meaning "tooth", i.e. "around the tooth".

The characteristic features of CP are detachment of the gingiva from the tooth (resulting in the formation of what is known as a "periodontal pocket"), loss of the periodontal ligament that attaches the gingiva to the root of the tooth and destruction of the alveolar bone that supports the tooth. Eventually, the bone is eroded to such an extent that the tooth falls out. This process occurs gradually and passes through several stages:

(a) Gingivitis. The gums are inflamed and swollen, but the depth of the gingival sulcus is only slightly increased compared to the disease-free state which is approximately 3 mm.

(b) Mild periodontitis. The gums are inflamed, and some periodontal ligament and alveolar bone have been destroyed. The gums have moved away

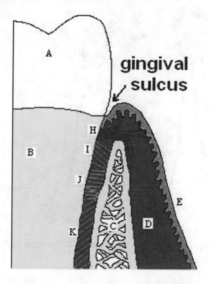

Fig. 21.7 The various components of the periodontium. **(A)** Crown of the tooth. **(B)** Root of the tooth, covered by cementum. **(C)** Alveolar bone. **(D)** Subepithelial connective tissue. **(E)** Oral epithelium. **(H–K)** The bundles of fibres that comprise the periodontal ligament. (Mikael Häggström/Public domain via Wikimedia Commons)

from the tooth surface to form a periodontal pocket which has a depth much greater than in (a).

(c) Moderate periodontitis. More alveolar bone and periodontal ligament have been lost, and the depth of the periodontal pocket has increased.

(d) Severe periodontitis. More alveolar bone and periodontal ligament have been lost, and the depth of the periodontal pocket has increased further (Fig. 21.8).

The main symptoms of CP are reddening of the gums, bleeding from the gingival crevice, gum tenderness, pain on chewing, detachment of the gingiva from the tooth (i.e. periodontal pocket formation), bad breath, recession of the gingivae and destruction of the bone and periodontal ligament which support the tooth (resulting in loosening of the tooth). Many of these symptoms can be seen in Fig. 21.9.

Measurement of the depth of the periodontal pocket with a probe (Fig. 21.10) is a useful means of determining the severity of the disease and whether or not it's progressing.

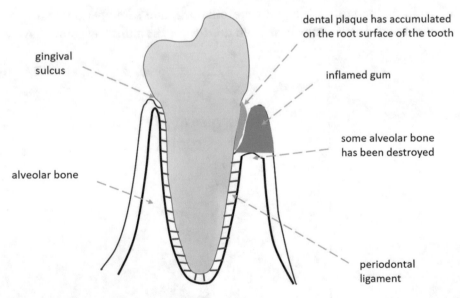

Fig. 21.8 Diagram showing the loss of alveolar bone and periodontal ligament that are associated with periodontitis

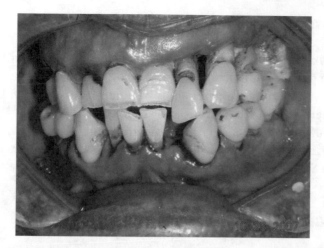

Fig. 21.9 A severe case of CP in a 60-year-old man. (AGUSTIN ZERON/Public domain. I, the copyright holder of this work, release this work into the public domain. This applies worldwide. In some countries this may not be legally possible; if so, I grant anyone the right to use this work for any purpose, without any conditions, unless such conditions are required by law. Via Wikimedia Commons)

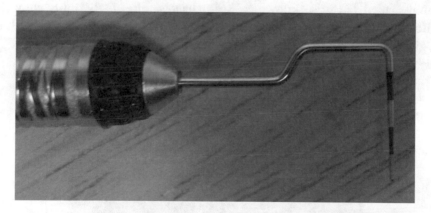

Fig. 21.10 A periodontal probe used to determine the depth of a periodontal pocket – note the depth markings on that part of the probe that is placed inside the periodontal pocket of the patient. (I, the copyright holder of this work, hereby release it into the public domain. This applies worldwide. If this is not legally possible, I grant any entity the right to use this work for any purpose, without any conditions, unless such conditions are required by law. Via Wikipedia)

How Common Is Periodontitis?

CP is the sixth most prevalent disease worldwide and affects approximately 743 million people – its global prevalence being 11.2%. The disease affects all age groups, but is most prevalent among adults and the elderly. In the USA approximately 35% of adults (30–90 years) are affected, while 45% of adults in the UK have the disease. The economic impact of CP is substantial, and it's been estimated that the loss of productivity associated with the disease amounts to $54 billion per year.

What Causes Chronic Periodontitis?

As with gingivitis, the cause of CP is dental plaque (Fig. 21.11).

Nourished by the GCF that floods into the gingival sulcus, the bacteria in plaque multiply, and so the plaque grows down along the tooth surface causing more inflammation and tissue destruction. The microbial communities in the plaque are again very complex and are dominated by Gram-negative anaerobic bacteria which can comprise as much as 75% of the microbiota. Spiral-shaped bacteria (known as spirochaetes, Fig. 21.12) also comprise much greater proportions of the microbiota than they do either in health or in gingivitis.

Fig. 21.11 Scanning electron micrograph of plaque showing a variety of different sizes and shapes of bacteria. (Credit: David Gregory & Debbie Marshall. CC BY 4.0. Wellcome Image Library)

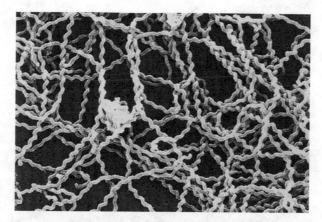

Fig. 21.12 Electron micrograph of a *Leptospira* species, a spiral-shaped organism that may be present in the periodontal pockets of those suffering from chronic periodontitis. (Janice Haney Carr, Centers for Disease Control and Prevention, USA)

Although the microbial communities associated with CP are very complex, a number of species are regularly detected in the periodontal pocket and are regarded as having an important role. Such species are known as "periodontopathogens" and include *Porphyromonas gingivalis, Tannerella forsythia, Treponema denticola, Filifactor alocis, Anaeroglobus geminatus, Eubacterium saphenum, Porphyromonas endodontalis* and *Prevotella denticola*. As with gingivitis, a detailed description of all these species is beyond the scope of this book. They can all survive and grow in the periodontal pocket environment which has a low oxygen content and a high content of proteins as the main nutrients. Furthermore, collectively they have a number of virulence factors that protect them from host defences in the pocket and also contribute to the destruction of the periodontal tissues. These include (i) molecules that enable them to stick to periodontal tissues; (ii) capsules that protect them from phagocytic cells; (iii) enzymes that can break down tissues, including alveolar bone; (iv) toxins that kill host defence cells; (v) enzymes that destroy the antibodies that have been produced against them; and (vi) components that induce an inflammatory response.

As well as being the major cause of tooth loss in adults, studies have suggested a possible link between CP and a number of other conditions including atherosclerotic cardiovascular disease, Alzheimer's disease, diabetes, adverse pregnancy outcome, respiratory disease, chronic kidney disease, rheumatoid arthritis, cognitive impairment, obesity, metabolic syndrome and cancer.

Box 21.1 Periodontitis and Alzheimer's Disease: Is There a Link?

There's some evidence that one of the bacteria responsible for periodontitis, *Porphyromonas gingivalis*, may also be involved in Alzheimer's disease (AD). An enzyme produced by this bacterium, known as gingipain, was found in the brain of more than 90% of patients with AD but in only 52% of patients without AD. Gingipain is an important virulence factor of the bacterium – it can break down a variety of human proteins and contributes to the tissue destruction that accompanies periodontitis. One of the proteins it can break down is a brain protein called Tau. A characteristic feature of the brain of an AD patient is the accumulation of tangles of Tau protein fragments. It's been suggested that damage to the gingival tissues during periodontitis allows *Por. gingivalis*, or its products (such as gingipain), to enter the bloodstream and eventually reach the brain. Once in the brain it could break down Tau protein and so initiate AD. This is very preliminary research, and the involvement of *Por. gingivalis* in AD hasn't been conclusively established.

Spiral-shaped oral bacteria (*Treponema* species) have also been detected more frequently in the brains of patients with AD than in the brains of those without the disease. Again, such bacteria may be able to gain access to the brain via damaged gingival tissues.

Although there's a long way to go before we can say that periodontitis can lead to AD, there's no harm in taking good care of your gums.

There are two possible ways in which CP may contribute to these other conditions: periodontopathogenic bacteria may be able to enter the bloodstream through the damaged oral tissues and then cause disease at other body sites, or the inflammatory response induced by periodontopathogenic bacteria may cause damage at sites beyond the oral cavity. However, it's important to appreciate that although there may be an association between CP and another disease, this doesn't necessarily mean that CP is the cause of that disease. This confusion arises because many of the risk factors for CP (age, smoking, obesity, socio-economic status, etc.) are the same as those for other conditions – this is known as "confounding". Just because two diseases show an association, this doesn't necessarily mean that one is the cause of the other. Most of the evidence in support of an association between CP and another disease have come from epidemiological studies. However, such studies often show that the association is a two-way relationship. As an example, epidemiological studies have shown that diabetes is a major risk factor for CP but have also shown that CP is a risk factor for diabetes.

Although CP may not be the cause of all of the conditions listed above, it's important to appreciate that patients with CP do have a greater risk of certain other diseases, in particular atherosclerotic cardiovascular disease, diabetes and adverse pregnancy outcome.

How Is Chronic Periodontitis Treated?

The main goals of treatment of CP are to control inflammation, stop disease progression and create conditions that will decrease the likelihood of a recurrence. Initial non-surgical therapy involves the professional removal of supragingival (i.e. above the gums) plaque and subgingival (i.e. below the gums) plaque and calculus by scaling. The surfaces of the tooth root are then smoothed (known as "root planing"). Healing of the tissues takes between 4 and 6 weeks following this treatment. Additional procedures may be undertaken if inflammation is still evident after re-evaluation. These include placing antibiotics (usually minocycline or doxycycline) or antiseptics (usually chlorhexidine) directly into the periodontal pocket. Another option is to use oral antibiotics – a mixture of amoxicillin and metronidazole has been shown to be very effective. However, the benefits of antibiotic use must be weighed against the potential risks such as resistance development, possible adverse reactions and side effects.

Box 21.2 Antimicrobial Photodynamic Therapy for Periodontitis

A number of studies have shown that antimicrobial photodynamic therapy (aPDT) is an effective alternative to antibiotics and antiseptics in the treatment of periodontitis. This involves the application of a light-activated antimicrobial agent into the periodontal pocket followed by irradiation of the pocket with red light from a low-power laser (Figure).

(continued)

Box 21.2 (continued)

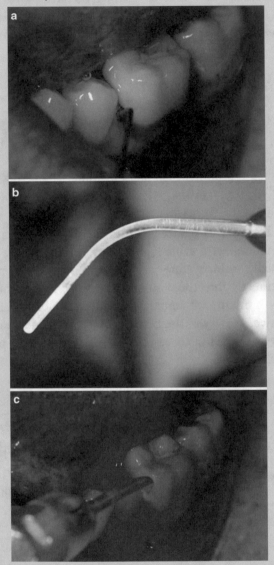

Figure The light-activated antimicrobial agent is placed in the periodontal pocket (a) and then irradiated for 2 minutes with red light (c) from a low-power laser (b)

This approach is being used in several countries throughout the world and has a number of important advantages over antibiotics and antiseptics: (i) it requires only a single application and is very rapid (a few minutes) in contrast to antibiotics or antiseptics which must be taken daily for several days or longer, (ii) microbes can't develop resistance to the treatment, (iii) it's effective against microbial biofilms and (iv) as well as being an antimicrobial, it also affects the immune response and can reduce inflammation.

Courtesy of Ondine Biomedical Inc., Vancouver, Canada.

Scaling and root planing with or without the use of antimicrobial agents is usually effective for treating most cases of CP. However, if you have deeper periodontal pockets (i.e. >4 mm), then some form of surgery may be necessary. One common surgical approach involves separating a section of the gum from the tooth to allow better access for scaling and root planing – this is known as "open flap debridement". The gum section is then replaced and stitched to the neighbouring gums, and the stitches are removed several days later. Other possible surgical procedures include bone grafting to replace lost alveolar bone and/or the use of soft tissue from the roof of the mouth to replace damaged gums.

How Can I Avoid Getting Periodontitis?

Preventive measures for CP are the same as for gingivitis and have been described above. The main risk factors for CP are increasing age, smoking, diabetes mellitus, stress, obesity, low socio-economic status, low educational level, being male and malnutrition.

Want to Know More?

American Academy of Periodontology. https://www.perio.org/consumer/types-gum-disease.html

American Dental Association. https://www.mouthhealthy.org/en/az-topics/g/gum-disease

British Society for Periodontology. https://www.bsperio.org.uk/periodontal-disease/sound-of-periodontitis.html

Centers for Disease Control and Prevention, USA. https://www.cdc.gov/oralhealth/conditions/periodontal-disease.html

Dental Health Foundation, Ireland. https://www.dentalhealth.ie/dentalhealth/causes/periodontaldisease.html

DermNet, New Zealand. https://www.dermnetnz.org/topics/gingivitis-and-periodontitis/

European Federation of Periodontology. https://www.efp.org/patients/what-is-periodontitis.html

Mayo Clinic, USA. https://www.mayoclinic.org/diseases-conditions/gingivitis/symptoms-causes/syc-20354453, https://www.mayoclinic.org/diseases-conditions/periodontitis/symptoms-causes/syc-20354473

National Health Service, UK. https://www.nhs.uk/conditions/gum-disease/symptoms/

National Institute for Clinical Care and Excellence (NICE), UK. Gingivitis and periodontitis, 2018. https://cks.nice.org.uk/gingivitis-and-periodontitis#!topicSummary

National Institute of Dental and Craniofacial Research, USA. https://www.nidcr.nih.gov/health-info/gum-disease/more-info

Bui FQ, Almeida-da-Silva CLC, Huynh B, Trinh A, Liu J, Woodward J, Asadi H, Ojcius DM. Association between periodontal pathogens and systemic disease. *Biomedical Journal* 2019 Feb;42(1):27-35. https://doi.org/10.1016/j.bj.2018.12.001. Epub 2019 Mar 2.

Campisi G, Pizzo G. Gingivitis. BMJ Best Practice, BMJ Publishing Group, 2019. https://bestpractice.bmj.com/topics/en-gb/620

Dahlen G, Basic A, Bylund J. Importance of virulence factors for the persistence of oral bacteria in the inflamed gingival crevice and in the pathogenesis of periodontal disease. *Journal of Clinical Medicine*. 2019 Aug 29;8(9). pii: E1339. https://doi.org/10.3390/jcm8091339.

Könönen E, Gursoy M, Gursoy UK. Periodontitis: a multifaceted disease of tooth-supporting tissues. *Journal of Clinical Medicine*. 2019 Jul 31;8(8). pii: E1135. https://doi.org/10.3390/jcm8081135.

Kumar PS. From focal sepsis to periodontal medicine: a century of exploring the role of the oral microbiome in systemic disease. *Journal of Physiology.* 2017 Jan 15;595(2):465-476. doi: https://doi.org/10.1113/JP272427. Epub 2016 Aug 28.

Mehrotra N, Singh S. Periodontitis. Treasure Island (FL): StatPearls Publishing LLC; 2019. https://www.ncbi.nlm.nih.gov/books/NBK541126/

Ng E, Lim LP. An overview of different interdental cleaning aids and their effectiveness. *Dentistry Journal (Basel)*. 2019 Jun 1;7(2). pii: E56. https://doi.org/10.3390/dj7020056.

Rathee M, Jain P. Gingivitis. Treasure Island (FL): StatPearls Publishing LLC; 2020. https://www.ncbi.nlm.nih.gov/books/NBK557422/

Stephen JM. Gingivitis. Medscape from WebMD, 2018. https://emedicine.medscape.com/article/763801-overview

22

Bad Breath

Abstract Bad breath (halitosis) is a common condition that affects about half of the adult population in many developed countries. It's due mainly to the production of offensive-smelling compounds by microbes in the mouth. The large, intricate surface of the tongue traps debris which is an excellent source of nutrients for microbes. The tongue provides an environment suitable for large microbial communities with high proportions of Gram-negative anaerobic bacteria. These anaerobes, as well as those present in dental plaque on the surface of teeth, can produce volatile sulphur and other malodorous compounds. Mechanical removal of plaque, together with gentle scraping of the tongue, reduces both the amount of debris and the population of oral microbes resulting in decreased production of malodorous compounds.

You have a date tonight and it's important. This is the one, a potential lifelong partner. You stand in front of the mirror doing your hair and lean in for a closer look. It's then that it hits you – a hint of bad breath. Oh, the horror. How can it be? Why tonight? You've brushed your teeth every day – it's not fair. You cup your hand over your mouth and nose and exhale. Yes, it's definitely there. It's dreadful, unbearable. I can't kiss anyone stinking like this. They'll turn away in disgust. Where's the breath freshener? Where's the chewing gum?

What Causes Halitosis?

Bad breath, or halitosis, is a term used to describe the presence of any unpleasant odour in the air that comes from the mouth. Halitosis is derived from "halitus" the Latin word for breath and the Greek word "osis" which simply denotes "a condition" or "pathological process". In 85–90% of cases, the smell is due to volatile compounds (VCs) produced by oral microbes, and these are mainly sulphur compounds, amines and fatty acids. The remaining cases of halitosis (often known as non-oral or extra-oral halitosis) are usually due to some medical disorder such as diabetes, kidney failure, liver failure, rheumatic fever, indigestion and cancer. Temporary episodes of bad breath may occur after eating certain foods such as onion, garlic, spiced food and durian fruit. These contain odorous compounds which are transferred via the blood stream to the lungs and then breathed out. This form of bad breath is known as "blood-borne halitosis" and may also occur following the use of certain drugs such as nitrates (used to treat angina), penicillamine (for rheumatoid arthritis) and disulfiram (for alcohol abuse). This chapter will focus on the most common cause of halitosis – that caused by oral microbes.

How Common Is Halitosis?

Estimates of the prevalence of the condition vary widely in different studies and in different countries. Globally, the prevalence has been estimated to be approximately 32%. In developed countries 8–50% of people consider themselves to suffer from recurrent episodes of halitosis. One of the problems in determining the prevalence of the condition is that it's often assessed and diagnosed on a very subjective basis – the examiner simply sniffs the exhaled air from the patient's mouth. Although attempts to standardise this have been made, no single, universally accepted method has been agreed on. Instruments capable of accurate measurement of the compounds responsible for bad breath are available, but these are expensive and aren't widely used. The condition affects people of all ages, and although it doesn't have serious health consequences, it can seriously undermine self-assuredness and social confidence, and in some cases this may lead to loneliness and then possibly other mental health impacts (Fig. 22.1).

Fig. 22.1 Advertisement for a mouthwash that was supposed to prevent bad breath. (Credit; Nesster. Attribution 2.0 Generic (CC BY 2.0))

Which Microbes Are Responsible for Halitosis?

As mentioned earlier, the VCs responsible for bad breath are produced by microbes living on the teeth and tongue. These microbes metabolise food debris, dead cells from the mouth as well as compounds present in saliva and generate VCs during the process. The accumulation of large numbers of microbes in the mouth due to poor oral hygiene will worsen the symptoms. The most important VCs contributing to halitosis are (i) volatile sulphur compounds (VSCs) such as hydrogen sulphide and methyl mercaptan; (ii) amines such as cadaverine, putrescine and trimethylamine; (iii) indole; and (iv) fatty acids such as butyric and isovaleric acids.

While a variety of oral microbes are able to produce one or more of these VCs, many Gram-negative anaerobic bacteria are able to produce a whole range of them including VSCs, amines and fatty acids. The most important of these bacteria are members of the genera *Porphyromonas, Prevotella, Treponema, Fusobacterium* and *Tannerella* (Fig. 22.2).

All of these bacteria can be found on the tongue and in the plaque that accumulates in the gap where the teeth meet the gums (see Chap. 21). The upper surfaces of the tongue, as well as its sides, are covered in small protrusions known as papillae, and these increase its surface area considerably (Fig. 22.3).

Because of its large surface area and its highly convoluted structure, the tongue can retain unexpectedly large quantities of food debris and dead epithelial cells, and these provide a plentiful supply of nutrients for odour-producing microbes (Fig. 22.4). It's been estimated that 60–70% of the VCs present in bad breath originate from the tongue.

How Is Halitosis Treated?

The most important thing to do first is to identify the cause of the halitosis so that treatment can then be correctly targeted. Treatment usually involves reducing the number of odour-causing microbes and removing the food debris and other sources of nutrients for these microbes. You can do this by regular and proper toothbrushing together with the use of inter-dental brushes and floss. You may have to ask the advice of dental professionals to either identify the cause of the halitosis and/or for advice on improving your oral hygiene habits. Gentle scraping of the tongue reduces the microbial communities that live there and has the additional benefit of removing food debris trapped between the papillae. You should carry this out on a regular basis each night before going to bed. A hard toothbrush and cold water, but no toothpaste, can be used or else a tongue scraper (Fig. 22.5). Interestingly, tongue scraping was popular in mouthwashes antiquity (Box 22.1) as well as in Europe in the early eighteenth century (Fig. 22.5).

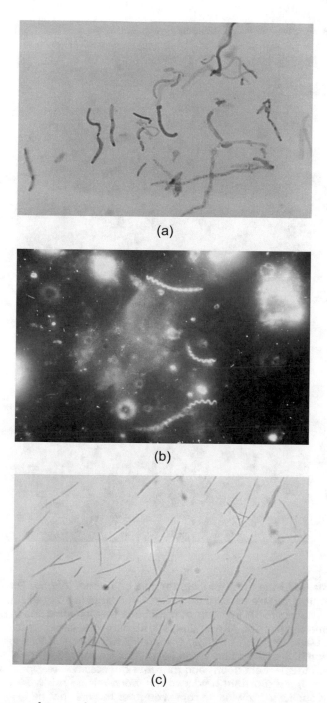

Fig. 22.2 Images of some of the microbes associated with halitosis. (a) Gram stain of a *Prevotella* species showing Gram-negative bacilli. (Dr. Holdeman, Centers for Disease Control and Prevention, USA). (b) Photomicrograph of a *Treponema* species. Note the long, corkscrew-like shape of the bacterium. (Susan Lindsley, Centers for Disease Control and Prevention, USA). (c) Gram stain of *Fusobacterium nucleatum* showing long, thin, tapering Gram-negative bacilli. (Dr. V. R. Dowell, Jr., Centers for Disease Control and Prevention, USA)

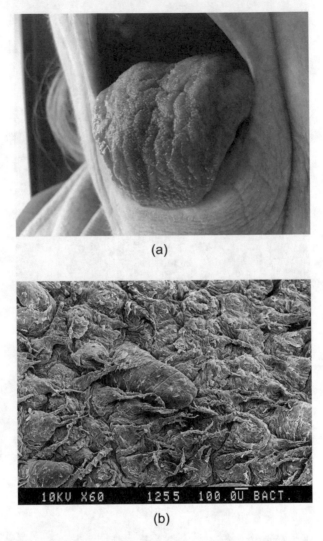

(a)

(b)

Fig. 22.3 Images of the surface of the tongue. (**a**) Human tongue. The tongue has a highly convoluted surface because it's covered in a large number of small protrusions (papillae) as well as smaller numbers of taste buds. (**b**) Magnified image of the tongue showing numerous papillae surrounding a taste bud – the larger structure in the centre. (Credit: David Gregory & Debbie Marshall. Attribution 4.0 International (CC BY 4.0)). (**c**) Magnified cross-section through the tongue showing the papillae and the gaps between them where debris and microbes can accumulate. (Jpogi at Wikipedia/Public domain. I, the copyright holder of this work, release this work into the public domain. This applies worldwide. In some countries this may not be legally possible; if so, I grant anyone the right to use this work for any purpose, without any conditions, unless such conditions are required by law. Via Wikimedia Commons). (**d**) Magnified view of a taste bud on the tongue surface. (Credit: David Gregory & Debbie Marshall. Attribution 4.0 International (CC BY 4.0))

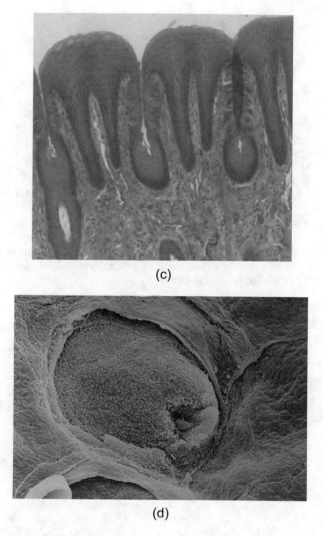

(c)

(d)

Fig. 22.3 (continued)

Fig. 22.4 Electron micrograph of the surface of the tongue. This shows the presence of large numbers of microbes in the form of a biofilm. (Credit: David Gregory & Debbie Marshall. CC BY 4.0. Wellcome Image Library)

Box 22.1 Tongue Cleaning and Scraping in Antiquity

Many ancient religions have emphasised the importance of having a clean mouth, including the tongue. The Hindus, for example, considered the mouth to be the gateway of the body, and, therefore, it was necessary to keep it very clean. Those who could afford it used tongue scrapers made of silver, gold, copper, tin or brass that had sharp curved edges. However, poorer Hindus used what was known as a "datana" or Indian toothbrush. This was made from a green twig and was about 8 inches long and as thick as the little finger. It was crushed and chewed at the end until it became a soft brush and used to clean the teeth. It was then bent into a V-shape and used to scrape the tongue twice a day.

Muslims are told in the Koran: "You shall clean your mouth for that is the way to praise God". For this purpose they used a brush made from wood and this was known as a "miswak". Wood from the arak tree (a small shrub – *Salvadora persica*) was soaked in water for 24 hours, and the ends were pounded with a hammer to form a brush which was used on the teeth and the tongue.

Generally, tongue scraping was not widely practised in ancient times in Western societies. Although the Romans were known to use iron tongue scrapers, it wasn't until the eighteenth century that such devices became popular among the upper classes of European society.

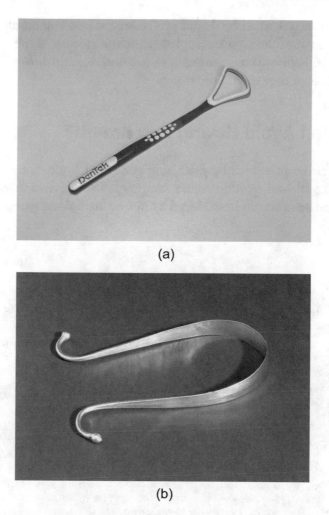

(a)

(b)

Fig. 22.5 Tongue scrapers. (a) A modern tongue scraper. (Niro5 at English Wikipedia. This work has been released into the public domain by its author, Niro5, at English Wikipedia. This applies worldwide. In some countries this may not be legally possible; if so, Niro5 grants anyone the right to use this work for any purpose, without any conditions, unless such conditions are required by law) (b) A silver tongue scraper from the early eighteenth century. (Credit: Science Museum, London. CC BY 4.0)

You might want to try supplementing these physical procedures by using mouthwashes containing antiseptics or products containing chemicals, such as zinc salts, that stop the smelly compounds being released into the breath. Chewing gum may help by increasing saliva production and perfumed products are also available to mask the odour.

If you suffer from gingivitis or periodontitis, this means that you have excessive plaque accumulation, and this can cause halitosis – these oral conditions can be treated by dental professionals. Less commonly, other diseases such as tonsillitis, gastroesophageal reflux disease, some cancers and some metabolic disorders are sometimes responsible for halitosis, and if these are present, then they should be treated.

How Can I Avoid Having Bad Breath?

Halitosis can be prevented by practising good oral hygiene, i.e. mechanical plaque removal and gentle tongue scraping. The prevention of transient blood-borne halitosis can be achieved by not eating onions, garlic and similar foodstuffs.

Box 22.2 Bad Breath and the Birth of an Emperor

Bad breath plays an important role in a famous Persian story ("The book of Darab") written in the twelfth century by Abu Taher Muhammad. King Darab, the ruler of Persia, had invaded Greece in about 500 BCE and defeated Philip II of Macedon. Darab demanded that he be given Philip's daughter, Nahid, because *"She is elegant as a cypress tree, her face as fresh as springtime. No one has ever seen any idol in China as lovely as she is, she outshines all others in her beauty. If the king sees her, she will please him: this cypress would be well placed in his garden"*. She was sent to Darab along with *".....ten camels carrying Greek brocade embroidered with jewels and gold, together with three hundred camel loads of carpets and necessities for the journey. .... sixty maidservants, each of them adorned with a diadem and earrings and carrying a golden goblet filled with jewels"* (Figure). Darab began living with her in his royal palace at the Persian capital in Pars. However: *"One night this lovely moon, arrayed in jewels and scents, lay sleeping beside the king. Suddenly she sighed deeply, and the king turned his head away, offended by the smell of her breath. This bad odour sickened him, and he frowned, wondering what could be done about it"*. His physicians advised the use of a herb called Sekandar, and this certainly cured the problem. But the king had lost all desire for Nahid and sent her back to Greece. However, Nahid was pregnant and eventually gave birth to a son who she called Sekander (after the herb) and, in order to avoid a scandal, she abandoned him on a mountain. He was found by an old woman who brought him up with the help of Aristotle, who happened to be living nearby. The boy was Alexander the Great.

Figure Nahid, daughter of Philip of Macedon, is presented to Darab. An illustration from "The tales of Darab: a medieval Persian prose romance". (Image courtesy of the British Library Board, London)

Want to Know More?

American Dental Association. https://www.ada.org/~/media/ADA/Publications/Files/for_the_dental_patient_sept_2012.ashx

Family Doctor, USA. https://familydoctor.org/condition/halitosis/

Mayo Clinic, USA. https://www.mayoclinic.org/diseases-conditions/bad-breath/symptoms-causes/syc-20350922

National Health Service, UK. https://www.nhs.uk/conditions/bad-breath/

National Institute for Clinical Care and Excellence (NICE), UK. Halitosis, 2019. https://cks.nice.org.uk/halitosis#!topicSummary

Oral Health Foundation, UK. https://www.dentalhealth.org/bad-breath?gclid=EAIaIQobChMIi5b-_Yj75QIVzLTtCh1RLQDOEAAYAiAAEgLB PD_BwE

Patient Info, UK. https://patient.info/doctor/halitosis

Royal Pharmaceutical Society, UK. https://www.pharmaceutical-journal.com/learning/learning-article/advising-patients-on-halitosis-and-oral-hygiene/20202477. article?firstPass=false

Bicak DA. A current approach to halitosis and oral malodour - a mini review. *Open Dentistry Journal*. 2018 Apr 30;12:322–330. https://doi.org/10.217 4/1874210601812010322. eCollection 2018. PMID: 29760825

Kapoor U, Sharma G, Juneja M, Nagpal A. Halitosis: current concepts on etiology, diagnosis and management. *European Journal of Dentistry*. 2016 Apr–Jun;10(2):292–300. https://doi.org/10.4103/1305-7456.178294. PMID: 27095913

Madhushankari GS, Yamunadevi A, Selvamani M, Mohan Kumar KP, Basandi PS. Halitosis – an overview: Part I – Classification, etiology, and pathophysiology of halitosis. *Journal of Pharmacy and Bioallied Sciences*. 2015 Aug;7(Suppl 2):S339–43. https://doi.org/10.4103/0975-7406.163441.

Patil S. Halitosis. Medscape from WebMD. 2016. https://emedicine.medscape.com/article/867570-overview

Porter SR, Fedele S. Halitosis. BMJ Best Practice, BMJ Publishing Group, 2017. https://bestpractice.bmj.com/topics/en-gb/1036

Tungare S, Paranjpe AG. Halitosis. Treasure Island (FL): StatPearls Publishing LLC; 2019. https://www.ncbi.nlm.nih.gov/books/NBK534859/

23

Oral Thrush

Abstract The yeast *Candida albicans* is responsible for a number of infections which are often called candidiasis. These affect several areas of the body including the skin and nails, gut, vagina and, less commonly, organs such as the heart if it invades the blood stream. In the mouth it's responsible for pseudomembranous candidiasis (oral thrush), erythematous candidiasis, denture stomatitis and angular cheilitis. Occasionally other species of *Candida* are involved. Although infants and the elderly are the main groups affected, oral fungal infections are also more common in those who are immunocompromised in some way and those undergoing radiotherapy or chemotherapy. Treatment involves the use of topical anti-fungal agents. In those who are immunocompromised, oral anti-fungal drugs may be necessary.

What Is Oral Candidiasis?

Oral thrush (or oral candidiasis) is an infection of the mouth caused by the yeast *Candida albicans* or other closely related yeasts. It has been known since antiquity (Box 23.1) and is the most common fungal infection of the mouth, most frequently affecting infants and older adults. The disease takes four main forms: pseudomembranous (whitish patches in the mouth), erythematous (red areas in the mouth), denture stomatitis (occurs underneath dentures) and angular cheilitis (affects the corners of the mouth).

© Springer Nature Switzerland AG 2021
M. Wilson, P. J. K. Wilson, *Close Encounters of the Microbial Kind*,
https://doi.org/10.1007/978-3-030-56978-5_23

Box 23.1 Oral Thrush: A Disease with a Long History

Oral candidiasis has been recognised since the time of Hippocrates who, in the fourth century BCE, described the disease in his book *Epidemics*. However, it wasn't until 1786 that the first funded research into the disease was undertaken – this was instigated by the Royal Society of Medicine in France. In 1839 a German surgeon, Bernhard von Langenbeck, described a fungus (which matches the features of *Candida albicans*) that he cultured from the mouth of a patient with typhus. However, he wrongly suggested that the fungus was responsible for typhus – this disease is actually caused by a bacterium. Then in 1841 the Swedish paediatrician Fredrik Theodor Berg (Figure) recognised that oral thrush was caused by a fungus. Subsequently, in 1853, the French biologist Charles-Philippe Robin named this fungus *Oidium albicans*. Since then it has been through several name changes (including *Mycoderma vini*, *Saccharomyces albicans* and *Monilia albicans*) until it was officially named *Candida albicans* in 1923.

Figure Portrait of Fredrik Theodor Berg by Hildegard Norberg (1880 CE). (Painting by Hildegard Norberg (1844–1917)/Public domain via Wikimedia Commons)

Pseudomembranous oral candidiasis is the form commonly known as "thrush" and is the type seen in about one third of cases of oral candidiasis (Fig. 23.1). It appears as creamy white or yellowish deposits (often known as "plaques") that stick to the lining of the mouth. If the deposits are removed by scraping, then the mucosa underneath is often red and/or bleeding. Although it can appear in any part of the mouth, it's most commonly seen on the palate, the lining of the cheeks, the inside of the lips and the tongue. Sometimes it spreads to the pharynx resulting in a condition known as oropharyngeal candidiasis. If you have a mild infection, you'll often feel nothing, but in more severe infections, you might experience a burning sensation in your mouth and/or changes in your taste perception.

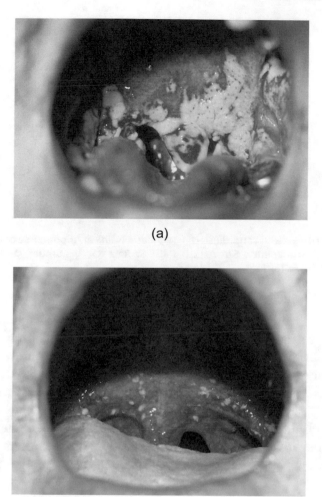

(a)

(b)

Fig. 23.1 Pseudomembranous oral candidiasis. (a) A patient suffering from pseudo-membranous oral candidiasis showing the characteristic white/yellowish plaques. (Centers for Disease Control and Prevention, USA). (b) Candidiasis affecting the tongue with patches on the upper palate and uvula (Klaus D. Peter, Gummersbach, Germany / CC BY 3.0 DE (https://creativecommons.org/licenses/by/3.0/de/deed.en))

The erythematous form of the infection appears as flat, red areas on the palate, the upper surface of the tongue or the lining of the cheek (Fig. 23.2). Symptoms include a burning sensation in the mouth, soreness of the lip and tongue and difficulties in swallowing.

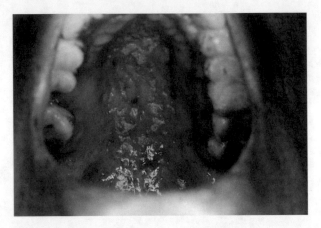

Fig. 23.2 Erythematous candidiasis of the palate. (Clinical Appearance of Oral Candida Infection and Therapeutic Strategies. Patil S, Rao RS, Majumdar B, Anil S. *Front Microbiol.* 2015; 6: 1391

Denture stomatitis is an inflammation of the part of the mouth that lies underneath a denture and is sometimes accompanied by a burning sensation. It often results from ill-fitting dentures, but wearing the dentures constantly or failing to clean them properly may also be responsible for the condition because this enables biofilms of *C. albicans* to form on the denture surface (Fig. 23.3).

Angular cheilitis (also known as angular stomatitis or perlèche) affects the corners of the lips and results in painful cracks, fissures or ulcers (Fig. 23.4). It often affects both corners of the mouth and *Staph. aureus* may also be present.

What Are the Risk Factors for These Conditions?

A number of factors are associated with an increased risk of developing oral candidiasis including:

- Immunosuppression due to HIV infection, diabetes, drugs such as steroids and malignancy.
- Reduced production of saliva due to antipsychotic drug use, Sjögren syndrome, radiotherapy or chemotherapy.

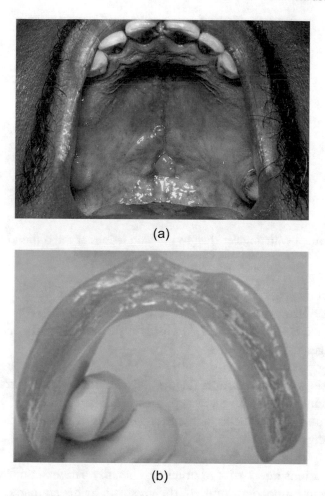

(a)

(b)

Fig. 23.3 Denture stomatitis. (**a**) Denture stomatitis of the palate that has developed beneath a denture. (Clinical Appearance of Oral Candida Infection and Therapeutic Strategies. Patil S, Rao RS, Majumdar B, Anil S. *Front Microbiol*. 2015; 6: 1391. (**b**) Formation of biofilms on the surface of a denture. (The effect of a commercial probiotic drink on oral microbiota in healthy complete denture wearers. Sutula J, Coulthwaite L, Thomas L, Verran J. *Microb. Ecol. Health Dis*. 2012; 23: https://doi.org/10.3402/mehd.v23i0.18404

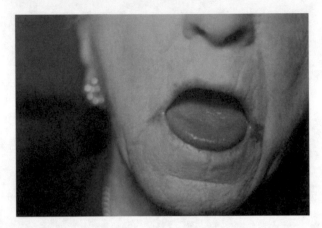

Fig. 23.4 Patient with angular cheilitis showing lesions at the corners of the mouth. (Centers for Disease Control and Prevention, USA)

- Poor oral hygiene.
- Presence of foreign bodies in the mouth such as dentures.
- Missing teeth – this increases skin folds at the corners of the mouth, which encourages angular cheilitis.
- Smoking.
- Antibiotic use – this disrupts the oral microbiota (which controls the yeast population) and so allows yeasts to grow unchecked.
- Poorly controlled diabetes increases sugar levels in saliva which encourages yeast overgrowth.

Complications rarely arise in otherwise healthy patients, but in those who are severely immunocompromised, the yeast can be disseminated throughout the body in the blood stream and cause infections at other locations. This can result in pneumonia, meningitis, endocarditis, osteomyelitis and peritoneal infections. Severe oesophageal infections can cause swallowing problems.

How Common Is It?

The disease has a high incidence in those who are immunocompromised in some way. It affects up to 50% of those undergoing some form of chemotherapy for cancer, up to 70% of patients receiving radiation therapy and up to 90% of HIV patients. In the otherwise healthy population, it's reported to affect 45% of neonates, 45–65% of children, 30–45% of adults, 50–65% of denture wearers and 65–88% of those residing in long-term care facilities.

What Happens During an Infection?

C. albicans is present in low numbers in the mouths of 45–65% of healthy infants and 30–55% of healthy adults where, generally, it does no harm. However, a number of factors (as listed above) can change the environment of the mouth, and the yeast can then take advantage and increase in numbers to such an extent that it can cause disease, i.e. it's an opportunistic pathogen (see Chap. 1).

C. albicans can attach itself to an epithelial cell by means of a number of molecules on its cell wall. Once this has happened, it starts to produce a protuberance (known as a "germ tube") which then grows into a long, filamentous structure (a hypha, Fig. 23.5) and produces enzymes that it uses to invade the cell. This damages the oral epithelium and causes an inflammatory response which eventually controls the infection. Infection is usually confined to the surface layers of the oral mucosa. The white deposits that accompany several forms of the infection contain a mixture of dead epithelial cells and fungal hyphae.

C. albicans is responsible for approximately 80% of cases, but other yeast species that may be involved include *C. glabrata*, *C. tropicalis*, *C. parapsilosis*, *C. guilliermondii*, *C. krusei* and *C. dubliniensis*.

How Is the Disease Diagnosed?

A diagnosis of oral candidiasis is usually made on the basis of an examination, and, if required, this can be confirmed by laboratory investigations. Hyphae of *C. albicans* can be easily seen with a microscope in a sample taken from the affected area. Samples may be grown in the laboratory in order to identify the yeast responsible and to determine its susceptibility to anti-fungal agents. This is especially useful for patients who are at risk of complications, fail to respond to therapy or have severe symptoms.

How Is the Disease Treated?

An important part of the treatment of oral candidiasis is addressing any underlying causes or identifiable risk factors.

If you have denture stomatitis, for example, it's important that you ask your dentist to check to see that your dentures fit properly. You should avoid wearing your dentures continually and should remove them at night. You should also regularly disinfect them – ideally every day. It's important that you practice meticulous oral hygiene, and this is essential for both treatment and prevention.

Mild to moderate cases can be treated with any of a number of topical anti-fungal agents which come in various forms – creams, gels, drops or solutions. These include clotrimazole, miconazole, amphotericin B, nystatin or gentian

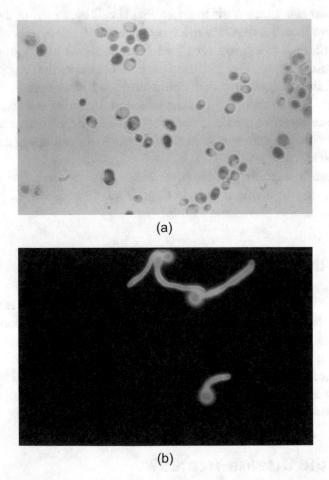

(a)

(b)

Fig. 23.5 Images of *C. albicans*. **(a)** The yeast form of *C. albicans* (×1200). (Dr. Lucille K. Georg, Centers for Disease Control and Prevention, USA). **(b)** Each of these yeast cells has produced a germ tube. (Dr. Brian Harrington, Centers for Disease Control and Prevention, USA). **(c)** Gram stain of a sample taken from a patient with candidiasis. This shows both the yeast and hyphal forms of *C. albicans* as well as epithelial cells. (Centers for Disease Control and Prevention, USA). **(d)** Colonies of *C. albicans* growing on an agar plate. (Dr. William Kaplan, Centers for Disease Control and Prevention, USA)

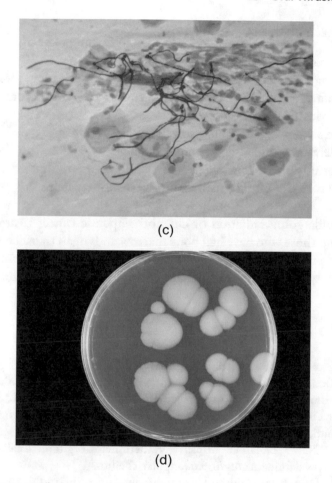

(c)

(d)

Fig. 23.5   (continued)

violet. For patients with severe disease or complications, particularly in those with immune problems, treatment with systemic anti-fungal agents is often necessary. Suitable drugs include fluconazole, itraconazole, ketoconazole and posaconazole.

Angular cheilitis often resolves by itself. If treatment is required then topical steroids, anti-fungal agents or antibiotics may be helpful.

How Can I Avoid Getting Oral Candidiasis?

There are a number of measures you can use to reduce the risks of developing the condition, and these include:

- Rinsing your mouth after meals.
- Brushing your teeth at least twice a day and flossing regularly.
- Having regular dental check-ups.
- Removing your dentures every night and regularly cleaning them.
- If you have few or no natural teeth, you should brush your gums and tongue with a soft brush twice a day.
- If you're a smoker, you should stop this habit.
- If you have any underlying condition, such as diabetes, you should make sure that this is well controlled.
- If you have a dry mouth, then artificial salivas may help.

Patients undergoing radiation or chemotherapeutic cancer treatment, HIV patients and those who have had bone marrow and organ transplants are often prescribed preventative anti-fungal medication to reduce the risk of fungal infections. Appropriate agents include fluconazole and itraconazole.

Want to Know More?

American Academy of Pediatrics. https://www.healthychildren.org/English/health-issues/conditions/infections/Pages/Thrush-and-Other-Candida-Infections.aspx

American Family Physician. https://www.aafp.org/afp/2007/0215/p501.html

Centers for Disease Control and Prevention, USA. https://www.cdc.gov/fungal/diseases/candidiasis/thrush/index.html

DermNet, New Zealand. https://www.dermnetnz.org/topics/oral-candidiasis/, https://www.dermnetnz.org/topics/angular-cheilitis/

KidsHealth, USA. https://kidshealth.org/en/parents/thrush.html

Mayo Clinic, USA. https://www.mayoclinic.org/diseases-conditions/oral-thrush/symptoms-causes/syc-20353533

National Health Service, UK. https://www.nhsinform.scot/illnesses-and-conditions/infections-and-poisoning/oral-thrush-in-adults

National Institute for Clinical Care and Excellence (NICE), UK. candida – oral, 2017. https://cks.nice.org.uk/candida-oral#!topicSummary

Oral Cancer Foundation, USA. https://oralcancerfoundation.org/dental/candida/

Patient Info, UK. https://patient.info/doctor/candidiasis

Primary Care Dermatology Society, UK. http://www.pcds.org.uk/clinical-guidance/angular-chelitis

Gleiznys A, Zdanavičienė E, Žilinskas J. *Candida albicans* importance to denture wearers. A literature review. *Stomatologija*. 2015;17(2):54–66.

Kumar M. Thrush. Medscape from WebMD, 2019. https://emedicine.medscape.com/article/969147-overview#a4

Ohshima T, Ikawa S, Kitano K, Maeda N. A proposal of remedies for oral diseases caused by Candida: a mini review. *Frontiers in Microbiology.* 2018 Jul 9;9:1522. https://doi.org/10.3389/fmicb.2018.01522. eCollection 2018.

Singh A, Verma R, Murari A, Agrawal A. Oral candidiasis: an overview. *Journal of Oral and Maxillofacial Pathology.* 2014 Sep;18(Suppl 1):S81–5. https://doi.org/10.4103/0973-029X.141325.

Taylor M, Raja A. Oral Candidiasis (Thrush). Treasure Island (FL): StatPearls Publishing LLC; 2020. https://www.ncbi.nlm.nih.gov/books/NBK545282/

Younai FS. Oral candidiasis. BMJ Best Practice, BMJ Publishing Group, 2019. https://bestpractice.bmj.com/topics/en-gb/106

24

Cold Sores

Abstract Oral herpes is a disease caused mainly by herpes simplex virus-1 (HSV-1) and primarily affects the face, mouth and throat. It causes pain, burning, tingling or itching at the infection site which is normally followed by clusters of blisters. After the first infection, the disease can recur as episodes of cold sores on the lips. It's usually mild and self-limiting, lasting 2–3 weeks, but can cause severe disease in those who are at risk such as the immunocompromised. It's a very common disease – 90% of the world's population are infected by the age of 40 years, and 40% of these will experience a recurrent infection. Infection is lifelong. Treatment includes medication for the relief of pain and fever – topical medications like anaesthetics and anti-virals may also be useful. Adequate fluid intake is important to avoid dehydration. Oral anti-viral agents are available for treating severe or frequently recurring infections.

What Is Oral Herpes?

The term "herpes" is used by the general public to refer to two broad types of condition – oral herpes and genital herpes. In the past these diseases were thought to be due to two distinct viruses – herpes simplex virus type 1 (HSV-1) in the case of oral infections while herpes simplex virus type 2 (HSV-2) was responsible for genital infections. However, it's now known that both viruses can cause infections in either area, although HSV-1 remains the most frequent cause of episodes affecting the mouth, being responsible for more than 90% of cases.

Herpes is a Greek word meaning "to creep" and refers to the slow spreading nature of the appearance of the lesions of the disease which were recognised more than 2000 years ago (Box 24.1). The existence of the virus was first suggested by J.B. Vidal in 1873, but it wasn't until 1941 that it was isolated – Margaret Smith (from the Washington University School of Medicine, St. Louis) grew the virus from a sample taken from a child with encephalitis.

HSV-1 primarily affects the face, mouth and throat resulting in a condition

Box 24.1 Oral Herpes and the Romans

The lip blistering associated with oral herpes has been a concern since Roman times. The second emperor of Rome, Tiberius (42 BCE–37 CE), banned kissing in public ceremonies in an attempt to curb the spread of oral blisters which were then rampant among Roman citizens. The first documented treatment for such blistering was cauterisation with a hot iron – this was practised by the Roman physician Aulus Cornelius Celsus (25 BCE–50 CE). In about 100 CE, the Roman physician, Herodotus, gave a detailed description of what was almost certainly a herpes infection. He noted that his patient had blisters on the lips as well as a fever, and he named this condition "herpes labialis". The Roman physician, Aelius Galenus (129–216 CE), noticed that although the blisters would sometimes disappear, any re-appearance was always on the same part of the lip.

Bust of the emperor Tiberius by an unknown Roman artist. (Digital image courtesy of the Getty's Open Content Program, the J. Paul Getty Museum, Los Angeles)

known as "oral herpes". In both children and adults, the first infection is usually not even recognised and may be without any symptoms at all

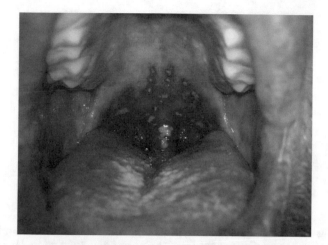

Fig. 24.1 Pharyngitis due to HSV-1 infection. (Klaus D. Peter, Gummersbach, Germany / CC BY 3.0 DE (https://creativecommons.org/licenses/by/3.0/de/deed.en))

(approximately two thirds of cases), but in 1–5-year-old children, it can give rise to fever, sore throat and gingivitis.

In those adults who do develop symptoms, the most frequent ones are pharyngitis (Fig. 24.1) and tonsillitis. Occasionally, the first infection affects the eyes and here it causes keratoconjunctivitis. The incubation period is 2–12 days, and the symptoms last for 2–3 weeks. In both children and adults, pain, burning, tingling or itching occurs at the infection site, and this is followed by the appearance of clusters of blisters. These soon break down and then look like small, shallow ulcers with a red base. The liquid from the blisters contains the virus particles and can spread infections to others. A few days later, they become crusted or scabbed. The sores may occur on the lips, the gums, the front of the tongue, the inside of the cheeks, the throat and the roof of the mouth (Fig. 24.2). They can be very painful and can make eating and drinking difficult.

After this first infection, the virus usually moves to a part of one of the nerves that supplies the affected area (typically the trigeminal nerve ganglion – Fig. 24.7) where it can remain dormant indefinitely or else become reactivated later to cause another infection. In 90% of cases of a recurrent infection, the lips are the sites affected – a condition known as "herpes labialis" or "cold sores" (Fig. 24.3). Prior to the appearance of the sores on the lip, the patient often experiences burning and tingling. Recurrent infection can be triggered by various factors such as exposure to sunlight, fatigue or stress.

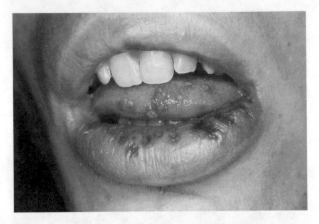

Fig. 24.2 Blisters on the tongue and lips of a patient suffering from an HSV-1 infection. (Robert E. Sumpter, Centers for Disease Control and Prevention, USA)

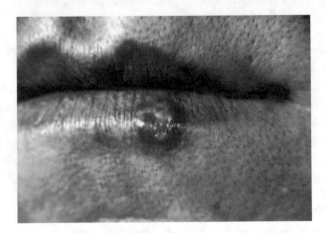

Fig. 24.3 This photograph is a close-up of the lips of a patient with a cold sore on the lower lip. (Dr. Herrmann, Centers for Disease Control and Prevention, USA)

As well as causing disease of the face and upper respiratory tract, HSV-1 can also infect the genitalia (see Chap. 29) mainly as a result of oral-genital contact.

How Is the Disease Transmitted?

HSV-1 is usually caught during childhood by direct contact with infected secretions which carry the virus particles, and these enter via breaks in the skin or mucous membranes. It can also be transmitted in saliva by kissing or by means of shared utensils or towels if these come into contact with the mouth or cuts/abrasions in skin. More rarely the eyes and nose can be a route of infection. Herpes infections are most contagious while you have symptoms. However, due to the long-term presence of the virus in the body, you can remain infective even when you aren't having a recurrence or experiencing symptoms; this is known as "asymptomatic shedding", which means you can infect someone else without even knowing you have the virus yourself. The virus can also be transmitted during oral-genital contact and sexual intercourse.

Although most (90%) of cases of oral herpes are due to HSV-1, the closely related virus HSV-2 can also occasionally be responsible – in such cases, transmission is mainly the result of oral-genital contact.

In most people, the infection is a mild, self-limiting illness. However, it can be a lot worse in people with certain skin diseases and can, rarely, cause severe or life-threatening complications in those at risk, particularly in immuno-compromised patients.

Are There Any Complications?

Possible complications of the infection include:

(i) Dehydration due to a low intake of fluids because of painful swallowing.

(ii) Herpetic whitlow, i.e. the appearance of small blisters on the hands, particularly the fingers (Fig. 24.4). This is often the result of thumb-sucking in infants.

(iii) Eye infection – known as herpetic keratoconjunctivitis. This can result from direct transfer of the virus by eye-rubbing.

(iv) Herpes gladiatorum – sores and blisters on the torso. This is seen mainly in those who engage in contact sports such as rugby and wrestling.

(v) Erythema multiforme – a hypersensitivity reaction that appears as bull's-eye-shaped spots on the legs, arms, hands and feet (Fig. 24.5). It generally last for no longer than 1–2 weeks.

(vi) Meningitis and encephalitis. These are rare complications. HSV-1 is responsible for only about 4% of viral meningitis cases. The number of cases of encephalitis due to HSV-1 has been estimated to be approximately one per million per year.

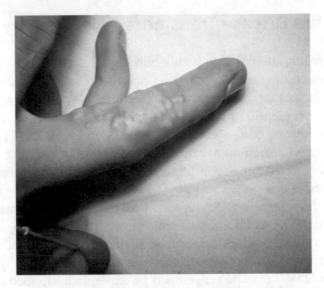

Fig. 24.4 Blisters on the finger due to HSV-1. (Dr. Thomas Sellers, Emory University, Centers for Disease Control and Prevention, USA)

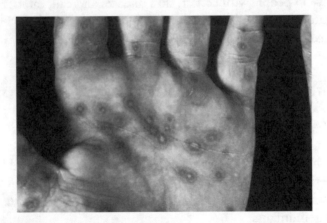

Fig. 24.5 Erythema multiforme of the hand showing the characteristic bull's-eye appearance of the lesions. (Dr. N.J. Fiumara, Centers for Disease Control and Prevention, USA)

(vii) Eczema herpeticum. The skin of people with pre-existing skin complaints, such as atopic dermatitis, is less well defended against infections. The HSV-1 virus spreads to large areas of the skin, and if not treated promptly, this condition can be life-threatening.

(viii) Neonatal herpes is a rare complication of genital herpes (see Chap. 29) but can cause significant disability and even death. Women with HSV can pass on the infection to their baby usually at the time of delivery, especially if it is their first infection. At most risk are those who are suffering their first infection in the last 6 weeks of pregnancy. This is because there isn't sufficient time for the protective maternal antibodies to pass into the foetus's circulation before delivery. Delivery by caesarean section is usually recommended if the infection has occurred for the first time during the last trimester of pregnancy.

How Common Is the Disease?

90% of the world's population will have been infected with HSV-1 by the age of 40 years, and 40% of these will experience a recurrent infection – recurrences typically happen 2–3 times a year. Between 20% and 40% of people will experience herpes labialis at some point in their lifetime. In the UK, herpes labialis accounts for about 1% of primary care consultations each year. During 2015–2016, the prevalence of HSV-1 in the USA was 47.8% among those aged 14–49 years.

What Happens During an Infection?

The HSV-1 virion is approximately 160 nm in diameter and contains a DNA molecule enclosed within an icosahedral protein structure known as the capsid (Fig. 24.6a). The capsid has a coat known as an envelope (Fig. 24.6a, b) which is formed from the membrane of the host cell in which the virion was produced. The region between the capsid and envelope, known as the tegument, contains a variety of proteins. The envelope has a number of protein spikes protruding from it, and these are used as adhesins to attach to host cells.

HSV-1 (and its close relative HSV-2) are members of the *Herpesviridae* virus family which also includes varicella zoster virus, Epstein-Barr virus and cytomegalovirus. They have three special properties that influence the way in which they cause infections in humans: (i) they can invade, and replicate within, the nervous system, (ii) they can remain dormant in nerve cell ganglia and (iii) they can be reactivated to cause recurrent infections.

The first stage in an infection involves binding of the virion to a host cell, and this is achieved by glycoproteins which form spikes on its surface. Once it has become attached, the envelope of the virion fuses with that of the host cell, and this allows the virion's capsid to enter the host cell. The capsid is transported to the host cell's nucleus where viral replication takes place. The

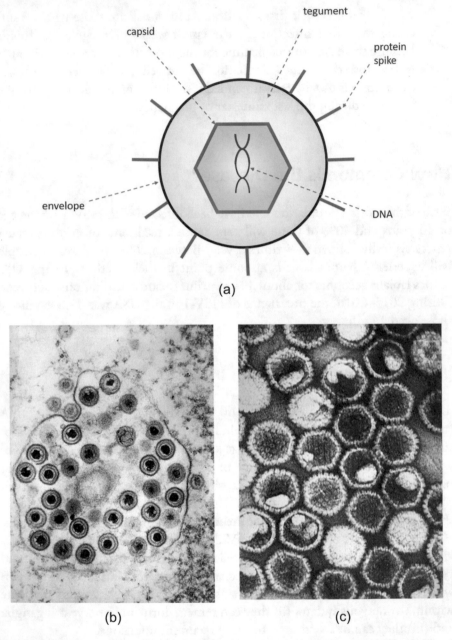

Fig. 24.6 The herpes simplex virus. (a) Diagram showing the structure of the virion of HSV-1. (b) Herpes simplex virions as seen through an electron microscope. The capsid surrounding the darkly stained DNA, as well as the envelope surrounding this, is clearly visible. (Dr. Fred Murphy and Sylvia Whitfield, Centers for Disease Control and Prevention, USA). (c) HSV virions seen at a higher magnification. (Dr. Fred Murphy and Sylvia Whitfield, Centers for Disease Control and Prevention, USA)

new virions are then released from the host cell. The whole process is similar to that involved in the replication of the varicella zoster virus (see Fig. 7.7 of Chap. 7).

Host cells that have been infected with the virus die and their contents are released and this results in inflammation and the formation of characteristic blisters. The fluid in these blisters contains large numbers of virions, cell debris and cells of the immune response. This first exposure to the virus results in antibody production that prevents further infections from happening later at different body sites. For example, if you have had oral HSV-1, you can't usually develop a genital HSV-1 infection in the future. It takes around 6 weeks from the first infection for these antibodies to provide effective protection.

The virus present in the skin and surrounding tissues can then invade more skin cells or invade a neighbouring nerve cell where it's transported via the axon to the nerve ganglion. HSV-1 affecting the mouth is transported via the trigeminal nerve to the trigeminal ganglion located on the inside of the skull close to the front of the ear. HSV-1 affecting the genital area is transported via the local sensory nerve to various dorsal root ganglia, most commonly the sacral ganglia, which are located just outside of the spinal cord in the lower back.

The virus then either replicates and travels back down the axon to cause ongoing infection or enters a latent state during which it isn't replicated – it can remain in this state indefinitely. However, certain triggers (such as UV light, stress, hormonal changes and fatigue) can re-activate it, and it then travels back along the nerve axon and, on reaching the mucosa, can cause a recurrence of the symptoms (Fig. 24.7). The most common symptom of a recurrent infection is the appearance of blisters or cold sores. However, this doesn't always happen and the virus may simply be shed. More than 90% of the population harbour HSV-1 in a latent form, but only approximately 25% of these experience repeated symptoms.

How Is the Disease Diagnosed?

The infection is usually diagnosed on the basis of the symptoms experienced by the patient. However, if the diagnosis is in doubt, the presence of the virus can be confirmed by analysis of samples taken from the blisters or sores. It's possible to grow the virus in the laboratory from such a sample, or else DNA from the virus can be detected by PCR (see Chap. 1). Testing for the presence of antibodies against the virus is also possible.

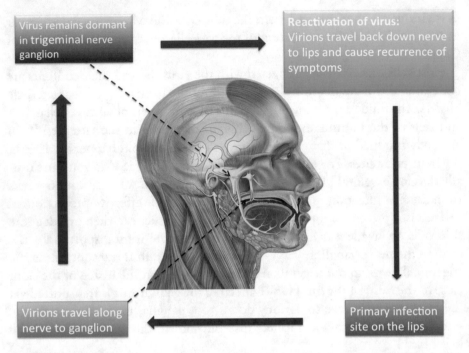

Fig. 24.7 Recurrence of symptoms is common in infections due to HSV-1. The trigeminal nerve and ganglion are shown in yellow on the central image. (Modified from Patrick J. Lynch, medical illustrator [CC BY 2.5 (https://creativecommons.org/licenses/by/2.5)] via Wikimedia Commons)

How Is It Treated?

As the disease is usually self-limiting, the use of systemic anti-viral agents isn't routinely recommended in those not at risk of serious complications. You should take paracetamol and/or ibuprofen to alleviate pain and fever, and it's important to ensure that your fluid intake is adequate so as to avoid dehydration. Topical anaesthetics may also be useful for pain control.

Although oral anti-viral agents aren't generally recommended for otherwise healthy adults with mild or moderate symptoms, they can be used for the treatment of severe infections. They are also used to prevent recurrences in those who are affected by frequent or persistent episodes. The most common agents are aciclovir, famciclovir and valaciclovir – these will shorten the duration and severity of the disease. Topical anti-viral agents are generally not particularly effective but may be used if you find them helpful for symptom control, especially if started early in the recurrence.

How Can I Avoid Getting Oral Herpes?

You can reduce the chances of acquiring HSV-1 by avoiding touching the saliva, skin or mucosal membranes of people who show signs of oral herpes. You should also avoid sharing their food, drinks, toothbrushes, drinking glasses and eating utensils. HSV-1 is also responsible for genital herpes (see Chap. 29) and so can be acquired during oral-genital contact – such activities are also able to transmit HSV-2, a less frequent cause of oral herpes.

Recurrence of the infection may be triggered by various factors such as sunlight. Avoiding triggers can reduce recurrence, for instance by using a sunscreen and/or a sunblock lip balm.

Want to Know More?

American Dental Association. https://ada.com/conditions/herpes-labialis/

American Family Physician. https://www.aafp.org/afp/2010/1101/p1075.html

DermNet, New Zealand. https://www.dermnetnz.org/cme/viral-infections/herpes-simplex/

Mayo Clinic, USA. https://www.mayoclinic.org/diseases-conditions/cold-sore/symptoms-causes/syc-20371017

National Health Service, UK. https://www.nhs.uk/conditions/cold-sores/

National Institute for Clinical Care and Excellence (NICE), UK. Herpes simplex – oral, 2016. https://cks.nice.org.uk/herpes-simplex-oral#!topicSummary

Patient Info, UK. https://patient.info/doctor/oral-herpes-simplex

Primary Care Dermatology Society, UK. http://www.pcds.org.uk/clinical-guidance/herpes-simplex

School of Medicine at Mount Sinai, USA. https://www.mountsinai.org/health-library/diseases-conditions/herpes-oral

World Health Organisation. https://www.who.int/news-room/fact-sheets/detail/herpes-simplex-virus

Ayoade FO. Herpes Simplex. Medscape from WebMD, 2018. https://emedicine.medscape.com/article/218580 overview

Lorenz BD. Herpes simplex virus infection. BMJ Best Practice, BMJ Publishing Group, 2019. https://bestpractice.bmj.com/topics/en-gb/53

Opstelten W, Neven AK, Eekhof J. Treatment and prevention of herpes labialis. *Canadian Family Physician*. 2008 Dec;54(12):1683-7. PMID: 19074705

Saleh D, Sharma S. Herpes Simplex Type 1. Treasure Island (FL): StatPearls Publishing LLC; 2019. https://www.ncbi.nlm.nih.gov/books/NBK482197/

Worrall G. Herpes labialis. *BMJ Clinical Evidence*. 2009 Sep 23;2009. pii: 1704. PMID: 21726482

Part VI

Infections of the Genito-Urinary System

25

Cystitis

Abstract Cystitis means inflammation of the bladder and has a variety of causes including infections, chemical irritants, radiotherapy and chemotherapy. In this chapter we will deal only with infective causes of cystitis.

Women are far more likely to get cystitis than men, and by the age of 24 years, about one third of women will have had at least one episode. Men under the age of 50 years rarely get cystitis. The main symptoms are pain during urination and in the lower abdomen as well as the need to urinate more often and urgently than normal. The urine produced is often dark, cloudy and offensive-smelling or contains blood. If untreated, the infection can spread up the urinary tract and affect the kidneys (pyelonephritis) which can make you seriously unwell, but, thankfully, this isn't common. In pregnant women untreated cystitis can result in pre-term delivery or miscarriage. In about 80% of cases, the infection is caused by *Escherichia coli*, a bacterium that lives in the gastrointestinal tract. A variety of other bacteria are responsible for the remaining cases. The problem also frequently accompanies urinary catheterisation, and such infections are among the most common healthcare-associated infections. Cystitis can get better by itself or can be successfully treated with a short course of an oral antibiotic such as nitrofurantoin or trimethoprim/sulfamethoxazole.

Have you ever passed blood in your urine? It really is one of the most unexpected and frightening surprises. This happened to me when I was on holiday one year and developed cystitis. It was probably brought on by dehydration due to the hot weather plus having long trips on buses meant I couldn't empty

M. Wilson, P. J. K. Wilson, *Close Encounters of the Microbial Kind*,
https://doi.org/10.1007/978-3-030-56978-5_25

my bladder as regularly as usual. For a few days, it had been a bit painful when I urinated so I did have some warning that something was amiss, but the most unusual symptom was seeing bright red blood in the toilet pan. My doctor asked me if my urine looked like rose wine. I'm not sure if it was the best vintage, but it certainly looked like that. She explained that the inflammation caused by an infection had led to blood leaking from my bladder into the urine. There are other, and more serious, causes of blood in the urine, but cystitis is a common one.

What Is Cystitis?

Acute cystitis is inflammation of the bladder (Fig. 25.1) and is most commonly due to a bacterial infection. The term is derived from "Kystis" the Greek word for bladder and the suffix "-itis" for inflammation.

The bladder is part of the urinary tract which also consists of the kidneys, ureter and urethra (Figs. 25.1 and 25.2). Its location in men and women is shown in Fig. 25.2. Infections can occur in any region of the urinary tract and are known collectively as urinary tract infections (UTIs). In addition to cystitis, UTIs include urethritis (affecting the urethra) and pyelonephritis (affecting the kidney). However, urethritis and pyelonephritis are much rarer than

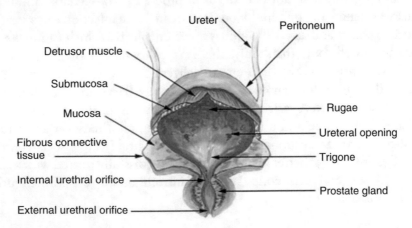

Fig. 25.1 The male bladder and associated structures. The bladder is a hollow, muscular organ which is supplied with urine by the kidneys via two ureters. The ureters open into the bladder through the ureteral openings. When empty, the bladder has many folds (known as rugae), and these enable it to expand as it fills with urine. The opening and closing of the internal and external urethral orifices are controlled by muscles. (US National Cancer Institute, Surveillance, Epidemiology and End Results (SEER) Program/ Public domain)

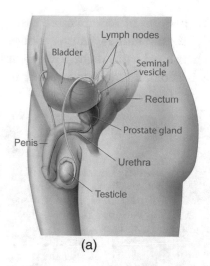

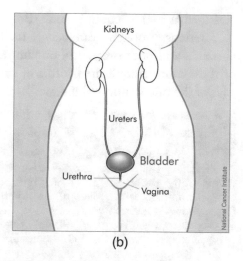

Fig. 25.2 The location of the bladder in (**a**) males, directly in front of the rectum, and (**b**) females, in front of the vagina and below the uterus. ((a) NIH Medical Arts, National Cancer Institute USA, Public Domain. (b) NIH Medical Arts, National Cancer Institute USA, Public Domain)

cystitis which accounts for 95% of all UTIs. "Acute cystitis" is the term used to describe a recent onset infection and these frequently affect young, sexually active women. As it affects the lower part of the urinary tract, it's also sometimes referred to as a "lower UTI".

The main symptoms of acute cystitis include (i) pain, burning or stinging during urination; (ii) needing to urinate more often and urgently than normal; (iii) feeling the need to urinate again soon after having urinated once; (iv) urine that's dark, cloudy and offensive-smelling or contains blood; (v) pain in the lower abdomen; and (vi) feeling generally unwell. If you also suffer vomiting or pain in the flank (on one side of the body, below the ribs), this is worrying because it suggests that the infection has spread to the kidney (i.e. pyelonephritis) which is a much more serious condition. Fortunately, pyelonephritis is relatively rare and affects <0.1% of the population each year.

Acute cystitis in healthy young women is classified as "uncomplicated cystitis". The infection is usually mild and can resolve without requiring medication. However, a number of factors can make cystitis more difficult to treat, and the condition is then referred to as "complicated cystitis". These factors include diabetes, pregnancy, a history of recurrent UTIs, impeded flow of urine, catheterisation, structural abnormalities of the urinary tract (such as polycystic kidneys, kidney stones or urine outflow obstruction) and immunodeficiency. Cystitis in men, which is very much rarer than in women, is also classified as being "complicated" (Box 25.1). In cases of "complicated cystitis", serious problems can occur, including sepsis, kidney failure, kidney abscesses and poor pregnancy outcomes such as low birth weight. Treatment of these more complicated forms of cystitis is generally more involved than that of the uncomplicated disease.

Box 25.1 Urinary Tract Infections in Men

UTIs rarely occur in men under 50 years of age, and when they do, they're generally associated with some other complicating factor such as prostatitis (inflammation of the prostate), orchitis (inflammation of the testicles), epididymitis (inflammation of the coiled tube at the back of the testicles), urethritis or some anatomical abnormality. As men age, they can develop various changes in the urinary tract which impair normal urination, and this can result in a UTI. The most common of these is enlargement of the prostate gland, this can cause the bladder not to empty fully meaning bacteria are not properly flushed out. The microbes responsible for UTIs in men are more varied than in women. Although, as in women, *E. coli* is the most prominent uropathogen, other organisms are frequently involved including species belonging to the genera *Klebsiella, Proteus, Providencia, Enterococcus* and *Staphylococcus*.

How Are We Protected from UTIs?

Like all regions of the body, the urinary tract has a number of defence systems that protect it from infection. The main ones are the following:

(i) The lining of the urethra and bladder is covered in mucus which prevents microbes from reaching the underlying epithelial cells.

(ii) The outer layer of epithelial cells is continually being shed which means that any microbes that have managed to penetrate the mucus and adhere to these cells are removed.

(iii) Urination flushes out any microbes present in the urethra and bladder.

(iv) Urine contains a chemical, urea, that can inhibit the growth of many microbes.

(v) Epithelial cells produce a variety of compounds known as antimicrobial peptides that can kill many microbes. These accumulate in the mucus overlying the epithelium and are also present in urine.

(vi) The surface of the urethra is covered in antibodies that prevent microbes from adhering to it.

Few microbes can overcome these defences, and consequently only a limited number of species are capable of causing UTIs – those that can do so are known as *uropathogens*.

How Common Is Acute Cystitis?

Acute cystitis is regarded as being the most common bacterial infection of humans. In the USA it results in more than ten million healthcare visits each year, and the associated costs have been estimated to be $2 billion per annum. Studies have shown that 11% of women in the USA suffer from it each year, and this includes only those who actually visit a physician – many sufferers don't. By the age of 24 years, it's been estimated that 33% of women will have experienced cystitis. In the UK, the number of visits to medical practitioners due to cystitis in 2007 was 0.7 million, and this resulted in prescriptions costing £3.6 million. A study carried out in England in 2014 found that 37% of females had suffered from cystitis in their lifetime and 28% had more than one episode. Overall, 11% reported one episode, and 6% more than one, during the previous year. On a global scale, data from 2001 showed that 150 million people suffer from cystitis each year and the direct healthcare cost associated with this was more than $6 billion.

Which Microbes Cause Cystitis?

In more than 80% of cases of uncomplicated cystitis, the microbe responsible for the infection is *Escherichia coli*.

Box 25.2 What's in a Name?

E. coli was first described by Theodor Escherich a German-Austrian paediatrician in 1886 – he called it *Bacterium coli commune* referring to the fact that it was a bacterium that was common in the colon of humans. However, the bacterium was given its current name in his honour in 1919.

Staphylococcus saprophyticus is a less-frequent cause of UTIs. The derivation of the genus name has already been described (Chap. 3). The species name comes from two Greek words – "sapros" which means "putrid" or "decaying" and "phyton" which means a "plant". Consequently *saprophyticus* means a plant that grows on decaying material, although we no longer regard bacteria as being plants.

Several species belonging to the genus *Proteus* also cause UTIs. The name is derived from the Greek god, Proteus, who was able to assume many different forms. Likewise, bacteria belonging to this genus can appear through the microscope as large bacilli, cocco-bacilli and filaments. They are also seen as single cells or in chains.

E. coli is a Gram-negative bacterium (Fig. 25.3) that lives in the gastrointestinal tract where it generally causes no problems. However, it can gain

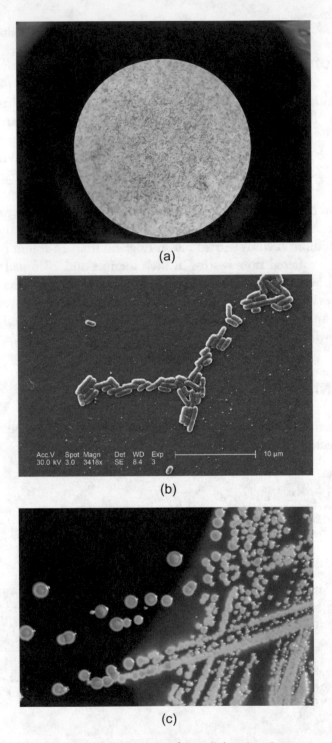

Fig. 25.3 Images of *E. coli*. (a) Gram stain of *E. coli* showing Gram-negative (i.e. red) bacilli (magnification ×100). (Sunil/CC BY (https://creativecommons.org/licenses/by/4.0)). (b) Digitally colourised electron microscopic image of a group of *E. coli* cells (magnification: ×3418). (Janice Haney Carr, Centers for Disease Control and Prevention, USA). (c) Colonies of *E. coli* growing on an agar plate. Each colony is approximately 2–3 mm in diameter. (Centers for Disease Control and Prevention, USA)

access to the urethra of healthy individuals and from there can ascend into the bladder which generally contains very few microbes. Here it gives rise to cystitis. This is known as "retrograde" or "ascending" infection and is the commonest route for microbes that cause UTIs. Sexual activity increases the risk of the disease because it can displace microbes from the vagina and perineum into the urethra. Many cases occur within 24 hours of sexual intercourse, and this has given rise to it being known as "honeymoon cystitis".

Another way in which microbes can infect the urinary tract is via the bloodstream. In this case, microbes that are causing an infection at another body site are carried in the bloodstream to the kidneys. This is known as a "descending" infection and occurs especially in those who are immunosuppressed.

Most (80%) of all UTIs occur in women, and the reasons for this are mainly anatomical. In women the distance from the anus to the urethra is approximately 5 cm, whereas in men it's 4–5 times greater – this means that it's far more difficult for microbes to move to the urethra from their normal habitat, the gastrointestinal tract. Another important factor is the length of the urethra. In women this is only approximately 4 cm, whereas in men it's approximately 5 times longer which is a much greater distance for such a small creature (*E. coli* is approximately 2 μm long) to traverse. This means that an *E. coli* cell has to travel a distance 20,000 times its length to reach the female bladder from the urethral opening but 100,000 times its length to reach the male bladder. *E. coli* swims in the mucus on the surface of the urethra using its whip-like flagella (Fig. 25.4) in order to reach the bladder.

What Happens During an Infection?

Once it has entered the bladder, *E. coli* penetrates the mucus that covers the surface of the epithelium and attaches itself to the underlying cells (Fig. 25.5) using specialised filaments called fimbriae or pili (Fig. 25.4).

After it's become attached, *E. coli* can invade the epithelial cells and this triggers inflammation and so our immune defence systems become activated. Unfortunately, however, the bacteria are now inside our own epithelial cells and so are protected from the phagocytes and antibodies we produce. The end result is that we suffer all of the adverse effects of bladder or kidney inflammation without actually killing the pathogen. We deal with this tricky situation by shedding the infected cells, and these are flushed away in our urine. In the remaining epithelial cells that aren't flushed away, the bacteria that re-emerge after these cells die are killed by our phagocytes with the help of antibodies. If you make a recovery eventually all of the bacteria are either killed or flushed away and the infection subsides. Administration of an appropriate antibiotic will, of course, help enormously and rapidly kills the infecting microbe.

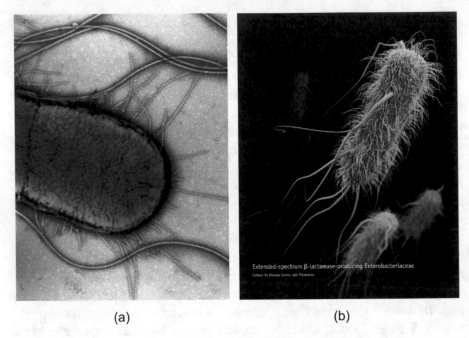

(a)   (b)

Fig. 25.4 Surface structures of *E. coli*. (a) *E. coli* as seen through an electron microscope. This shows the long, whip-like flagella that are used to propel the bacterium and the shorter, hair-like fimbriae (or pili) that are used to adhere to surfaces. (Credit: David Gregory & Debbie Marshall. CC BY 4.0). (b) Three-dimensional computer-generated image of a group of *E. coli* based on electron microscopic images. The image shows the flagella and fimbriae (pili) on the surface of the bacteria. (Alissa Eckert, Jennifer Oosthuizen and James Archer, Centers for Disease Control and Prevention, USA)

E. coli has a number of virulence factors in addition to its flagella and pili: (a) It produces several toxins that can kill epithelial cells so releasing their contents which are an important source of nutrients for them, (b) it has a capsule which protects it from our phagocytic defence cells, and (c) it produces proteins that can remove iron from our cells – iron is essential for the growth of all bacteria.

Although *E. coli* is responsible for most cases of uncomplicated cystitis, other bacteria that may cause the disease include *Staphylococcus saprophyticus* (responsible for 10–15% of cases) and species belonging to the genera *Klebsiella*, *Proteus* and *Enterobacter*.

In the case of complicated cystitis, *E. coli* is again the organism most likely to be involved but accounts for fewer than 40% of cases. A much wider range of microbes are responsible, and these include species belonging to the genera *Enterobacter*, *Proteus*, *Klebsiella*, *Citrobacter*, *Serratia*, *Pseudomonas*, *Enterococcus*, *Staphylococcus* and *Candida*. Complicated cystitis occurs frequently following a medical procedure that is often performed in hospitals – catheterisation (Box 25.3).

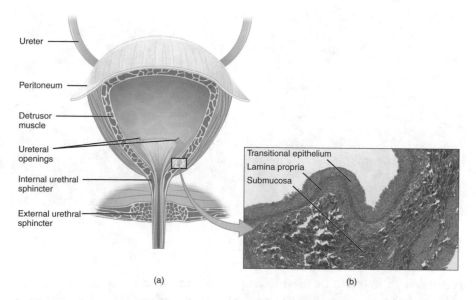

(a)   (b)

Fig. 25.5 Cross-section through the wall of the bladder showing the cells of the epithelium that line the inside of the bladder. This epithelium is known as a "transitional epithelium" because it consists of layers of epithelial cells that enable the bladder to expand (as it fills with urine) and contracts (as it empties). It is covered in a protective layer of mucus that prevents many microbes from adhering to it. (OpenStax College/CC BY (https://creativecommons.org/licenses/by/3.0))

Box 25.3 Catheterisation and Cystitis

Between 15% and 25% of hospitalised patients will have a urethral catheter inserted at some point during their stay in hospital, and more than 30 million bladder catheters are used annually in the USA.

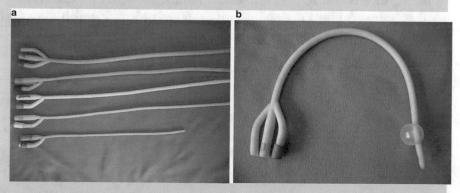

Urinary catheter. (a) Urinary catheters of different sizes. (Saltanat ebli/CC0). (b) Following insertion through the urethra and into the bladder, the catheter is kept in place by a small balloon which is inflated after being inserted. Urine collects in the bladder until its level reaches the opening in the top of the catheter. The urine then drains out through the opening and into a collecting bag. (Saltanat ebli/CC0)

Microbes in the urethra can adhere to the outer surface of the catheter where they grow and reproduce and this, ultimately, results in the formation of a biofilm (see Chap. 20) on the catheter surface. Microbes from this biofilm can then enter the bladder and cause an infection – this is known as a catheter-associated UTI (CAUTI). Almost all patients who are catheterised for longer than 1 week will experience a CAUTI.

CAUTIs have long been recognised as being the most common healthcare-associated infection. It has been estimated that approximately 500,000 CAUTIs occur each year in the USA and each of these incurs an additional cost (in 2007) of $749–1007. In the UK in 2016 there were 43,000–61,000 CAUTIs, and this resulted in 1300–1700 deaths. The economic burden associated with these infections amounted to £150–331 million.

How Is Cystitis Diagnosed?

A number of tests can be used to help determine whether or not you have a UTI, and these are carried out on a "mid-stream urine (MSU)" specimen. A MSU specimen is regarded as being representative of the urine that's actually present in the bladder, and this is what's needed for analysis. A MSU avoids collection of the first and last parts of the urine produced during urination. The first part of the urine stream will contain microbes from the urethra, and this isn't what's wanted for analysis and should be discarded.

The appearance of your urine is helpful in diagnosis because healthy urine is usually clear. Any cloudiness, the presence of a red colour (due to blood, rather than having just consumed beetroot!) or unusual smell could indicate a UTI. The urine sample is usually also tested using a simple dipstick test (Box 25.4) which can give rapid clues about the likelihood of a UTI. An advantage of the dipstick is that it gives immediate results. However, it's disadvantages are that it doesn't give a definite diagnosis, it can't identify which microbe is present and it doesn't suggest any treatment options.

Box 25.4 Urinalysis Using a Dipstick

A simple test that can be carried out on a urine sample involves dipping a thin plastic or paper strip coated in panels of certain chemicals that change colour if the urine contains any of a number of indicators of infection (or other problems). The main ones include:

(i)   The presence of protein (proteinuria) – implies a UTI or a malfunction of the kidneys.
(ii) An abnormal pH. The pH of urine is usually slightly acidic; a high pH (i.e. alkaline) is suggestive of a UTI.

(iii)   The presence of sugar – indicative of high blood sugar levels and possible diabetes.

(iv)   The presence of nitrites – suggestive of a UTI. Nitrites are produced from nitrates (normally present in urine) by some bacteria.

(v)   The presence of leukocyte esterase – suggestive of a UTI. Leukocyte esterase is an enzyme produced by white blood cells which are absent from healthy urine but are present during a UTI.

(vi)   The presence of blood. This could indicate a UTI or something more serious such as kidney disease, kidney stones, cancer of the kidney or bladder.

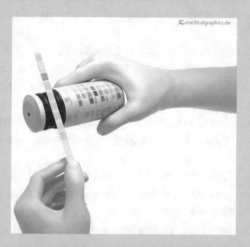

Using a urine test strip. The test strip has been dipped in the patient's urine, and the results are read by comparing the colour changes obtained with those shown printed on the container. (www.medicalgraphics.de CC BY-ND 3.0. The images shown in the section "free pictures" are available under the Creative Commons license "Attribution-No Derivative Works". The illustrations can be used free of charge also for commercial purposes)

A more useful test, however, involves microbiological analysis to answer the following questions: (i) Are there any microbes present? (ii) If there are, how many are there? (iii) Which microbe is it? (iv) To which antibiotics is it sensitive? This analysis is carried out in a microbiology laboratory and takes about 48 hours. The presence of a few microbes in a MSU doesn't necessarily mean that the person has a UTI – it could merely show that the sample wasn't a true MSU or has been contaminated in some way. It's generally agreed that if only 1 type of microbe is present and that there are more than 100,000 live bacteria per ml of urine, then there is a high likelihood of a UTI.

How Is Cystitis Treated?

Treatment of cystitis and other UTIs involves eliminating the infecting microbe. This can happen spontaneously and is helped by drinking plenty of fluids. Alternatively, a systemic antibiotic can be taken.

In the case of uncomplicated acute cystitis, the antibiotic of choice is usually nitrofurantoin for 3–5 days, trimethoprim/sulfamethoxazole for 3 days or trimethoprim alone. A useful alternative may be fosfomycin. However, as is the case with all antibiotic prescribing, alternative choices may be necessary depending on the resistance and sensitivity patterns of uropathogens in the local area.

In complicated cystitis or upper UTIs, where any of a broader range of uropathogens may be responsible for the infection, a broad-spectrum antibiotic such as a fluoroquinolone is usually the first choice antibiotic because this is effective against many types of bacteria. A cephalosporin (such as cefpodoxime) could be a suitable alternative in patients allergic to fluoroquinolones or where resistance to these agents is high. Possible adverse effects should, of course, always be considered.

In all types of UTI, therapy may have to be modified once the results of the microbiological analysis become available and the antibiotic susceptibility of the uropathogen has been determined.

If cystitis is left untreated, the uropathogen can make its way up towards the kidneys and cause pyelonephritis which is a much more serious condition and may require hospitalisation. In pregnant women cystitis should be treated promptly because delay or complications can result in poor pregnancy outcomes including pre-term delivery or miscarriage.

How Can I Avoid Getting Cystitis?

A few simple measures can help to prevent cystitis in women, and these include (i) prompt urination after sexual intercourse – this flushes out microbes that may have been introduced into the urethra, (ii) wiping the anus from front to back to avoid transferring uropathogens to the urethra, (iii) keeping well hydrated which produces plenty of urine to flush the renal tract and (iv) avoiding the use of spermicides as these have been found to increase the risk of UTIs.

Probiotics have been heavily promoted as a way of reducing UTIs, but there's little evidence to support their use. In contrast, there is some evidence of the efficacy of D-mannose (a sugar) in preventing and treating UTIs. D-mannose can bind to the pili of *E. coli* which prevents it from adhering to epithelial cells of the urinary tract. However, further robust clinical studies are needed to substantiate these initial promising findings. Components of cranberry juice have also been shown to block bacteria adhering to epithelial cells in laboratory studies, and some, but not all, clinical trials have found that it's moderately effective in preventing UTIs (however, see Box 25.5). In those who suffer from recurrent UTIs, taking an appropriate antibiotic to prevent an infection (antibiotic prophylaxis) may be necessary, and this could be taken daily or after sexual intercourse.

Box 25.5 Cranberries for the Prevention of Urinary Tract Infections

For many years there's been great interest in the possibility of using cranberry juice (*Vaccinium macrocarpon*) for preventing UTIs, particularly in women. Cranberries contain benzoic acid which is excreted in the urine as hippuric acid, and this can prevent the growth of some of the bacteria responsible for UTIs. However, it's been calculated that you'd need to consume more than 4 L of cranberry juice a day to produce sufficient hippuric acid to exert any antibacterial effect. Cranberries also contain proanthocyanidin, and in the laboratory this has been shown to prevent *E. coli* from sticking to the mucosa of the urinary tract. However, in 2012 an analysis of 24 clinical studies involving 4473 participants showed that cranberry products had no effect on the occurrence of UTIs. This result applied overall as well as to the various subgroups involved in the studies – women with recurrent UTIs, older people, pregnant women or children with recurrent UTIs.

Cranberries. (Courtesy of Cjboffoli published under CC BY 3.0)

Want to Know More?

American Family Physician. https://www.aafp.org/afp/2009/0315/p503.html

American Urological Association. https://www.urologyhealth.org/urologic-conditions/urinary-tract-infections-in-adults

Bladder and Bowel Community, UK. https://www.bladderandbowel.org/bladder/bladder-conditions-and-symptoms/bacterial-cystitis/

Bladder Health, UK. https://bladderhealthuk.org/bladder-conditions/cystitis/bacterial-cystitis

Centers for Disease Control and Prevention, USA. https://www.cdc.gov/antibiotic-use/community/for-patients/common-illnesses/uti.html

Department of Health, Western Australia. https://healthywa.wa.gov.au/Articles/A_E/Cystitis

Mayo Clinic, USA. https://www.mayoclinic.org/diseases-conditions/cystitis/symptoms-causes/syc-20371306

National Health Service, UK. https://www.nhs.uk/conditions/cystitis/

National Institute for Clinical Care and Excellence (NICE), UK. Urinary tract infection (lower) – women, 2019. https://cks.nice.org.uk/urinary-tract-infection-lower-women#!topicSummary

National Institute for Clinical Care and Excellence (NICE), UK. Urinary tract infection (lower) – men, 2018. https://cks.nice.org.uk/urinary-tract-infection-lower-men#!topicSummary

Patient Info, UK. https://patient.info/womens-health/lower-urinary-tract-symptoms-in-women-luts/cystitis-in-women

Abou Heidar NF, Degheili JA, Yacoubian AA, Khauli RB. Management of urinary tract infection in women: A practical approach for everyday practice. *Urology Annals*. 2019 Oct–Dec;11(4):339–346. https://doi.org/10.4103/UA.UA_104_19. PMID: 31649450

Benton TJ. Urinary tract infections in men. BMJ Best Practice, BMJ Publishing Group, 2019. https://bestpractice.bmj.com/topics/en-gb/76

Brusch JL. Urinary Tract Infection (UTI) and Cystitis (Bladder Infection) in Females. Medscape from WebMD, 2020a. https://emedicine.medscape.com/article/233101-overview

Brusch JL. Urinary Tract Infection (UTI) in males. Medscape from WebMD, 2020b. https://emedicine.medscape.com/article/231574-overview

Lala V, Minter DA. Acute Cystitis. Treasure Island (FL): StatPearls Publishing LLC; 2020. https://www.ncbi.nlm.nih.gov/books/NBK459322/

Lee UJ. Urinary tract infections in women. BMJ Best Practice, BMJ Publishing Group, 2019. https://bestpractice.bmj.com/topics/en-gb/77

Rowe TA, Juthani-Mehta M. Diagnosis and management of urinary tract infection in older adults.. *Infectious Disease Clinics of North America*. 2014 Mar;28(1):75–89. https://doi.org/10.1016/j.idc.2013.10.004. Epub 2013 Dec 8.

Storme O, Tirán Saucedo J, Garcia-Mora A, Dehesa-Dávila M, Naber KG. Risk factors and predisposing conditions for urinary tract infection. *Therapeutic Advances in Urology*. 2019 May 2;11:1756287218814382. https://doi.org/10.1177/1756287218814382. eCollection 2019 Jan-Dec. PMID: 31105772

26

Vaginitis

Abstract Infections are one of the causes of inflammation of the vagina (vaginitis); they are commonly due to bacteria (bacterial vaginosis), fungi (vaginal candidiasis) or protozoa (trichomoniasis). The main symptoms include irritation of the genital area, itching, inflammation of the labia and perineum, a vaginal discharge, pain during intercourse and discomfort during urination.

Bacterial vaginosis (BV) is the most commonly reported vaginal infection among women of childbearing age. However, approximately 40% of women with BV have no symptoms. BV increases the risk of acquiring sexually transmitted infections, and, if present in pregnancy, is associated with preterm birth, premature rupture of the membranes and a low birth weight neonate. The disease results from a disruption of the normal vaginal microbiota in which the proportion of lactobacilli decreases, while that of Gram-negative anaerobes increases. The condition responds to treatment with topical or oral antibiotics.

Vaginal candidiasis (VC) is the second most common type of infective vaginitis and occurs most frequently among young women of childbearing age. It's usually caused by *Candida albicans*, but other *Candida* species may be responsible. Oral or topical anti-fungal agents are effective in treating VC.

Trichomoniasis is caused by *Trichomonas vaginalis* which is a protozoan and is the most prevalent, non-viral, sexually transmitted infection although it's uncommon in the UK. Up to 70% of infected women have no symptoms. It increases the risk of acquiring other sexually transmitted infections and can result in infertility. If acquired during pregnancy, it can lead to pre-term birth, premature rupture of the membranes and a low birth weight neonate. It's treated with oral metronidazole or tinidazole.

What Is Vaginitis?

Vaginitis means inflammation affecting the vagina (Fig. 26.1); usually the vulva is also involved (Fig. 26.2). Confusingly, as well as being called vaginitis, the condition is known by a number of other names including vulvovaginitis, bacterial vaginosis, thrush (when the microbe responsible is a yeast), vulvitis and trichomoniasis (when the microbe responsible is a protozoan).

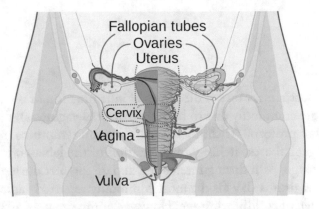

Fig. 26.1 Vaginitis is an inflammation of the vagina and has a variety of causes. (CDC, Mysid/Public domain)

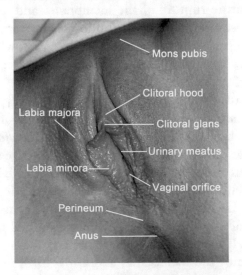

Fig. 26.2 The external female genitalia (the vulva). (Vagina039.jpg: Londoner500 derivative work: Lamilli/Public domain)

Vaginitis is frequently the result of a microbial infection, but the condition may also be caused by chemical or physical irritants or by a hormone deficiency (known as atrophic vaginitis). In order to infect the vagina, a microbe must be able to overcome the body's antimicrobial defence systems that operate there. The vaginal epithelium is covered with a layer of mucus which traps microbes and prevents them from reaching the underlying cells. It also contains high proportions of a range of antimicrobial compounds produced by epithelial cells, and these can kill many microbes that are trapped there. The outer cells of the epithelium are, of course, continually being shed, and this will remove any microbes that have managed to breach the mucus layer. The vaginal epithelium contains high levels of a polysaccharide, glycogen, which is continually being used as an energy source for the constituent cells. This produces high concentrations of acetic and lactic acids which make the vagina a very acidic region – it's pH is usually as low as 4 which means that it's 1000 times more acidic than pure water. Many microbes are unable to survive in such an acidic environment. Specific sensors (known as Toll-like receptors – see Chap. 1) in the epithelium can recognise pathogenic microbes, and when they do so, they release chemical messengers (cytokines) that trigger an immune response to deal with these unwanted organisms.

What Are the Symptoms?

Common symptoms of vaginitis include irritation of the genital area, itching, inflammation around the labia and perineal areas, a discharge from the vagina, pain during intercourse and discomfort during urination. However, many cases can be without any symptoms at all. It can affect girls and women of all ages and is the most common gynaecological condition seen in primary care, accounting for ten million medical practitioner visits per year in the USA. In the UK it results in 300,000 GP consultations each year.

Which Microbes Cause the Disease?

Interestingly, the condition can result from infection with a variety of different types of microbes – bacteria, fungi and protozoa. The most common causes of vaginitis are (i) bacteria (40–45% of cases) and the condition is then known as bacterial vaginosis, (ii) the yeast *Candida albicans* (20–25%) and the condition is known as thrush or vaginal candidiasis and (iii) a protozoan *Trichomonas vaginalis* (15–20% in the USA, but is uncommon in the UK) and the condition is known as trichomoniasis.

Box 26.1 What's in a Name?

The name of the yeast *Candida albicans* is rather odd because *Candida* comes from the Latin word "candidus", meaning white, while *albicans* is the present participle of the Latin word "albicō" which means becoming white – so the name means "white becoming white". The link with whiteness is because the fungus was first identified as being associated with oral candidiasis in which the tongue often becomes covered in a white coating. Although the fungus was discovered in 1839, it didn't receive its current name officially until 1954.

Trichomonas vaginalis was discovered in 1839 by the French bacteriologist Alfred François Donné. Its name is derived from the Greek word "thrix" which means hair (referring to the flagella that can be easily seen through a microscope) and the Greek "monas" which means a single unit – referring to the fact that it exists as a single cell. The species name, *vaginalis*, refers to its normal habitat – the vagina.

What Is Bacterial Vaginosis?

Bacterial vaginosis (BV), the most frequent infective form of vaginitis, is often accompanied by a white or greyish discharge with a fish-like odour and pain on urination. Neither soreness nor itching is common. However, as many as 40–50% of those affected don't have any symptoms.

How Is BV Diagnosed?

The disease can be diagnosed on the basis of the following indicators: (i) the vaginal fluid has a pH greater than 4.5 (Fig. 26.3), (ii) the detection of a fishy odour on adding 10% potassium hydroxide (KOH) to the fluid, (iii) the presence of vaginal epithelial cells covered with bacteria (known as "clue cells" – Fig. 26.4) and (iv) a homogeneous milky vaginal discharge. These are known as the Amsel criteria (after Dr. Richard Amsel who devised this method of diagnosis), and if three of these four indicators are present, the patient is likely to have BV.

Low-risk cases with characteristic symptoms that are unlikely to be attributable to other causes may be treated without having further investigations.

How Common Is BV?

BV is the most commonly reported vaginal infection among women of childbearing age. A number of risk factors have been identified, and women are

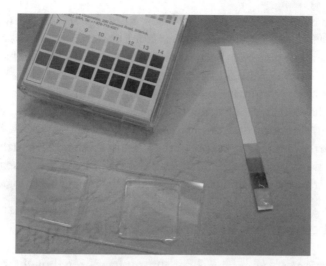

Fig. 26.3 In BV, the pH of the vaginal fluid or discharge is often greater than 4.5. The pH can be tested using indicator papers. In this case the pH of the fluid is between 8 and 9 and so is suggestive of a diagnosis of BV. (Mikael Häggström/CC0 via Wikimedia Commons)

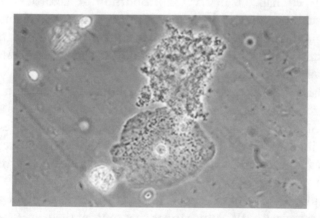

Fig. 26.4 Photomicrograph of a vaginal sample showing a normal vaginal epithelial cell (the lower of the two) and another (the upper one) with its exterior covered in bacteria giving it a roughened, stippled appearance known as a "clue cell". The presence of clue cells is one of the diagnostic criteria of bacterial vaginosis. (Image courtesy of M. Rein, Centers for Disease Control and Prevention, USA)

more likely to suffer from the disease if they (i) smoke, (ii) have a low socio-economic status, (iii) douche regularly, (iv) have a high number of sexual partners, (v) are commercial sex workers, (vi) frequently engage in vaginal intercourse, (vii) have recently used antibiotics, (viii) use an intrauterine contraceptive device and (ix) are of black ethnicity. In the USA, 21 million women (29%) aged 14–49 years suffer from bacterial vaginosis, and an estimated 7.4 million new cases occur each year. In the UK, studies have reported a prevalence ranging from 5 to 12%.

If you have BV, you may also develop any of a number of complications, even though you may not actually display any symptoms of the disease. Possible complications that can arise in pregnant patients with BV include pre-term birth, premature rupture of the membranes and low birth weight. In those who aren't pregnant, BV increases the risk of acquiring sexually transmitted infections (STIs) and complications following gynaecological surgery.

Which Microbes Are Responsible for BV?

Over the years a very wide range of organisms have been thought to be responsible for BV. However, after many studies it's become clear that no single bacterial species, or cluster of species, is present in all cases. It's now agreed that BV arises as a result of a disturbance (known as a dysbiosis) of the vaginal microbiota (see Chap. 1). In BV the proportion of lactobacilli is decreased (they are usually present in high proportions in the vaginal microbiota of healthy women), and there is an increase in the proportions of various Gram-negative bacteria such as *Gardnerella vaginalis*, *Prevotella* species, *Porphyromonas* species and *Mobiluncus* species (Fig. 26.5).

A diagnosis of BV can be made in the laboratory by examining a Gram stain of a sample of the discharge (Fig. 26.6): BV is suggested if there's a low proportion of large Gram-positive bacilli (i.e. *Lactobacillus* species) accompanied by an increase in the proportions of small Gram-negative bacilli (i.e. *Gardnerella*, *Porphyromonas*, *Prevotella*) and curved Gram-negative bacilli (i.e. *Mobiluncus*).

It's beyond the scope of this book to describe in detail all of the bacteria that are associated with BV. Suffice to say that most of them are Gram-negative anaerobic bacilli.

Fig. 26.5 (continued) and Prevention, USA). **(b)** Gram stain of a *Prevotella* species showing Gram-negative bacilli. (Dr. Holdeman, Centers for Disease Control and Prevention, USA). **(c)** Vaginal sample that has been stained to show the presence of *Mobiluncus* species – these appear as red-stained curved bacilli. (More than meets the eye: associations of vaginal bacteria with gram stain morphotypes using molecular phylogenetic analysis. Srinivasan S, Morgan MT, Liu C, Matsen FA, Hoffman NG, Fiedler TL, Agnew KJ, Marrazzo JM, Fredricks DN. *PLoS One*. 2013 Oct 24;8(10):e78633

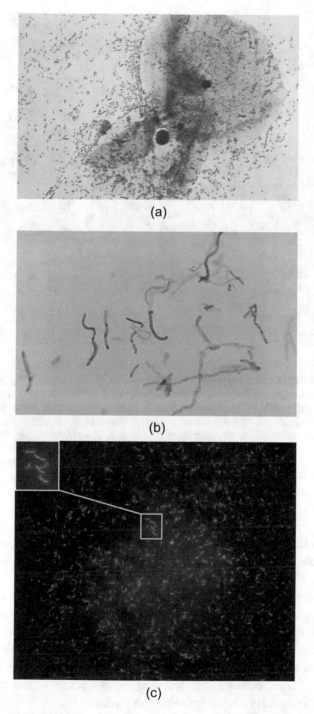

Fig. 26.5 Images of some of the bacteria responsible for bacterial vaginosis. (a) Gram stain showing *Gardnerella vaginalis* (small Gram-negative bacilli) and vaginal epithelial cells (considerably larger than the bacteria). (Joe Miller, Centers for Disease Control

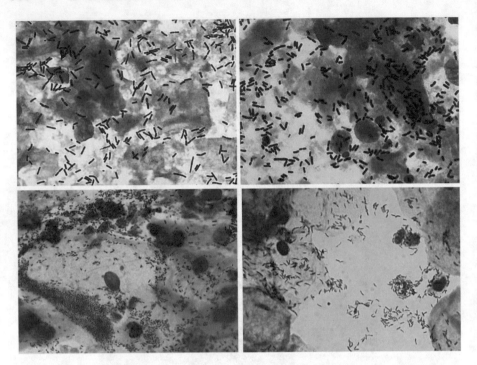

Fig. 26.6 Use of the Gram stain of vaginal samples in diagnosing bacterial vaginosis. The top two images show samples from two healthy individuals. They show the presence of large numbers of large Gram-positive bacilli (i.e. *Lactobacillus* species) but very few Gram-negative bacilli. The bottom two images show samples from two individuals with bacterial vaginosis. In both cases, no large Gram-positive bacilli can be seen, but large numbers of small Gram-negative bacilli (i.e. *Porphyromonas*, *Prevotella* and *Mobiluncus*) are present. (Comparison between Gram stain and culture for the characterization of vaginal microflora: definition of a distinct grade that resembles grade I microflora and revised categorization of grade I microflora. Verhelst, R. et al. *BMC Microbiology* (2005) 5, 61

Although the way in which disruption of the microbiota results in an infection is poorly understood, the microbes associated with BV collectively display a range of virulence factors that are likely to be involved in the disease process. These include (i) toxins that kill epithelial cells; (ii) enzymes that break down the mucus coating of the epithelium (this results in the characteristic vaginal discharge); (iii) enzymes that break down antibodies, thereby inhibiting the immune defence system; (iv) enzymes that break down tissue proteins such as collagen; and (v) compounds that induce inflammation. The fishy odour of the discharge that accompanies BV is due to the production of amines by the bacteria associated with the disease.

How Is BV Treated?

If you have BV but aren't pregnant and don't have any symptoms, then treatment isn't usually required, unless you are about to have some surgical procedure. If treatment is required, then either oral or topical (vaginal) metronidazole or clindamycin is the usual antibiotic used. Both of these are effective against anaerobic bacteria which are considered to be the main bacteria responsible. For persistent or recurrent infection, intravaginal metronidazole gel can be considered as a preventative measure, but advice should first be sought from a specialist as this may be an "off-label use" of the drug. If you are pregnant, then it's best to consult your obstetrician for advice.

How Can I Avoid Getting BV?

If you refrain from douching and from using irritants such as strong soaps or bubble baths, then you're less likely to get BV. Hormonal contraception, the use of male condoms and having a circumcised partner all reduce the risk of BV. Smoking and a copper IUD can increase rates of BV as does receiving oral sex.

BV is not an STI but being sexually active and recently changing your sexual partner increases the risk of developing the disease. It's probably correct to consider BV a sexually associated infection. The route of transmission isn't clear and there is no direct male equivalent. The exact trigger isn't known, but anything that reduces vaginal acidity will favour abnormal bacterial growth and result in a dysbiosis of the vaginal microbiota. The infection is more prevalent in lesbian couples than heterosexuals, but it has also been detected in virgins.

What Is Vaginal Candidiasis?

As the name implies, vaginal candidiasis (VC) is a vaginal infection with *Candida* species. In 80–90% of cases, the causative organism is *C. albicans*, but other species, particularly *C. glabrata*, may also be responsible. These microbes are present in the vagina of as many as 50% of healthy, symptom-free individuals. *C. albicans* is a fungus that can exist as ovoid yeast cells or can produce a long protuberance (known as a germ tube) which eventually grows to form a long filament (Fig. 26.7). The hyphae are able to penetrate epithelial

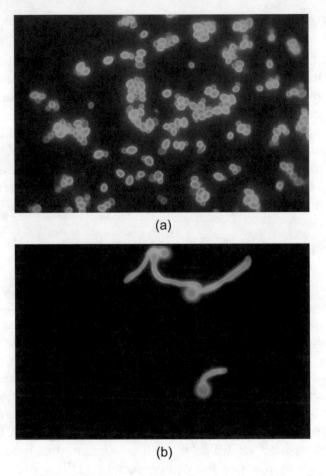

Fig. 26.7 Images of *Candida albicans*. (**a**) Yeast form of the organism. (Image courtesy of Maxine Jalbert and Dr. Leo Kaufman, Centers for Disease Control and Prevention, USA). (**b**) Hyphal form of the organism. (Image courtesy of Dr. Brian Harrington, Centers for Disease Control and Prevention, USA and the Mercy Hospital, Toledo, OH). (**c**) Gram stain of a vaginal smear showing *C. albicans*. Both the yeast form of the organisms and filamentous hyphae are visible. (Image courtesy of Dr. Stuart Brown, Centers for Disease Control and Prevention, USA). (**d**) Colonies growing on an agar plate. (Dr. William Kaplan, Centers for Disease Control and Prevention, USA)

cells and so damage the vaginal epithelium. The fungus produces enzymes that can break down vaginal tissues as well as the antibodies that are produced against it. It also induces inflammation and has proteins in its wall that prevent it from being engulfed by phagocytic cells.

(c)

(d)

Fig. 26.7 (continued)

What Are the Symptoms of VC?

Common symptoms of VC are genital itching or soreness, pain during sexual intercourse, pain or discomfort when urinating and a thick, white, odourless vaginal discharge with a "cottage cheese-like" appearance. VC is the most common type of vaginal infection after BV and occurs most frequently among young women of childbearing age.

How Common Is VC?

Approximately 75% of women will have at least one episode of vaginal candidiasis during their life, while 40%–45% will have two or more episodes. An estimated 1.4 million outpatient visits for vaginal candidiasis occur annually in the USA, while in the UK the number is approximately 333,000. Worldwide, recurrent vaginal candidiasis affects about 138 million women annually; 372 million women are affected by recurrent vulvovaginal candidiasis over their lifetime.

What Are the Risk Factors for VC?

A number of factors that predispose to VC have been identified, and these include pregnancy, chemical irritants and douching, diabetes mellitus, HIV or other immunocompromised states, increased oestrogen levels and the use of broad-spectrum antibiotics.

How Is VC Diagnosed?

Diagnosis in low-risk, mild cases is usually made on the basis of the symptoms without the need of investigations, but, if required, a sample of any vaginal discharge you are producing can be examined under the microscope to look for yeast and hyphal forms of the organism. The yeast form of *C. albicans* is found in the vagina of VC-free women, whereas both forms are present in those with VC. The pH of the discharge can also be checked – in VC the discharge has a pH of 4.5 or less.

Complications of VC

Women with VC, especially those with frequent recurrences, often experience loss of confidence and self-esteem. They may also be unable to carry on with their normal physical activities and have difficulties with their sexual life. VC is not an STI, although it can be triggered by sexual activity. Male partners can develop penile irritation, but this is rare.

Treatment failure is quite common, and around 5% of women suffer recurrent episodes of VC the cause of which is poorly understood. The inflamed areas can become secondarily infected by bacteria. Serious complications are very rare but can occur, mostly in debilitated and hospitalised patients. These include candidaemia where the yeast enters the bloodstream and causes widespread infection.

How Is VC Treated?

If you have VC, you should avoid chemical irritants such as soaps and perfumed products and wear loose-fitting non-synthetic clothes. Topical or oral anti-fungal agents such as clotrimazole or fluconazole can be used to treat and prevent the disease. Unfortunately, the optimal treatment for recurrent cases hasn't yet been defined, and you may have to try several different regimes. Evidence doesn't support the use of probiotics, oral or vaginal lactobacillus, a change in diet or tea tree oil. The use of probiotics to treat VC has been heavily promoted by a number of companies, but a recent meta-analysis of clinical studies has concluded that "there is insufficient evidence for the use of probiotics as adjuvants to conventional antifungal medicines or used alone for the treatment of VC in non-pregnant women".

How Can I Avoid Getting VC?

Any predisposing factors (listed above) should be managed or avoided including vaginal douching. The use of perfumed "feminine hygiene" products and close-fitting synthetic clothes should also be avoided. Vaginal dryness during sexual intercourse can trigger an episode, and some spermicidal jellies may increase susceptibility. Poorly controlled diabetic patients should try to improve their sugar levels and even non-diabetics with recurrent episodes tend to have high normal circulating sugar levels.

What Is Trichomoniasis?

Trichomoniasis is a disease of the genital tract caused by the protozoan *Trichomonas vaginalis*. As well as causing vaginitis in women, it can also infect men although usually this doesn't cause many symptoms. Occasionally, infected men experience pain or itching of the penis as well as pain during intercourse and urination.

The disease is approximately ten times more common in women than in men – approximately 5.3% of women are affected by trichomoniasis. An estimated 270 million cases of trichomoniasis occur each year worldwide. In the USA the annual number of cases is 3.7 million, while in the UK it is only approximately 6000. Women suffering from trichomoniasis usually produce a green, yellow or white, frothy and odorous vaginal discharge (Fig. 26.8).

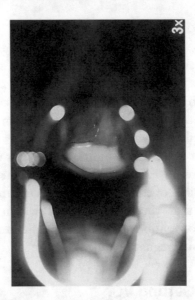

Fig. 26.8 A copious vaginal discharge is usually found in women suffering from tricho-moniasis. (Image courtesy of Public Health Image Library, USA)

Other symptoms include vaginal itching, pain during intercourse and pain during urination. However, over 70% of sufferers may have no symptoms.

A number of complications may arise in pregnant women with the disease including pre-term birth, premature rupture of the membranes and low birth weight. There is also a risk of maternal sepsis in the post-natal period. Men with trichomoniasis may suffer from prostatitis, increased risk of HIV, prostate cancer and infertility.

What Kind of Microbe Is *T. vaginalis*?

T. vaginalis is a parasite that colonises the urogenital tract of humans. It is unicellular (9 x 7 μm in size – much larger than bacteria), can exhibit a variety of shapes, has a characteristic undulating membrane and moves by means of five flagella (Fig. 26.9).

The microbe grows best in the absence of oxygen, but can survive in the presence of low concentrations of the gas, and it reproduces by splitting in two (binary fission) every 8–12 h. It can survive outside the body in urine, semen and swimming pool water, although it is primarily regarded as an STI and there are few reports of other routes of spread. On reaching the vaginal epithelium, it attaches to one of the cells and kills it by releasing a number of toxins. It also induces inflammation and produces enzymes that can degrade components of human tissues as well as antibodies. An important virulence

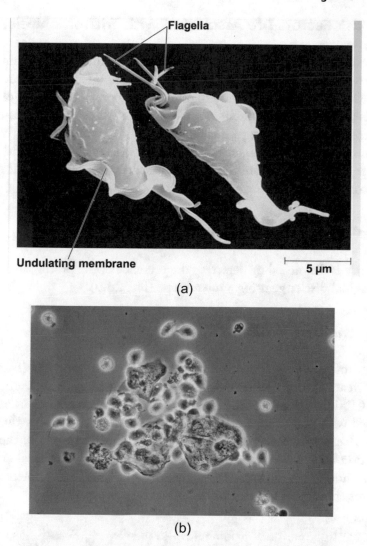

Fig. 26.9 *Trichomonas vaginalis*. (a) Electron micrograph of *T. vaginalis*. Fickleandfreckled This file is licensed under the Creative Commons Attribution 2.0 Generic license via Wikimedia Commons. (b) A sample of a vaginal discharge from a woman with trichomoniasis. A large number of *T. vaginalis* cells (generally pear-shaped with flagella) are visible as well as a small number of (larger) epithelial cells. (Joe Miller, Centers for Disease Control and Prevention, USA)

factor is its ability to phagocytose (engulf and destroy) a range of human cells including epithelial cells, white blood cells and erythrocytes. This provides nutrients for the microbe but, obviously, results in considerable damage to its human host and partially disables the antimicrobial defences that are being used against it.

What Risk Factors Are Associated with Trichomoniasis?

The disease can be transmitted during vaginal, anal and oral sex, and it's the most prevalent, non-viral STI. Risk factors for the infection include having multiple sexual partners, a history of other STIs, a previous episode of trichomoniasis and having sex without a condom. Infection can also be passed from a mother to her baby during a vaginal delivery.

In women, trichomoniasis increases the risk of cervical cancer and acquiring other STIs, furthermore the chronic inflammation of the genitourinary tract accompanying the disease may lead to infertility.

How Is Trichomoniasis Diagnosed?

The disease is diagnosed by detecting the presence of the protozoan in a sample of vaginal discharge using a microscope (Fig. 26.9).

How Is the Disease Treated?

A variety of treatments have been tried throughout the ages (Box 26.2). Ideally, treatment of trichomoniasis is best provided by a specialist Sexual Health Clinic. Both pregnant and non-pregnant women (as well as men) can be treated with oral metronidazole. All sexual partners, including those from up to 4 weeks before diagnosis, should also be treated to reduce the risk of re-infection and further spread in the community. The individual and these contacts should also be screened for other STIs. Pregnant women with no symptoms should have treatment as advised by their specialist.

Box 26.2 Treatment of Trichomoniasis in the Past

A variety of approaches have been used in the past to treat this disease. In the early twentieth century, compounds of arsenic were used and achieved some success. Despite their toxicity, some practitioners recommended the insertion into the vagina of as many as 70 tablets containing these compounds each week. Deaths from the use of such toxic compounds were reported well into the second half of the twentieth century. A number of other substances have also been tried and even ultraviolet irradiation. The success of the sulphonamides in the treatment of gonorrhoea prompted their use for trichomoniasis, but they were found to be ineffective. The introduction of metronidazole (Flagyl) in the late 1950s for treating the disease was a major breakthrough and continues to be highly effective.

Want to Know More?

American College of Gynecologists and Obstetricians. https://www.acog.org/Patients/FAQs/Vaginitis?IsMobileSet=false

American Family Physician. https://www.aafp.org/afp/2000/0901/p1095.html

British Association for Sexual Health and HIV. https://www.bashh.org/documents/4413.pdf, https://www.bashh.org/documents/UK%20national%20guideline%20on%20the%20management%20of%20TV%20%202014.pdf

Centers for Disease Control and Prevention, USA. https://www.cdc.gov/std/bv/default.htm

DermNet, New Zealand. https://www.dermnetnz.org/topics/vaginitis/

Mayo Clinic, USA. https://www.mayoclinic.org/diseases-conditions/vaginitis/symptoms-causes/syc-20354707

National Health Service, UK. https://www.nhs.uk/conditions/vaginitis/

National Institute for Clinical Care and Excellence (NICE), UK. Trichomoniasis, 2019. https://cks.nice.org.uk/trichomoniasis#!topicSummary

National Institute for Clinical Care and Excellence (NICE), UK. Bacterial vaginosis, 2018. https://cks.nice.org.uk/bacterial-vaginosis#!topicSummary

National Institute for Clinical Care and Excellence (NICE), UK. Candida – female genital, 2017. https://cks.nice.org.uk/candida-female-genital#!topicSummary

Patient Info, UK. https://patient.info/sexual-health/vaginal-discharge-female-discharge

Girerd PH. Bacterial Vaginosis. Medscape from WebMD, 2018. https://emedicine.medscape.com/article/254342-overview

Gor HB. Vaginitis. Medscape from WebMD, 2018. https://emedicine.medscape.com/article/257141-overview

Hay P. Bacterial vaginosis. F1000Research. 2017 Sep 27;6:1761. https://doi.org/10.12688/f1000research.11417.1. eCollection 2017. PMID: 29043070

Hildebrand JP, Kansagor AT. Vaginitis. Treasure Island (FL): StatPearls Publishing LLC; 2020. https://www.ncbi.nlm.nih.gov/books/NBK470302/

Illanes DS. Vaginitis. BMJ Best Practice, BMJ Publishing Group, 2019. https://bestpractice.bmj.com/topics/en-gb/75

Jeanmonod R, Jeanmonod D. Vaginal Candidiasis (Vulvovaginal Candidiasis). Treasure Island (FL): StatPearls Publishing LLC; 2020. https://www.ncbi.nlm.nih.gov/books/NBK459317/

Krapf JM. Vulvovaginitis. Medscape from WebMD, 2018. https://emedicine.medscape.com/article/2188931-overview#a1

Meites E. Trichomoniasis: the "neglected" sexually transmitted disease. Infectious Disease Clinics of North America. 2013 Dec;27(4):755–64. https://doi.org/10.1016/j.idc.2013.06.003. Epub 2013 Oct 25.

Nasioudis D, Linhares IM, Ledger WJ, Witkin SS. Bacterial vaginosis: a critical analysis of current knowledge. BJOG: An International Journal of Obstetrics and Gynaecology. 2017 Jan;124(1):61–69. https://doi.org/10.1111/1471-0528.14209. Epub 2016 Jul 11. PMID: 27396541

Paladine HL, Desai UA. Vaginitis: diagnosis and treatment. *American Family Physician.* 2018 Mar 1;97(5):321–329. PMID: 29671516

Peters BM, Yano J, Noverr MC, Fidel PL Jr. Candida vaginitis: when opportunism knocks, the host responds. *PLoS Pathogens.* 2014 Apr 3;10(4):e1003965. https://doi.org/10.1371/journal.ppat.1003965. eCollection 2014 Apr.

Schumann JA, Plasner S. Trichomoniasis. Treasure Island (FL): StatPearls Publishing LLC; 2019. https://www.ncbi.nlm.nih.gov/books/NBK534826/

Smith DS. Trichomoniasis. Medscape from WebMD, 2020. https://emedicine.medscape.com/article/230617-overview

27

Chlamydia

Abstract Infections of the reproductive system by *Chlamydia trachomatis* are among the most common sexually transmitted infections – those most at risk are young, sexually active adults. In women the infection usually occurs in the cervix resulting in a variety of symptoms, but up to 70% of those affected are asymptomatic. Lack of symptoms allows the infection to spread unnoticed, and untreated chlamydia infection can result in pelvic inflammatory disease which has serious consequences such as infertility and an increased risk of ectopic pregnancy. The infection is also associated with an increased risk of spontaneous abortion, stillbirth, preterm delivery and cervical cancer.

In men the infection usually results in urethritis, but up to 90% don't experience any symptoms. If untreated, it can progress to cause epididymitis or epididymo-orchitis and can result in infertility.

The disease is transmitted by sexual contact with the penis, vagina, mouth or anus of an infected partner. It can also be passed on from an infected mother to her baby primarily at delivery. Oral antibiotics are used to treat those infected and their sexual contacts. The risk of acquiring the infection can be reduced by using condoms, limiting the number of sex partners, undergoing regular screening and avoiding douching.

What Is Chlamydia?

Infections of the genital tract by *Chlamydia trachomatis* are among the most frequently encountered sexually transmitted infections (STIs) worldwide – such an infection is often known simply as "chlamydia". In women the infection usually occurs in the passage through the cervix – the endocervical canal – while in men the urethra is the main site affected (Fig. 27.1). Infected individuals very often experience no symptoms and so the disease is often referred to as being a "silent" infection.

What Are the Main Symptoms?

Only approximately 30% of women infected with chlamydia develop symptoms, and these include pain when urinating, vaginal discharge, pain in the abdomen or pelvic area, pain during intercourse, bleeding after sex and bleeding between periods. The lack of symptoms in most women is a great worry because if the infection isn't treated promptly the microbe can spread further into the womb, fallopian tubes or pelvis. If this occurs, you may experience fever, chills, muscle pain, nausea, vomiting and pelvic or abdominal pain. Spread of the microbe to these upper regions of the reproductive system can result in pelvic inflammatory disease (PID) which has serious consequences including reduced fertility or infertility, persistent pelvic pain and an increased risk of ectopic pregnancy (where a fertilised egg implants itself outside the womb).

The incubation period for the disease in men is variable but is typically 5–10 days following exposure. As many as 90% of men who are infected are asymptomatic. If you do have symptoms, then the most likely ones you'll experience will be pain when you urinate, a white, cloudy or watery discharge from the tip of your penis, burning or itching in your urethra and pain in your testicles. It's the second most common cause of urethritis (after gonorrhoea) in men. If left untreated, the disease can progress to cause painful inflammation of the epididymis (epididymitis) and, sometimes, of the testicles as well (epididymo-orchitis). This can result in infertility. *Chlam. trachomatis* is one of the most frequent pathogens in epididymitis among sexually active men <35 years of age.

Fig. 27.1 (continued) staff (2014). Medical gallery of Blausen Medical 2014;. WikiJournal of Medicine 1 (2). https://doi.org/10.15347/wjm/2014.010. ISSN 2002-4436. / CC BY (https://creativecommons.org/licenses/by/3.0)). **(b)** Male reproductive system showing the main sites affected by chlamydia – the urethra, epididymis and testicle. (NIH Medical Arts, National Cancer Institute, USA. Public Domain)

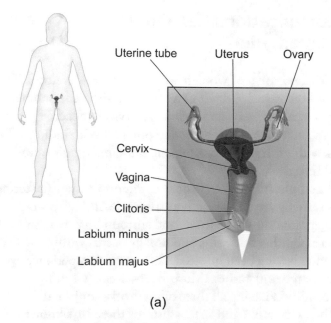

(a)

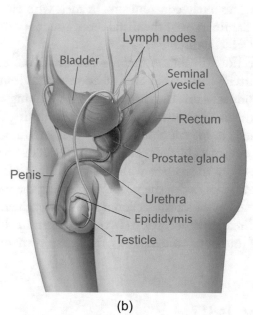

(b)

Fig. 27.1 Diagrams of the male and female reproductive systems. **(a)** Diagram of the female reproductive system. The main site affected by chlamydia is the endocervical canal – the hollow passage through the cervix that opens into the vagina. (BruceBlaus. When using this image in external sources, it can be cited as:Blausen.com

What Complications Can Occur If I Get Chlamydia?

Between 10% and 40% of women with chlamydia develop PID. The infection has also been associated with an increased risk of spontaneous abortion, stillbirth, preterm delivery and possibly cervical cancer. People with chlamydia are also at a higher risk of having other STIs including HIV.

Oral sex with an infected person can cause pharyngitis, and anal sex can cause rectal infections.

Another complication that is more common in men (affecting 3–8% of patients) is sexually acquired reactive arthritis (SARA). This is a painful, non-infective inflammation of the joints and surrounding structures, triggered by an infection at another body site, and usually occurs within the first few weeks or months after having been infected. You may also experience eye symptoms, urethral discharge and rashes. More rarely, Reiter's syndrome can develop, which is a condition having all three of urethritis, conjunctivitis and a reactive arthritis. The exact cause and mechanism of these isn't known, but infection and immune factors are likely contributors. In men who engage in receptive anal intercourse, inflammation of the distal rectal mucosa can occur and may be accompanied by a rectal discharge.

Non-genital infections in adults include conjunctivitis and pneumonia. Trachoma is a recurrent chlamydial eye infection which leads to scarring of the eye lids and blindness– this is mostly seen in developing countries.

The organism can also be transmitted to a newborn during delivery, and this most commonly results in infection of the baby's eyes and lungs. Conjunctivitis that develops 5–12 days after birth is the most common of these and occurs in 18–44% of neonates born to mothers with chlamydia. If this is left untreated, it can result in blindness. The risk of the neonate acquiring the infection is greatly reduced if it is delivered by caesarean section rather than vaginally.

Severe and life-threatening infections have been seen in the form of sepsis and myocarditis. If untreated, the disease can persist in many women for up to 4 years.

How Common Is It?

Approximately 4.2% of women and 2.7% of men are affected by chlamydia worldwide, and in 2016 about 127 million new cases occurred globally. Those most at risk are young, sexually active adults, and approximately 75% of cases occur in those under 25 years of age. In young people between 15 and 24 years, infection rates are 1.5–10%. More than 200,000 new cases were diagnosed in England in 2017, and these accounted for 48% of all new STIs. Approximately two thirds of those affected were women. In the USA about 1.7 million cases were reported in 2017 (528.8 cases per 100,000 population), and approximately two thirds of these were in women.

What Risk Factors Are Associated with the Disease?

The disease is transmitted by unprotected sexual contact with the penis, vagina, mouth or anus of an infected partner. It can also be passed on from an infected mother to her baby by contact with infected discharge. Risk factors for chlamydia in both men and women include being sexually active before the age of 25 years, multiple sex partners within the preceding year, not using a condom consistently, a history of prior STIs and low socio-economic status.

What Microbe Causes Chlamydia?

The organism responsible for chlamydia is *Chlamydia trachomatis*. This is a Gram-negative bacterium that can only grow and reproduce inside a human cell and is therefore known as an "obligate intracellular parasite". It has a complex life cycle (Fig. 27.2) which involves two forms of the organism – an elementary body (outside human cells) and a reticulate body (inside cells).

The elementary body (EB) has the appearance of a Gram-negative bacterium but is much smaller, being only 200–400 nm in diameter. This form can survive outside the human body (but can't grow or reproduce there) and can initiate an infection once it comes into contact with a susceptible host cell such as an epithelial cell of the human reproductive system. Once inside the epithelial cell, the EB is converted within about 2 hours into a reticulate body (RB) which is much larger (600–1500 nm in diameter). The RB uses nutrients from the host cell to grow and reproduce, and this results in the formation of a large structure known as a "cytoplasmic inclusion" or "inclusion body" (Fig. 27.3). Once the supply of nutrients has been exhausted, the RBs within the inclusion are transformed into a large number of EBs. The cell then bursts releasing large numbers (more than 500) of infectious EBs which can then infect other epithelial cells. The whole life cycle lasts 2–3 days.

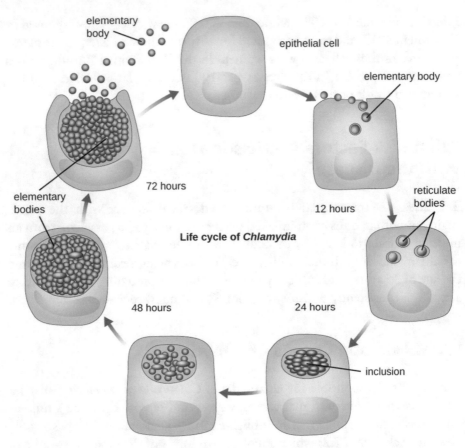

Fig. 27.2 The life cycle of *Chlam. trachomatis*. (CNX OpenStax [CC BY 4.0 (https://creativecommons.org/licenses/by/4.0)])

Fig. 27.3 (continued) Bergamini C, Fato R, Cavallini C, Donati M, Nardini P, Foschi C, Cevenini R. *BMC Res Notes.* 2014 Apr 11;7:230. (**c**). An inclusion body of *Chlam. trachomatis* inside an epithelial cell as viewed through an electron microscope. A number of RBs can be seen in the spherical inclusion body below the nucleus of the cell. Bar = 4 μm. (Skilton RJ, Cutcliffe LT, Barlow D, Wang Y, Salim O, et al. (2009) Penicillin Induced Persistence in Chlamydia trachomatis: High Quality Time Lapse Video Analysis of the Developmental Cycle. PLoS ONE 4(11): e7723. https://doi.org/10.1371/journal.pone.0007723. Copyright:ß2009 Skilton et al.

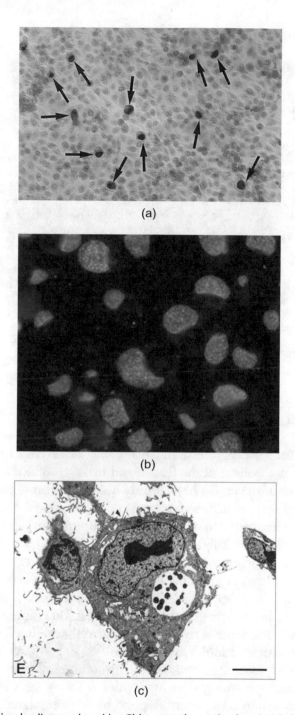

Fig. 27.3 Inclusion bodies produced by *Chlam. trachomatis* when inside human cells. **(a)** Human cells some of which have inclusion bodies of *Chlam. trachomatis* – these are darker staining and are arrowed (magnification ×50). (Dr. E. Arum and Dr. N. Jacobs, Centers for Disease Control and Prevention, USA). **(b)** Inclusion bodies of *Chlam. trachomatis* stained green within human cells (stained red). (Infection of human monocytes by *Chlamydia pneumoniae* and *Chlamydia trachomatis*: an in vitro comparative study. Marangoni A,

Box 27.1 What's in a Name?

The name of the genus, *Chlamydia*, is derived from the Greek word "khlamu-dion" which means a small cloak. The species name comes from the Greek word "trakhōma" which means roughness. The organism was discovered in 1907 by Ludwig Halberstaedter and Stanislaus von Prowazek who observed small granules (known as "inclusions") inside the cells present in scrapings they had taken from the conjunctiva of patients suffering from an eye infection known as trachoma. They concluded that this must be the microbe responsible for the disease, and because these inclusions were draped, like a cloak, around the nucleus of the infected cell, they proposed the name *Chlamydozoa* from the Greek word for a cloak. The microbe was first thought to be a protozoan, then a virus, and then it was shown to be a rather odd type of bacterium – one that could only live within another cell. Such organisms are termed "obligate intracellular parasites". Trachoma is a serious eye disease caused by *Chlam. trachomatis* and is responsible for 1.9 million cases of blindness worldwide. Characteristically it causes roughening of the inside of the eyelid – hence its name from the Greek word for roughness. It affects more than 80 million people mainly in Africa, Asia and Central and South America. *Chlam. trachomatis* therefore derives its name from the effects it has on human cells and one of the diseases that it causes.

What Happens During an Infection?

In order to cause a genital infection, *Chlam. trachomatis* has to evade the antimicrobial defence systems of the female and male reproductive systems. The defence systems of the vagina have already been described (Chap. 26). In the cervix an additional defence mechanism is the presence of a mucociliary escalator similar to that present in some regions of the respiratory system (described in Chap. 1). This continually sweeps incoming microbes back into the vagina and away from the uterus.

In men, the urethra is protected in the same way as described for the urethra of women (Chap. 25). However, the urethra is also the passageway for semen as well as urine. Semen has high levels of a wide variety of antimicrobial compounds, and some of these will remain in the urethra following ejaculation and will provide additional protection.

Once *Chlam. trachomatis* reaches the genitourinary system of men or women, the pathogen-recognition molecules (TLRs – see Chap. 1) on the surfaces of the various cells present respond by releasing signalling molecules that start an inflammatory response which attracts cells of the immune defences to the site of the infection. The first stage of the infectious process is the attachment of the EB to an epithelial cell. This is achieved by molecules on its surface known as adhesins. These are important virulence factors that bind to molecules (known as receptors) on the surface of epithelial cells and induce the uptake of the EB by the cell.

Many human cells respond to microbial invasion by committing suicide – this process is known as apoptosis. Apoptosis is an effective means of ensuring that the host cell is not going to provide a safe environment for the invader or, in the case of viruses and intracellular parasites like *Chlam. trachomatis*, be used to produce more of the microbial invader. *Chlam. trachomatis*, however, can inhibit apoptosis, and this is an important virulence factor as it ensures its survival and reproduction. Living inside a human cell also enables the microbe to evade many of the defence systems – phagocytes, antibodies etc. – that are trying to destroy it.

How Is Chlamydia Diagnosed?

The infection can't be diagnosed by the techniques usually used for bacterial infections, i.e. by growing the organism from a sample taken from the infected site. This is because *Chlam. trachomatis* can't easily be grown in the laboratory – it does not grow on agar plates like the majority of bacteria. However, it can be grown on human cells in the laboratory – a technique known as "cell culture". But this is expensive, difficult to perform and slow.

The most widely used diagnostic approach is based on detecting the presence of DNA from *Chlam. trachomatis* in an appropriate sample. Such tests are very rapid and have been described in Chap. 1. Suitable samples for analysis from both men and women are swabs from the affected area (endocervix or vagina for women, urethra for men) or a first-catch urine. A first-catch urine sample is one taken immediately after urination has begun as this is most likely to contain microbes from the urethra where *Chlam. trachomatis* is most likely to be found. Oral, eye or anorectal swabs may also be examined.

How Is Chlamydia Treated?

Antibiotics are an effective treatment for the infection and these should be started as soon as possible – sometimes even prior to knowing any test results, if other indications suggest a high likelihood of the disease. Delaying treatment increases the risk of serious complications such as infertility. Ideally you should seek treatment at a specialist STI clinic and be screened for other STI infections (Box 27.2). Such clinics will also give advice and support with regard to tracing sexual contacts as well as how to notify your partner.

The antibiotics of first choice for uncomplicated infections are usually azithromycin or doxycycline. Alternatively, erythromycin, ofloxacin or levofloxacin may be used. Treatment in pregnancy should be guided by a specialist. All sexual contacts, within the last 60 days, of those infected should be advised to seek full STI testing and treatment.

Box 27.2 Screening for Sexually Transmitted Infections

In the USA, the Centers for Disease Control and Prevention has made the following recommendations regarding testing for STIs:

- *All adults and adolescents from ages 13 to 64* should be tested at least once for HIV.
- *All sexually active women* younger than 25 years should be tested for gonorrhoea and chlamydia every year. Women 25 years and older with risk factors such as new or multiple sex partners or a sex partner who has an STD should also be tested for gonorrhoea and chlamydia every year.
- *All pregnant women* should be tested for syphilis, HIV and hepatitis B starting early in pregnancy. At-risk pregnant women should also be tested for chlamydia and gonorrhoea starting early in pregnancy. Testing should be repeated as needed to protect the health of mothers and their infants.
- *All sexually active gay and bisexual men* should be tested at least once a year for syphilis, chlamydia and gonorrhoea. Those who have multiple or anonymous partners should be tested more frequently for STDs (i.e. at 3- to 6-month intervals).
- *Sexually active gay and bisexual men* may benefit from more frequent HIV testing (e.g. every 3–6 months).
- *Anyone who has unsafe sex or shares injection drug equipment* should get tested for HIV at least once a year.

How Can I Avoid Getting Chlamydia?

The most effective way of avoiding the disease is to refrain from any sexual activity with other people. However, if you do intend to engage in sexual activities with someone else then you should (i) use a condom (Box 27.3), (ii) limit the number of sexual partners you have, (iii) undergo regular screening for the disease and (iv) avoid douching, if you are a woman, because this reduces the vaginal microbiota and so decreases its effectiveness as a barrier to infection.

Chlamydia is such an important health concern that many countries have introduced screening programmes to try to identify people with the infection as early as possible. In England, for example, any sexually active person under the age of 25 is encouraged to have a test every year and when they change their partner. Chlamydia home testing kits are simple and often only involve collecting a urine sample or self-taken swab which is then posted to the laboratory.

Box 27.3 Consistent and Correct Condom Use: Guidance from the Centers for Disease Control and Prevention, USA

The level of protection provided by condoms depends on the particular STI as these vary in the way they are transmitted. Male condoms may not cover all infected areas or areas that could become infected. Consequently, they provide greater protection against those STIs that are transmitted only by genital fluids such as chlamydia, gonorrhoea, trichomoniasis and HIV infection. However, they are less effective against those STIs that are transmitted primarily by skin-to-skin contact because the condom may not cover all of the infected area – these include genital herpes, human papillomavirus [HPV] infection, syphilis and chancroid.

In order to achieve maximum protection from the use of condoms, it's essential to use them consistently and correctly. The failure of condoms to protect against transmission of a STI is usually the result of inconsistent or incorrect use, rather than product failure. Incorrect condom use diminishes their protective effect because it can result in condom breakage, slippage or leakage. It's important to use a condom throughout the entire sex act, from the start of sexual contact until after ejaculation.

How to Use a Condom Consistently and Correctly

- Use a new condom for every act of vaginal, anal and oral sex throughout the entire sex act. Before any genital contact, put the condom on the tip of the erect penis with the rolled side out.
- If the condom doesn't have a reservoir tip, pinch the tip enough to leave a half-inch space for semen to collect. While holding the tip, unroll the condom all the way to the base of the erect penis.
- After ejaculation and before the penis gets soft, grip the rim of the condom and carefully withdraw. Then gently pull the condom off the penis, making sure that semen doesn't spill out.
- Wrap the condom in a tissue, and throw it in the trash where others won't handle it.
- If you feel the condom break at any point during sexual activity, stop immediately, withdraw, remove the broken condom and put on a new condom.
- Ensure that adequate lubrication is used during vaginal and anal sex – this might require water-based lubricants. Oil-based lubricants (such as petroleum jelly, mineral oil, massage oils, body lotions and cooking oil) shouldn't be used because they can weaken latex so causing breakage.

Want to Know More?

Centers for Disease Control and Prevention, USA. https://www.cdc.gov/std/chlamydia/stdfact-chlamydia.htm

European Centre for Disease Control and Prevention. https://www.ecdc.europa.eu/sites/portal/files/media/en/publications/Publications/chlamydia-control-europe-guidance.pdf

Family Planning Association, UK. https://www.sexwise.fpa.org.uk/stis/chlamydia

Healthtalk, UK. http://www.healthtalk.org/young-peoples-experiences/sexual-health/chlamydia

Mayo Clinic, USA. https://www.mayoclinic.org/diseases-conditions/chlamydia/diagnosis-treatment/drc-20355355

National Health Service, UK. https://www.nhs.uk/conditions/chlamydia/

National Institute for Clinical Care and Excellence (NICE), UK. Chlamydia – uncomplicated genital, 2019. https://cks.nice.org.uk/chlamydia-uncomplicated-genital#!topicSummary

Patient Info, UK. https://patient.info/sexual-health/sexually-transmitted-infections-leaflet/chlamydia

World Health Organisation. https://apps.who.int/iris/bitstream/handle/10665/246165/9789241549714-eng.pdf;jsessionid=2DABAEC1DFFB87CAF0A86BBEE599B1AC?sequence=1

Gitsels A, Sanders N, Vanrompay D. Chlamydial infection from outside to inside. *Frontiers in Microbiology*. 2019 Oct 9;10:2329. https://doi.org/10.3389/fmicb.2019.02329. eCollection 2019.

Mohseni M, Sung S, Takov V. Chlamydia. Treasure Island (FL): StatPearls Publishing LLC; 2020. https://www.ncbi.nlm.nih.gov/books/NBK537286/

O'Connell CM, Ferone ME. *Chlamydia trachomatis* genital infections. *Microbial Cell*. 2016 Sep 5;3(9):390-403. doi: https://doi.org/10.15698/mic2016.09.525.

Qureshi S. Chlamydia (Chlamydial Genitourinary Infections). Medscape from WebMD, 2018. https://emedicine.medscape.com/article/214823-overview

Rompalo A. Genital tract chlamydia infection. BMJ Best Practice, BMJ Publishing Group, 2019. https://bestpractice.bmj.com/topics/en-gb/52

28

Gonorrhoea

Abstract Gonorrhoea is a common sexually transmitted infection that affects mainly the cervix in women and the urethra in men. However, other sites may be involved including the pharynx and rectum. In women the main symptom is a vaginal discharge, but many patients are asymptomatic and the infection can spread to the upper regions of the reproductive tract resulting in pelvic inflammatory disease. Most men with the disease have a urethral discharge – the disease may spread to neighbouring tissues resulting in epididymitis and prostatitis. The disease is caused by the bacterium *Neisseria gonorrhoeae* and is transmitted via infected secretions – these can come from any infected site (penis, vagina, pharynx, rectum) and be transferred to any mucosal surface. Newborn babies can acquire the disease from their mother during a vaginal delivery, and this can result in conjunctivitis, arthritis and meningitis. In both men and women, the bacterium can occasionally spread to other parts of the body resulting in a skin rash, joint pains and fever. Treatment involves using systemic antibiotics, but *N. gonorrhoeae* is now resistant to many of these, and so a combination of two antibiotics is generally necessary. Treatment of sexual contacts of infected patients is also important to reduce the spread of the disease. The risk of getting gonorrhoea can be reduced by having fewer sexual partners and using condoms during penetrative sex.

© Springer Nature Switzerland AG 2021
M. Wilson, P. J. K. Wilson, *Close Encounters of the Microbial Kind*,
https://doi.org/10.1007/978-3-030-56978-5_28

What Is Gonorrhoea?

Gonorrhoea is a common sexually transmitted infection (STI) caused by the bacterium *Neisseria gonorrhoeae*. Infected men usually have a urethral discharge whereas about half of infected women have no symptoms. The main site of infection is the genitourinary tract, but it can also affect the rectum or pharynx in those who participate in anal or oral sex, respectively.

> **Box 28.1 What's in a Name?**
>
> The disease gonorrhoea was named by Galen, a Greek physician living in the second century CE, who thought that the purulent discharge from the urethra was semen. The name comes from the Greek words "gonos" and "rhoe" which mean "seed" and "flow", respectively. The organism responsible for the disease, Neisseria gonorrhoeae, was discovered in 1878 by the German physician Albert Ludwig Sigesmund Neisser who saw the organism in samples of pus from patients suffering from the disease. Albert Neisser called the organism he saw "Micrococcus der gonorrhoe". It was first grown in the laboratory in 1882 by the German bacteriologists Leistikow and Loeffler. The genus name of the organism, Neisseria, was adopted in 1885 in honour of Neisser. The species name, gonorrhoea, is derived from the disease it causes.

The main symptoms of the various forms of the disease in men and women are summarised in Table 28.1.

The most common site of infection in women is the cervix (around 80–90%), followed by the urethra (80%), rectum (40%) and pharynx (10–20%). If symptoms develop, they often appear within 10 days of infection. Infection of the cervix results in inflammation (known as

Table 28.1 Symptoms of gonorrhoea

Site of infection	Symptoms and approximate frequency of their occurrence	
	Men	Women
Urethra	• Discharge (>80%). • Dysuria (>50%). • Asymptomatic (<10%).	Dysuria (10–15%)
Rectum	• Usually asymptomatic. • Anal discharge (12%). • Perianal/anal pain, pruritus or bleeding (7%).	Usually asymptomatic
Pharynx	Usually asymptomatic (>90%)	Usually asymptomatic (>90%)
Endocervix	–	• Frequently asymptomatic (up to 50%); • Increased or altered vaginal discharge (up to 50%). • Lower abdominal pain (up to 25%).

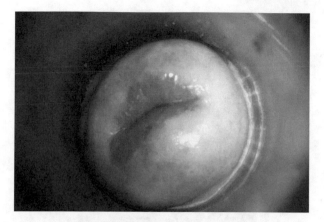

Fig. 28.1 Cervix of a woman with gonorrhoea showing the inflammation that is characteristic of the disease. (Image courtesy of Centers for Disease Control and Prevention, USA)

"endo-cervicitis" – Fig. 28.1), and in some cases it can spread to the upper regions of the genital tract and affect the uterus, fallopian tubes and ovaries – this is known as pelvic inflammatory disease. The most common symptom of gonorrhoea in women is a vaginal discharge which is usually described as thin, purulent, and mildly odorous. Patients may also experience pain during sexual intercourse or abnormal vaginal bleeding.

In men the main site of infection is the urethra, and this results in urethritis. The initial symptom is a burning feeling during urination and a watery discharge. A few days later, the discharge usually becomes more profuse and purulent (i.e. contains pus – Fig. 28.2), and, occasionally, this may contain blood. Some patients also experience inflammation of the epididymis (the coiled tube at the back of the testicle) which can be painful and results in a red, swollen scrotum.

The disease can be acquired during any form of sexual contact that enables the transfer of infected secretions from one mucous membrane to another. This may occur during any penetrative sex that involves a mucosa-lined orifice, i.e. the vagina, anus or throat. The probability of transmission for a single unprotected heterosexual contact is around 58% for male-to-female and 23% for female-to-male transmission. The incubation period is usually between 2 and 5 days but may be as long as 10 days.

Neonates can acquire the disease usually following exposure to infected cervical discharge during birth. Symptoms usually develop 2–5 days after birth and include conjunctivitis (known as "ophthalmia neonatorum" – Fig. 28.3), arthritis and meningitis.

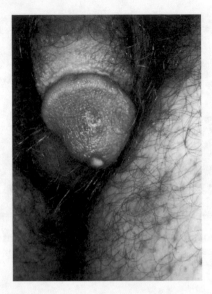

Fig. 28.2 Purulent discharge from the penis of a patient with gonorrhoea. (Dr. Gavin Hart, Centers for Disease Control and Prevention, USA)

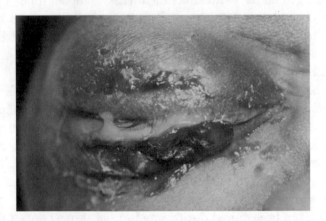

Fig. 28.3 The eye of an infant suffering from ophthalmia neonatorum. (J. Pledger, Centers for Disease Control and Prevention, USA)

What Are the Risks of Getting Gonorrhoea?

The main risk factors for acquiring the disease are:

- Young age.
- History of previous STIs.
- Co-existent STIs.

- New or multiple sexual partners.
- Recent sexual activity abroad.
- Certain sexual activities such as anal intercourse and frequent oral sex.
- Inconsistent condom use.
- History of drug use or commercial sex work.
- Male same-sex intercourse.

Are There Any Complications?

A variety of complications can arise from the disease, and some of these are not uncommon. In women, pelvic inflammatory disease (PID) can develop in up to one third of patients. This can result in chronic pelvic pain (around 40% of those with PID), tubal infertility (around 10% of those with PID) and ectopic pregnancy (around 10% of those with PID). Other, less frequent, complications in women include abscesses, premature labour and miscarriage.

In men, the disease may spread to various neighbouring tissues resulting in acute epididymitis, prostatitis, seminal vesiculitis, penile lymphangitis, peri-urethral abscess and infection of Tyson's and Cowper's glands. It can result in infertility.

In both men and women, the bacterium can occasionally (around 1–3% of cases) spread to other parts of the body resulting in more widespread infections including (i) skin lesions, (ii) joint pain, arthritis and tenosynovitis, (iii) meningitis and (iv) endocarditis or myocarditis. The most common features of such disseminated gonococcal infections (DGIs), which can be fatal, include skin rash (around 75% of cases – Fig. 28.4), tenosynovitis (around 68%), fever (around 60%), joint pains (around 52%) and arthritis (around 48%).

Like many of the STIs, having gonorrhoea increases the risk of both transmitting and acquiring HIV infection. This is why full STI screening should normally include all of the STIs.

How Common Is the Disease?

Gonorrhoea is the second most common bacterial STI worldwide. The WHO estimated that there were 78 million new cases of gonorrhoea in 2012 among those aged 15–49 years, with a global incidence rate of 19 per 1000 women and 24 per 1000 men.

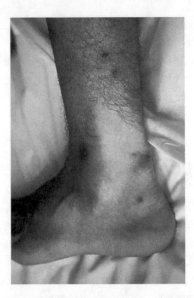

Fig. 28.4 View of a patient's left lower leg and foot, revealing the presence of a number of lesions caused by the wider dissemination of *N. gonorrhoeae*. (Dr. S. E. Thompson, VDCD; J. Pledger, Centers for Disease Control and Prevention, USA)

In the USA, gonorrhoea is the second most commonly reported notifiable disease with an estimated 820,000 new gonococcal infections reported each year. In 2016, the incidence rate of the disease was 146 cases per 100,000. The rates of reported cases are highest in the age group 20–24 years with 705 and 685 cases per 100,000 in men and women, respectively, in 2017. One report has estimated that the annual cost of gonorrhoea and its complications in the USA is $162.1 million. In the UK in 2015, there were 45,140 new cases of gonorrhoea, and this represented an 11% increase on 2014. The overall rate of diagnosis was approximately 76 cases per 100,000 population, but in the 20–24 year age group, it was 270 cases per 100,000 population. The highest rates of the disease were among the young – 56% of cases occurred in those aged 15–24 years.

What Causes Gonorrhoea?

The bacterium responsible for the disease, *N. gonorrhoeae* (also referred to as "the gonococcus"), is a Gram-negative coccus (usually the individual cells are kidney bean-shaped) that often appears in pairs when viewed through the microscope (Fig. 28.5).

The bacterium prefers an oxygen-containing environment and is described as being a fastidious organism because it needs a wide range of nutrients in order to grow. Humans are the only natural host for the gonococcus, and it can't survive for long outside the human body. Approximately 250 cells of the organism are required to initiate an infection in an individual. However, as

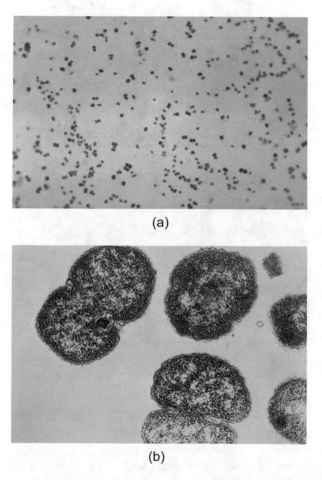

(a)

(b)

Fig. 28.5 Images of *N. gonorrhoeae*. (a) Gram-stain of *N. gonorrhoea* showing Gram-negative cocci many of which are in pairs. (Renelle Woodall, Centers for Disease Control and Prevention, USA). (b) Electron micrograph of *N. gonorrhoeae* showing single cells and pairs of cells. (Joe Miller, Centers for Disease Control and Prevention, USA). (c). A three-dimensional computer-generated image of *N. gonorrhoeae*. Note the hair-like appendages (known as pili or fimbriae) which are involved in adhesion of the bacterium to cells and other surfaces. The artistic recreation was based upon scanning electron microscopic imagery. (James Archer, Centers for Disease Control and Prevention, USA). (d) Colonies of *N. gonorrhoeae* growing on an agar plate. (W. Jerry Brown, Centers for Disease Control and Prevention, USA)

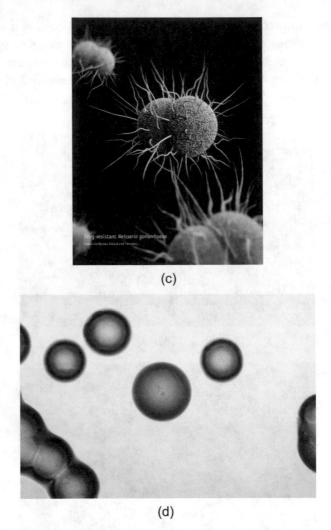

(c)

(d)

Fig. 28.5 (continued)

many as one third of those individuals who are exposed to the gonococcus don't become infected.

What Happens during an Infection?

Once it reaches a mucosal surface, the gonococcus attaches to epithelial cells using proteins in its cell as well as hair-like structures known as pili or fimbriae (Fig. 28.5). Within 24–48 hours, it invades and then exits the epithelial cell or manages to pass between adjacent cells and so penetrates through to the underlying tissues. This, of course, induces an inflammatory response, and neutrophils soon arrive on the scene. Neutrophils can phagocytose the bacteria, but gonococci are able to survive within these cells (Fig. 28.6).

Invasion of epithelial cells and the subsequent inflammatory response cause detachment of regions of the mucosa which, along with gonococci-containing neutrophils, are released from the infected individual in the form of a purulent discharge. The organism can be passed to another person if they come into contact with the pus or discharge as it can survive inside the neutrophils. The main virulence factors of the organisms are listed in Table 28.2.

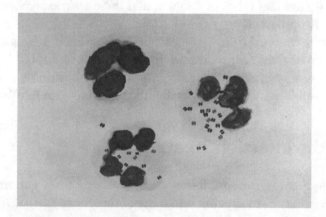

Fig. 28.6 Gram-stained photomicrograph of a urethral exudate from a patient with gonorrhoea showing the presence of both intracellular and extracellular Gram-negative diplococci. (Dr. Norman Jacobs, Centers for Disease Control and Prevention, USA)

Table 28.2 Virulence factors of *N. gonorrhoeae* and their function

Virulence factor	Function
Pili	Involved in adhesion to, and invasion of, epithelial cells
Proteins in cell wall	Involved in adhesion to, and invasion of, epithelial cells
Lipopolysaccharide (LPS)	Induces the release of inflammatory cytokines
Antigenic variation	The gonococcus is able to continually change the chemical composition of its pili and LPS which means that antibodies produced against these antigens quickly become redundant
Porins (molecules in the cell wall)	Enable survival of gonococci within neutrophils

How Is Gonorrhoea Diagnosed?

Diagnosis of gonorrhoea is based on the disease symptoms together with laboratory analysis of samples taken from any infected sites such as the urethra, vagina, throat, rectum and eyes. Microscopic examination of the sample is a rapid and simple diagnostic approach. The sample is Gram stained and searched for the presence of neutrophils and Gram-negative diplococci (see Fig. 28.6). An attempt can also be made to grow the gonococcus from the sample. Although culture and microscopic examination of samples have for many years been the main laboratory tests performed, nucleic acid amplification tests (NAATs – see Chap. 1) are increasingly being used to detect the presence of gonococcal DNA in samples, and these are much more rapid than conventional approaches.

Culture of gonococci from samples has the significant advantage that the organism can be tested for its sensitivity to a range of antibiotics, and this information is very useful given that gonococci are becoming resistant to a number of antibiotics. NAATs generally don't provide such information.

How Is the Infection Treated?

Prior to the discovery of antibiotics, those suffering from gonorrhoea were exposed to a variety of treatments, and some of these are described in Box 28.2.

Box 28.2 Historical Treatments for Gonorrhoea

Throughout the nineteenth century, one of the most popular treatments for the disease involved the use of a type of pepper from Indonesia known as cubebs. The unripe fruit of the plant was dried and powdered and administered in a drink. However, it had an awful taste and caused bowel irritation.

Another popular remedy at the time was balsam of copaiba which was the resin from the *Copaifera* tree found in South America. In 1859, 151,000 pounds of copaiba balsam were imported into Great Britain for use mainly in the treatment of gonorrhoea.

Once Neisser had established in 1878 that the disease was caused by a bacterium, research for a specific antibacterial drug began. Initially compounds of arsenic, antimony, bismuth, gold and mercury were tried. These were taken orally, injected intravenously or syringed directly into the urethra.

Heat treatment was a popular remedy for the treatment of many diseases and so was used in the early twentieth century for the treatment of gonorrhoea. Then in 1932 researchers at the University of Rochester in New York discovered that the gonococcus could be killed by 2 hours of exposure to 41.5°–42.0 °C, and this gave huge impetus to thermal therapy. A popular procedure involved a fever cabinet in which the patient (apart from their head) was enclosed for 4–6 hours at a temperature above 41 °C. This was repeated every 3 days on 4–6 occasions. It was unpleasant – but it was successful in more than 60% of cases. Attempts to treat the disease by the local application of heat (to the vagina or rectum) were also made, with some success.

In 1937 the introduction of sulphonamides revolutionised the treatment of the disease and replaced thermal therapy. However within a few years, resistance to this antimicrobial had emerged (see Box 28.3).

The main goals of treatment are to reduce the impact of the disease on the individual and to reduce transmission of the infection within the community. Specialist sexual health services are best placed to do this, and you should attend one of these to ensure best care. Screening for other STIs should be performed and partners and other recent sexual contacts informed and treated as appropriate. Sexual partners usually start treatment while waiting for their test results.

Antibiotics are an essential part of treatment, but the growing resistance of gonococci to many of these is causing significant problems. Antibiotics such as sulphonamides, penicillins, fluoroquinolones, oral cephalosporins and tetracyclines are now no longer recommended on their own as first-line agents because of the emergence of resistant strains of gonococci (Box 28.3). If you have gonorrhoea, then there is a high chance that you will also be infected with *Chlamydia trachomatis*; consequently therapy with two antibiotics that are effective against both organisms is usually recommended. This approach also usefully delays the development of drug resistance. Suitable combinations include ceftriaxone plus oral azithromycin or ceftriaxone with doxycycline. In all cases, it's vital to perform tests to determine the antimicrobial sensitivity of the particular strain of gonococcus that is infecting you so that an appropriate choice of antibiotics can be made.

Neonates with suspected eye infections should always have samples taken for laboratory analysis. Specialist input while waiting for the results should be considered, especially in those with severe symptoms or rapid onset, or if the baby is distressed. Early treatment reduces the risk of sight-threatening complications. Oral, intravenous or intramuscular antibiotics are recommended as topical treatment is not sufficient, and usually treatment for both chlamydia and gonorrhoea is given until the test results are known.

Box 28.3 *N. gonorrhoeae*: The Emergence of Another Superbug

The sulphonamides were among the first antimicrobial agents to be used as an effective treatment for gonorrhoea – these were introduced in the mid-1930s. However, by the 1940s there was widespread resistance to these drugs. Fortunately, at that time penicillin was becoming more widely available and was a very effective replacement. *N. gonorrhoeae* then gradually became resistant to penicillin which meant that higher doses were required for treating the disease. Resistance eventually reached such a high level that by 1989 penicillin was no longer of any use. Resistance also developed to other antibiotics that had been introduced in the meantime. By 1977 there was widespread resistance to erythromycin, and in 1986 tetracycline was no longer recommended for treating the disease. In the 1980s the fluoroquinolones were being used widely for treatment, but, again, the development of resistance to these drugs meant that by 1990 they were no longer recommended. In the 1990s the cephalosporins cefixime and ceftriaxone became the main antibiotics used for gonorrhoea in most countries. However, reports of treatment failure using cefixime emerged in Japan and Europe in the 2010s. In 2013 the Centers for Disease Control and Prevention in the USA classified antibiotic-resistant *N. gonorrhoeae* as being an urgent threat to public health. Most authorities currently recommend dual therapy with ceftriaxone or cefixime in combination with azithromycin for treating gonorrhoea. But recently, in 2018, there were several reports of treatment failure using this drug combination. Fortunately, a UK patient infected with a strain resistant to this drug combination was successfully treated with another antibiotic, ertapenem. All the indications are that we're running out of options for treating diseases caused by this important pathogen.

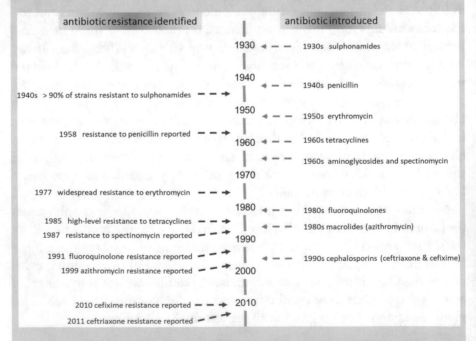

antibiotic resistance identified		antibiotic introduced
	1930	1930s sulphonamides
	1940	1940s penicillin
1940s > 90% of strains resistant to sulphonamides		
	1950	1950s erythromycin
1958 resistance to penicillin reported	1960	1960s tetracyclines
		1960s aminoglycosides and spectinomycin
	1970	
1977 widespread resistance to erythromycin		
	1980	1980s fluoroquinolones
1985 high-level resistance to tetracyclines		1980s macrolides (azithromycin)
1987 resistance to spectinomycin reported	1990	
1991 fluoroquinolone resistance reported		1990s cephalosporins (ceftriaxone & cefixime)
1999 azithromycin resistance reported	2000	
2010 cefixime resistance reported	2010	
2011 ceftriaxone resistance reported		

Timeline showing the emergence of resistance to each new antibiotic following its introduction into clinical practice

What Is the Prognosis?

When the correct treatment for a recently acquired infection is started promptly and completed as advised, the prognosis is good with a full recovery being usual. Tests to ensure treatment has been successful should always be performed. If your symptoms continue despite correct treatment, then this is more likely to be due to re-infection than persistence of the original infection. In severe cases of PID, intravenous antibiotics may be necessary. Patients with DGIs or other serious complications should usually be admitted to hospital especially if they are unwell, or else specialist assessment and input should be sought.

How Can I Avoid Getting Gonorrhoea?

Currently there are no vaccines available for gonorrhoea. Safe sex practices will reduce your risk of contracting gonorrhoea such as using condoms during any penetrative sex. Other risk-reducing measures include delaying the commencement of sexual activity, practising monogamy and reducing the number of sexual partners. Adequate treatment of infected patients is an important preventive measure as it helps to reduce transmission of the disease within the community. The partners of infected patients should also be evaluated and treated. To avoid re-infection, sex partners should abstain from sexual intercourse until they and their partner(s) have completed their treatment.

Want to Know More?

Centers for Disease Control and Prevention, USA. https://www.cdc.gov/std/gonorrhea/default.htm
DermNet, New Zealand. https://www.dermnetnz.org/topics/gonorrhoea/
European Centre for Disease Prevention and Control. https://www.ecdc.europa.eu/en/gonorrhoea/facts
Family Planning Organisation, UK. https://www.sexwise.fpa.org.uk/stis/gonorrhoea
Mayo Clinic, USA. https://www.mayoclinic.org/diseases-conditions/gonorrhea/symptoms-causes/syc-20351774
National Health Service, UK. https://www.nhs.uk/conditions/gonorrhoea/
National Institute for Clinical Care and Excellence (NICE), UK. Gonorrhoea, 2019. https://cks.nice.org.uk/gonorrhoea#!topicSummary

Patient Info, UK. https://patient.info/sexual-health/sexually-transmitted-infections-leaflet/gonorrhoea

Terrence Higgins Trust, UK. https://www.tht.org.uk/hiv-and-sexual-health/sexual-health/stis/gonorrhoea

World Health Organisation. https://www.who.int/reproductivehealth/publications/rtis/gonorrhoea-treatment-guidelines/en/

Abraha M, Egli-Gany D, Low N. Epidemiological, behavioural, and clinical factors associated with antimicrobial-resistant gonorrhoea: a review.. *F1000Research*. 2018 Mar 27;7:400. https://doi.org/10.12688/f1000research.13600.1. eCollection 2018.

Creighton S. Gonorrhoea. *British Medical Journal Clinical Evidence* 2014 Feb 21;2014. pii: 1604. PMID: 24559849

Hill SA, Masters TL, Wachter J. Gonorrhea - an evolving disease of the new millennium. *Microbial Cell*. 2016 Sep 5;3(9):371-389. https://doi.org/10.15698/mic2016.09.524.

Morris S. Gonorrhoea infection. BMJ Best Practice, BMJ Publishing Group, 2019. https://bestpractice.bmj.com/topics/en-gb/51

Suay-García B, Pérez-Gracia MT. Future prospects for *Neisseria gonorrhoeae* treatment. *Antibiotics (Basel)*. 2018 Jun 15;7(2). pii: E49. https://doi.org/10.3390/antibiotics7020049.

Wong B. Gonorrhea. Medscape from WebMD, 2018. https://emedicine.medscape.com/article/218059-overview

Young A, Wray AA. Urethritis. Treasure Island (FL): StatPearls Publishing LLC; 2020. https://www.ncbi.nlm.nih.gov/books/NBK537282/

29

Genital Herpes

Abstract Genital herpes is an infection of the genitalia due to herpes simplex virus-1 (HSV-1) or herpes simplex-2 (HSV-2). Only about 20% of those affected have symptoms – blisters in the genital area accompanied by a burning or itching sensation and perhaps fever and headaches. In women the blisters can appear on the vulva, perineum, buttocks, cervix and vagina, while in men they can be on the penis, scrotum, groin and thighs and buttocks and around the anus. However, rashes may occur anywhere on the skin.

Following infection, the virus becomes dormant in nerve cells and can be reactivated in the future – it's therefore a lifelong condition. As many as 90% of those infected with HSV-2 have a recurrence with symptoms within 1 year, whereas recurrences are less common in those infected with HSV-1. Recurrences generally become less frequent with time.

It's a highly contagious disease and is acquired by direct contact with the mucosal surfaces or skin of infected individuals or via infected secretions. HSV-2 is usually transmitted during vaginal or anal intercourse, while HSV-1 is generally transferred from the mouth to the genital area. Spread can also happen from one site to another in the same person (auto-inoculation).

The disease is usually diagnosed by finding HSV DNA in a sample of fluid taken from the blistering rash or sore. Treatment involves anti-viral agents (acyclovir, valacyclovir or famciclovir) which reduce symptoms and also the likelihood of virus transmission. Analgesics for pain relief can be helpful as well as a topical anaesthetic when the pain is severe. The risks of acquiring the infection can be reduced by having few sexual partners, not having sex with someone who has herpes blisters or sores on their genitals, not receiving oral sex from someone with a cold sore and using condoms for penetrative sex.

What Is Genital Herpes?

Genital herpes can be caused by either the herpes simplex virus-2 (HSV-2) or HSV-1. Historically, most cases of genital herpes have been due to HSV-2, but the proportion due to HSV-1 has increased in recent years. The proportion of cases due to each of the viruses varies in different countries; in the UK the majority of new primary genital infections are now caused by HSV-1 rather than by HSV-2.

Primary genital infection with either virus usually goes unnoticed, although symptoms do occur in approximately one fifth of cases. Because of this, most people don't know they have the disease, especially considering that often infections are caught from people who themselves have no symptoms at the time – this is known as asymptomatic viral shedding.

In those who do experience symptoms, painful blisters appear on the skin and/or mucosa of the genitals about 4–7 days after sexual contact with an infected individual, and this may be accompanied by a burning or itching sensation. The blisters then burst to produce painful red sores that can last for up to 3 weeks. In women these lesions may appear on the vulva, perineum, buttocks, cervix and vagina (Fig. 29.1). The symptoms of the disease were first described in 1736 by the French physician Jean Astruc (Box 29.1).

Fig. 29.1 Appearance of typical lesions of genital herpes due to HSV-2 in a woman. The blisters have a reddish appearance on the skin, but a rather white-brownish appearance on the vaginal epithelium as shown by the black arrow. (Mikael Häggström, uploaded with permission from patient/CC0)

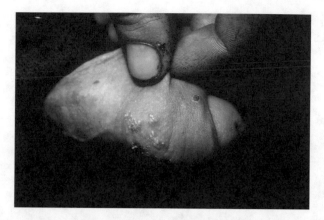

Fig. 29.2 Appearance of lesions of genital herpes due to HSV-2 on the penis. (Dr. Dancewicz, Centers for Disease Control and Prevention, USA)

In men the lesions may appear on the head (glans) or shaft of the penis, the scrotum, groin and thighs and buttocks and around the anus (Fig. 29.2).

Other symptoms in women include fever, headache, muscle pain (in the back and legs), constipation, genital pain, pain on urination and a vaginal discharge. In men, fever and headache may be accompanied by urethritis, as well as a urethral discharge, and this results in painful urination. Primary episodes of the infection generally last for between 2 and 4 weeks.

Within the first year after the first infection with HSV-2, 70–90% of patients experience a symptomatic recurrence. One third of these have frequent (at least 6 times a year) recurrences. In contrast, recurrence is much less frequent (20–50%) in those infected with HSV-1. Symptoms may include genital pain or tingling or shooting pains in the legs, hips or buttocks. This is followed by the appearance of the genital rash, either hours or 1–2 days later. Compared to the first infection, recurrences are much less severe and last for a shorter period of time (6–48 hours). Fever and other symptoms associated with the initial infection are much less common.

Box 29.1 First Description of the Symptoms of Genital Herpes

Jean Astruc (1684–1766) was a French physician who was Professor of Medicine at the universities of Montpelier and Paris as well as being physician to the King of France. Among the many medical books he wrote, the most famous was on venereal diseases *De morbis venereis libri sex* which was published in 1736. In this he described, for the first time, the symptoms of genital herpes.

JEAN ASTRUC,

Portrait of Jean Astruc. (Image courtesy of the National Library of Medicine, USA. Public Domain)

The disease was well known to nineteenth-century physicians specialising in venereal diseases who also noticed its frequent association with gonorrhoea and syphilis. However, it wasn't until the early twentieth century that it was recognised as being caused by a microbe, thanks to the research of Gruter and Lowenstein in Germany.

Are There Any Complications?

Possible complications of genital herpes include (i) acute urinary retention – particularly in women, (ii) meningitis, (iii) encephalitis, (iv) hepatitis, (v) neonatal infection, (vi) pelvic inflammatory disease, (vii) pneumonitis and (viii) secondary infection of the rash by microbes such as *Candida albicans*.

The virus can pass from mother to child when it's in the womb (rarely), during delivery or after delivery, and this may result in a serious life-threatening infection in the newborn (Fig. 29.3). The greatest risk of transmission is

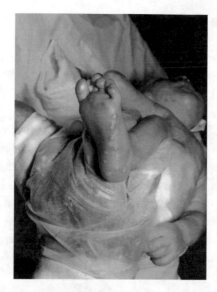

Fig. 29.3 Herpes simplex ulcerations on the surface of an infant's foot. Women, who acquire genital herpes during pregnancy, can transmit the virus to their babies. Untreated, HSV infections in newborns can result in intellectual disability and death. (Image courtesy of Judith Faulk, Centers for Disease Control and Prevention, USA)

when the mother acquires a primary infection when she is close to labour or within the last trimester. In contrast, an infection prior to pregnancy results in the mother producing protective antibodies, and these will pass to the foetus in the womb during a subsequent pregnancy. The risk of transmission to the infant is 20–50% for those women who have recently acquired a primary infection but is less than 1% in those with a recurrent infection.

Infection with HSV-2 increases the risk of both acquiring and passing on HIV infection.

What Are the Risk Factors Associated with the Disease?

The main risk factors for genital herpes include (i) being young (15–24 years old) and sexually active, (ii) a previous history of STIs, (iii) a high number of lifetime sexual partners, (iv) more than one partner in the last year or a recent new partner, (v) early age of first sexual intercourse, (vi) unprotected sexual intercourse, (vii) male same-sex intercourse, (viii) female sexual partners of males who engage in same-sex intercourse, (ix) having HIV infection and (x) being immunocompromised.

How Big Is the Problem?

Genital herpes is a global issue, and 417 million people worldwide were estimated to be suffering from the infection in 2012. With regard to genital infection with HSV-1, in 2012 it was estimated that 140 million people aged 15–49 years were affected worldwide.

Each year in the USA, 776,000 people are diagnosed with new genital herpes infections. 12% of those aged 14 to 49 years have HSV-2 infection, but the prevalence of genital herpes is thought to be greater than this because an increasing number of infections are caused by HSV-1. In the USA, Canada and many European countries, during the last decade, at least half of the first episodes of genital herpes were caused by HSV-1. In the USA, HSV-2 infection is more common among women than among men; the proportions of 14–49-year-olds infected during 2015–2016 were 16% and 8% in women and men, respectively.

In the UK, genital herpes is one of the most common STIs with up to 23% of adults being affected – the proportions of cases due to HSV-1 and HSV-2 are similar. In 2015 in England, 30,658 new cases of genital herpes were diagnosed – 41% were in those aged 15–24 years and 92% were in heterosexual men and women.

What Happens During an Infection?

The structure of the HSV-2 virus is very similar to that of HSV-1 (see Chap. 25). The way in which it attaches to, and replicates within, human cells (Fig. 29.4) is also very similar to HSV-1 and is described in Chap. 25.

Genital herpes, of course, affects the epithelial cells of the genital mucosa where it causes inflammation and cell death at the site of the infection and this results in the characteristic lesions of the disease. It can also enter through tiny splits in the skin which are often present in areas prone to trauma such as elbows, knees and hands. Following initial infection, the virus enters a latent state within certain nerve cells. During genital herpes it's in the dorsal root ganglion at the base of the spine where the virus is able to persist indefinitely. Reactivation of the virus may be induced by trauma (such as surgery), ultraviolet light, smoking, drinking alcohol, stress and fatigue. It then travels back along the nerve, and, on reaching the mucosa or skin, it causes the usual rash again.

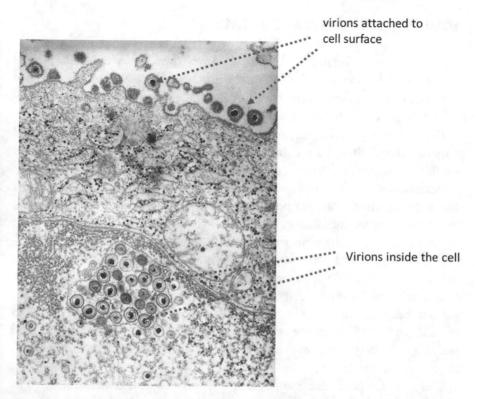

virions attached to
cell surface

Virions inside the cell

Fig. 29.4 Electron micrograph showing herpes simplex virions attached to, and within, a human epithelial cell. (Dr. Fred Murphy; Sylvia Whitfield, Centers for Disease Control and Prevention, USA)

How Is Genital Herpes Diagnosed?

A diagnosis of genital herpes on the basis of a clinical examination alone may be very difficult because (i) many cases are asymptomatic, (ii) the symptoms of the disease may be similar to those of other STIs such as syphilis and (iii) the disease may cause atypical symptoms that occur at unusual sites such as the thighs or the buttocks. If you are suffering from the characteristic symptoms of the disease, then diagnosis can be more straightforward, and suspicion should lead to testing for confirmation.

Diagnosis in the laboratory requires testing of a sample of blister fluid taken from the rash. The sample is either tested for the presence of herpes viral DNA by PCR or else an attempt is made to grow the virus. Results are obtained more rapidly using PCR (a few hours as opposed to 2–5 days for viral culture), and this approach is also more sensitive.

How Is Genital Herpes Treated?

If you have genital herpes, it's best that you're cared for by Specialist Sexual Health Services so you can be screened for other STIs and given appropriate support and counselling. It can be upsetting to be told you have an STI, especially if you are in a monogamous relationship. It's important that the complexities of the infection including asymptomatic shedding are explained properly. These clinics have expertise in advising how to notify your partner and in tracing previous sexual contacts.

Treatment of the disease involves the use of anti-viral agents. These can't eliminate the virus from the body but do prevent it from being replicated. There's no cure for the disease, but anti-viral medication reduces symptoms and transmission of the virus to others. The most widely used drugs are aciclovir, valaciclovir and famciclovir. Aciclovir reduces viral shedding, decreases new blister formation and reduces the duration of genital symptoms. Valaciclovir can be used for treating both primary and recurrent infections and has been shown to reduce the transmission of genital herpes from an infected partner to an uninfected partner. It also reduces asymptomatic viral shedding. Topical therapy with anti-viral agents is of minimal benefit and isn't usually recommended.

Analgesics for pain relief may be helpful, and a topical anaesthetic, such as lidocaine, may be applied if you experience severe pain during urination. Saline bathing of the affected areas is also soothing. You should postpone having sexual intercourse until after your treatment has been completed.

Certain at-risk groups, such as pregnant women and those who are immunocompromised, may require specialist input. Admission to hospital may be appropriate for complications such as urinary retention or severe secondary infection or if you are very unwell.

How Can I Avoid Getting Genital Herpes?

The most effective way of avoiding genital herpes is not to engage in vaginal, anal or oral sex. For those who are sexually active, HSV is primarily passed on during any activities that involve direct skin-to-skin contact with the blisters and sores that it causes. However, as said previously, the virus can also be spread during shedding from someone who shows no sign of any symptoms. Asymptomatic shedding occurs with both oral and genital herpes, but to cause an infection, direct contact has to occur with the affected area. The virus will be present in the saliva of someone with oral herpes, and the infection can be passed on if you come into direct contact with this. However the virus doesn't survive for long outside the body and isn't transmitted in water – for example, a bath or swimming pool. HSV can also cause infections of the fingers and, less commonly, of the toes. Such infections, known as herpetic whitlow (see Chap. 24), are highly contagious, and it's also possible to spread the infection to other parts of your own body such as the genitals.

If you are sexually active, the risk of acquiring the disease is reduced if you (i) are in a long-term, mutually monogamous, relationship with a partner who is known to not be infected, (ii) don't have sex with someone who has blisters or sores on his or her genitals, (iii) don't receive oral sex from somebody with a cold sore, (iv) use a condom every time you have sex and (v) don't share sex toys.

With regard to condom use, it's important to be aware that condoms don't fully protect against herpes transmission because blisters or sores are often present at skin sites not covered by the condom. Also, the virus can be released from areas of the skin that don't have a visible herpes sore.

Men who are circumcised have been shown to have a reduced risk of acquiring HSV-2 although it doesn't reduce the risk of them transmitting the virus to female sexual partners.

Want to Know More?

American College of Obstetricians and Gynecologists. https://www.acog.org/Patients/FAQs/Genital-Herpes?IsMobileSet=false

American Sexual Health Association. http://www.ashasexualhealth.org/stdsstis/herpes/signs-symptoms/

Centers for Disease Control and Prevention, USA. https://www.cdc.gov/std/herpes/stdfact-herpes.htm

Family Planning Association, UK. https://www.sexwise.fpa.org.uk/stis/genital-herpes

Herpes Viruses Association, UK. https://herpes.org.uk/frequently-asked-questions/?gclid=EAIaIQobChMIqbLh4pPY5QIVRLDtCh0ToAtjEAAYAiAAEgI9g_D_BwE

Mayo Clinic, USA. https://www.mayoclinic.org/diseases-conditions/genital-herpes/symptoms-causes/syc-20356161

National Health Service, UK. https://www.nhs.uk/conditions/genital-herpes/

National Institute for Clinical Care and Excellence (NICE), UK. Herpes simplex – genital, 2017. https://cks.nice.org.uk/herpes-simplex-genital#!topicSummary

Patient Info, UK. https://patient.info/sexual-health/sexually-transmitted-infections-leaflet/genital-herpes

Ayoade FO. Herpes Simplex. Medscape from WebMD, 2018. https://emedicine.medscape.com/article/218580-overview

Groves MJ. Genital herpes: a review. *American Family Physician*. 2016 Jun 1;93(11):928-34.PMID: 27281837

Jaishankar D, Shukla D. Genital herpes: insights into sexually transmitted infectious disease. *Microbial Cell*. 2016 Jun 27;3(9):438-450. https://doi.org/10.15698/mic2016.09.528. PMID: 28357380

Lorenz BD. Herpes simplex virus infection. BMJ Best Practice, BMJ Publishing Group, 2019. https://bestpractice.bmj.com/topics/en-gb/53

Mathew J, Sapra A. Herpes Simplex Type 2. Treasure Island (FL): StatPearls Publishing LLC; 2020. https://www.ncbi.nlm.nih.gov/books/NBK554427/

Sauerbrei A. Optimal management of genital herpes: current perspectives. *Infection and Drug Resistance*. 2016 Jun 13;9:129-41. https://doi.org/10.2147/IDR.S96164. eCollection 2016.

WHO guidelines for the treatment of genital herpes simplex virus. Geneva: World Health Organization; 2016. https://www.who.int/reproductivehealth/publications/rtis/genital-HSV-treatment-guidelines/en/

30

Genital Warts

Abstract Genital warts are grey- or skin-coloured, fleshy growths on the genitals or around the anus which may be itchy or inflamed. They result from infection with the human papilloma virus (HPV) although many of those infected don't have any symptoms. Genital warts are mostly caused by low-risk strains of HPV. Most infections clear without any treatment within 2 years, but some remain longer term or indefinitely. High-risk strains of the virus can cause cancer of the cervix, penis, anus and oropharynx. HPV is transmitted during vaginal, anal and oral sex as well as genital to genital contact and the sharing of sex toys. It's the most common viral sexually transmitted infection, and its prevalence is greatest in persons aged 17–33 years. The warts themselves can be treated at home by the topical application of a variety of drugs, but this doesn't eradicate the viral infection which can remain after treatment. Alternatively, they can be removed by a clinician using a variety of procedures. Condoms are useful in reducing transmission of the virus. Vaccination against some types is possible and has resulted in dramatic reductions in its prevalence in those countries in which it has been introduced.

What Are Genital Warts?

Genital warts (sometimes referred to as "condylomata acuminata") arise as a result of infection with the human papilloma virus (HPV – see Chap. 9) and appear as small, fleshy growths, bumps or skin changes on the genitals or around the anus. They may appear as single, isolated bumps (Fig. 30.1a), in

M. Wilson, P. J. K. Wilson, *Close Encounters of the Microbial Kind*,
https://doi.org/10.1007/978-3-030-56978-5_30

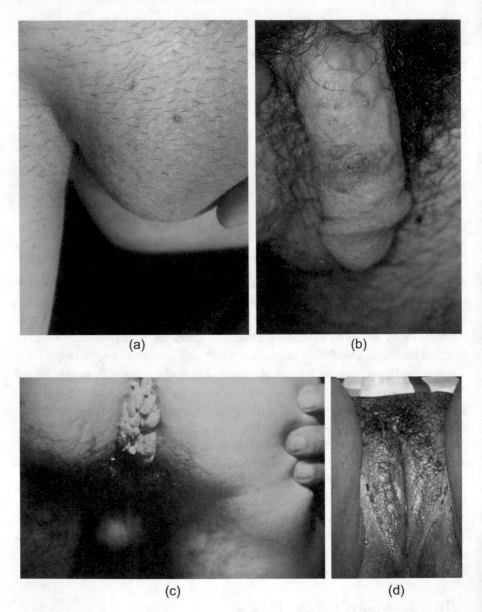

Fig. 30.1 Appearance of genital warts. (**a**) A single wart on the testicle. (Bikepunk2. This file is made available under the Creative Commons CC0 1.0 Universal Public Domain Dedication. Via Wikimedia Commons). (**b**) Multiple warts on the penis. (Dr. M.F. Rein, Centers for Disease Control and Prevention, USA). (**c**) Clusters of warts on the anus. (Dr. Wiesner, Centers for Disease Control and Prevention, USA). (**d**) Warts on the female external genitalia – mainly on the labia majora. (Dr. M.F. Rein, Centers for Disease Control and Prevention, USA)

groups (Fig. 30.1b) or as clusters with a cauliflower-like appearance (Fig. 30.1c). The warts are usually, but not always, painless, and in some cases they may be itchy and/or inflamed. They range in size from a few millimetres to several centimetres, although they're generally 1–3 mm in diameter. They vary in colour and may be white, skin-coloured, red or much darker, and they may feel hard or soft. In women, the most common places for warts to develop are on the vulva, on the cervix, inside the vagina, around or inside the anus and on the upper thighs. In men they may appear anywhere on the penis, on the scrotum, inside the urethra, around or inside the anus and on the upper thighs. When warts are present inside the urethra, this can result in bleeding and difficulty in urinating. Less commonly, warts may also appear on the lips or mouth – this usually only occurs in those who are immunocompromised such as HIV patients. Many people infected with HPV don't have any symptoms but can still pass on the virus to others.

Are There Any Complications?

The warts themselves usually don't pose a serious threat to health, but their appearance can be unpleasant and can result in psychological distress because of shame and embarrassment. Anogenital warts can sometimes cause physical problems because of their location. For example, if they are in the urethra, they can obstruct urination.

Any individual can be infected with several types of HPV at one time, and although the low-risk HPVs can cause visible warts, the high-risk HPVs generally don't cause any symptoms. Some high-risk HPV types can cause cancer after many years, and these are responsible for around 6% of all human cancers.

Cervical carcinoma can occur after many years in some women who are infected with particular strains of the virus – HPV 16 and HPV 18. However, this can be prevented by vaccination. In England the cervical screening programme has been changed to include high-risk HPV testing.

Penile cancer can occur with high-risk strains of HPV, and, in men who have sex with men, anal cancer is a possible complication – approximately 85% of anal cancer cases worldwide are attributable to HPV infection. Oropharyngeal cancer is another risk for those who engage in oral sex. In the USA every year approximately 19,400 women and 12,100 men are affected by cancers caused by HPV.

Some HPV types can also affect the mouth, larynx, nasal cavity and eyes causing conjunctivitis.

Infection very early in life may result in a rare condition called respiratory or laryngeal papillomatosis where small swellings develop in the airways or voice box, causing problems with the voice, swallowing or breathing.

How Is the Infection Transmitted?

Genital warts is predominantly a sexually transmitted infection (STI) and the HPV is transferred to a sexual partner during direct skin-to-skin contact or exposure to secretions during vaginal sex, anal sex, non-penetrative genital to genital contact, the sharing of sex toys and oral sex. In general, two thirds of individuals who have sexual contact with a partner who has genital warts develop symptoms within 3 months. You may be infected with the virus, and transmit it to someone else, even if you don't actually have any visible warts – this situation can last for many months or even years. Asymptomatic transmission is a very important way in which the virus is spread.

Nonsexual transmission of the virus is also possible but is much less common. This can occur when HPV on the skin of an infected individual, or on contaminated objects (fomites), comes into contact with the skin of an individual that has been damaged in some way. The HPV can penetrate through visible or microscopic injuries on the skin surface.

It's also possible for you to transfer HPV from one of your body sites, such as the hands, to your genital area. However, this isn't very common because the different viral strains have a preference for different body locations.

The virus can be passed from mother to baby, and, although there are many possible routes for this, most transfer is thought to occur during delivery.

How Common Is the Disease?

HPV infection is a very common STI, and it's been estimated that about 80% of sexually active people will become infected at some point in their lives. Globally, it's the most common viral STI. The prevalence of the disease is greatest in those aged 17–33 years, with a peak incidence in those aged 20–24 years. In the USA, 14 million new infections occur each year, and about 79 million people are thought to have an active infection at any given time. In the UK, genital warts are the most common viral STI, accounting for almost 16% of all STIs in 2015. The number of new cases of the disease in

England in 2015 was 68,310, i.e. 126 per 100,000 population. In England, it was most common in the 20–24 age group, with a rate of 649 per 100,000 population. The disease is a considerable financial burden on health services – in England in 2008, the annual expenditure associated with HPV infection was estimated as being £16.8 million.

Which Viruses Are Responsible for Genital Warts?

There are more than 180 different types of HPV, but only about 40 of these are able to cause genital warts. Approximately 90% of cases of anal and genital warts are due to two types – HPV 6 and HPV 11 – and these have a low risk of serious complications such as cancer.

Two types, HPV 16 and HPV 18, are associated with a high risk of cancer, and another 11 types are associated with a moderate risk. More than 70% of cervical, vaginal and penile cancers are caused by HPV 16 or HPV 18.

The way in which HPV causes an infection has been described in Chap. 9.

Diagnosis of the disease is usually made on the basis of the appearance and location of the characteristic warts. Biopsies aren't usually necessary but are an option if confirmation is required.

What Are the Risk Factors?

Genital warts is a common STI and can be avoided by limiting unprotected sexual activity and the number of sexual partners. Smoking tobacco increases rates of persisting HPV infection and the cancers associated with this. Male circumcision reduces infection in men by around 35%.

How Is the Disease Treated?

If you suffer from genital warts, then it's very important that you, as well as your current sexual partners, are screened to rule out other STIs. Ideally, you should seek treatment in a specialist sexual health clinic.

Treatment is aimed at removing or destroying the warts, especially if they are causing physical or cosmetic impact or distress, but this isn't likely to eradicate the infection. Not treating is an option, as around 50% of untreated warts will resolve within 12 months and most HPV infections resolve spontaneously, with around 95% of people clearing HPV within 2 years of genital infection. It can be unclear what the ideal treatment is, and many options have considerable failure and relapse rates. The location and size of the warts, as well as patient preferences, guide the best treatment choice.

A number of topical treatments are available that you can use at home, but, unfortunately, these can cause adverse effects. Podophyllotoxin is an agent that stops replication of HPV-infected cells. It's usually used to treat clusters of small warts and is applied as a liquid to the affected site twice a day for several days. The treatment is then stopped for 4 days. Most patients need four to five cycles of treatment. The drug shouldn't be used during pregnancy. Local inflammation, burning, itching and pain are common side effects.

Imiquimod is a drug that stimulates the immune system, making it more effective against HPV, and is recommended for the treatment of larger warts. It needs to be applied as a cream in various regimes usually for several weeks. Side effects include itching, redness, burning and pain. Recurrence rates are relatively low when compared with other treatments. The drug shouldn't be used during pregnancy.

Polyphenon E is an extract of green tea that has a variety of useful effects. It is immunostimulatory, reduces viral replication and suppresses tumours, although how it achieves these effects is not completely understood. Side effects include itching, redness, irritation and pain.

In addition to these self-applied treatments, a number of other options are available which include:

- Cryotherapy – this destroys the warts by freezing with liquid nitrogen. It's effective, with few adverse effects but can be painful and usually requires multiple treatments. It's often used to treat multiple small warts, particularly those that develop on the shaft of the penis or on, or near, the vulva. Side effects include skin irritation, blistering and pain at the site of treatment.
- Trichloroacetic acid destroys warts by breaking down the proteins inside them and is useful for treating small, hard warts. Recurrence rates are high.
- Surgical removal for large or difficult locations can eliminate warts in a single visit. It can cause scarring, so it may not be suitable for very large warts.
- Laser surgery is used to treat large genital warts that can't be treated using other methods because they're difficult to access, e.g. warts deep inside the anus or urethra.
- Photodynamic therapy (PDT) involves the application of a photosensitising agent that is activated by laser light of a particular wavelength (described

further in Chaps. 3, 15 and 21). This results in the production of reactive molecules such as singlet oxygen that destroy the warts.

There is no definite cure for genital warts and recurrence isn't uncommon. Many patients either fail to respond to treatment or the disease recurs following an adequate response. Recurrence rates exceed 50% after 1 year although these can be higher in those who are immunocompromised.

How Can I Avoid Getting Genital Warts?

Using condoms helps to prevent some types of some types of the spread of genital warts. However, they aren't 100% effective because the skin around the genital area is often infected and this isn't covered by the condom.

A number of vaccines can prevent some types of HPV infection; however, their availability, as well as the immunisation schedules employed, vary from country to country. The impact of vaccination in those countries where it has been introduced has been very significant. For example, within 6 years of HPV vaccination availability, infections due to HPV types 6/11/16/18 among Australian women (aged 18–24) decreased by 86%. In 2011, genital wart prevalence had decreased by up to 92.6% among HPV vaccine-eligible Australian females aged <21 years.

Want to Know More?

American Family Physician. https://www.aafp.org/afp/2004/1215/p2335.html
Family Planning Association, UK. https://www.sexwise.fpa.org.uk/stis/genital-warts
Mayo Clinic, USA. https://www.mayoclinic.org/diseases-conditions/genital-warts/symptoms-causes/syc-20355234
National Health Service, UK. https://www.nhs.uk/conditions/genital-warts/
https://www.nhsinform.scot/illnesses-and-conditions/sexual-and-reproductive/genital-warts
National Institute for Clinical Care and Excellence (NICE), UK. Warts – anogenital, 2017. https://cks.nice.org.uk/warts-anogenital#!topicSummary
Patient Info, UK. https://patient.info/sexual-health/sexually-transmitted-infections-leaflet/anogenital-warts
Terrence Higgins Trust, UK. https://www.tht.org.uk/hiv-and-sexual-health/sexual-health/stis/genital-warts-and-hpv?gclid=EAIaIQobChMIntXYj7jY5QIVSrDtCh0FWgBrEAAYBCAAEgI9cPD_BwE

US Department of Health and Human Services. https://www.womenshealth.gov/a-z-topics/genital-warts

Ghadishah D. Genital Warts. Medscape from WebMD, 2018. https://emedicine.medscape.com/article/763014-overview

Leslie SW, Sajjad H, Kumar S. Genital Warts. Treasure Island (FL): StatPearls Publishing LLC; 2020. https://www.ncbi.nlm.nih.gov/books/NBK441884/

Mejilla A, Li E, Sadowski CA. Human papilloma virus (HPV) vaccination: Questions and answers. *Canadian Pharmacists Journal (Ott).* 2017 Jul 12;150(5):306-315. doi: https://doi.org/10.1177/1715163517712534. eCollection 2017 Sep-Oct.

Mendoza N, Tyring SK. Genital warts. BMJ Best Practice, BMJ Publishing Group, 2019. https://bestpractice.bmj.com/topics/en-gb/228

Scheinfeld N. Update on the treatment of genital warts. *Dermatology Online Journal.* 2013 Jun 15;19(6):18559. PMID: 24011309

Yuan J, Ni G, Wang T, Mounsey K, Cavezza S, Pan X, Liu X. Genital warts treatment: beyond imiquimod. *Human Vaccines and Immunotherapeutics.* 2018 Jul 3;14(7):1815-1819. https://doi.org/10.1080/21645515.2018.1445947. Epub 2018 Apr 9.

31

Syphilis

Abstract Syphilis is a sexually transmitted infection caused by the bacterium *Treponema pallidum*. It's mostly transmitted by direct person-to-person sexual contact with an infected individual, but also from mother to baby, usually before birth, and rarely via infected blood products. Individuals infected with *T. pallidum* typically follow a disease course with four stages – primary, secondary, latent and tertiary. These stages can span more than 10 years. Diagnosis on the basis of clinical symptoms is difficult because these are varied and can be similar to those of other infections – syphilis has been called the "great imitator". Because *T. pallidum* can't be grown in the laboratory, diagnosis is usually based on detecting the presence of antibodies to the organism in the patient's blood or the organism can be identified under the microscope from samples taken from the patient. The disease can be treated with penicillin (or doxycycline in those who are allergic to penicillin) which is effective in all stages of the disease. It's also effective in congenital syphilis and for treating pregnant women with the disease. Condom use reduces the sexual transmission of syphilis although only if the sores or rashes are properly covered. Treatment is best provided by specialist sexual health clinics and should include contact tracing and partner notification; specialist follow up should be offered. Treatment and testing should be given to the sexual partners of someone who has syphilis.

© Springer Nature Switzerland AG 2021
M. Wilson, P. J. K. Wilson, *Close Encounters of the Microbial Kind*,
https://doi.org/10.1007/978-3-030-56978-5_31

What Is Syphilis?

Syphilis is a sexually transmitted infection (STI) which is caused by the bacterium *Treponema pallidum* which is an organism found only in humans and some non-human primates. The risk of acquiring syphilis after having unprotected sexual intercourse with someone with primary or secondary syphilis is between 10% and 60%. The symptoms of the disease are varied, and, because these can be similar to those of other infections, it's been called the "great imitator" or "great imposter". Transmission is usually by direct person-to-person sexual contact with an infected person, but it can also occur via blood product transfusion and, occasionally, through breaks in the skin that come into contact with the syphilis ulcers or sores of an infected individual. It can also be passed on from mother to child via several routes: across the placenta to the foetus, from the birth canal and during breastfeeding if the mother has a lesion on her breast. Infection with *T. pallidum* typically follows a disease course that's classified into four stages which can span more than 10 years – primary, secondary, latent and tertiary stages.

Box 31.1 What's in a Name?

The name of the genus comes from two Greek words – "trepo" and "nema" which mean "to turn" and "a thread", respectively. This is appropriate for a long, thin spiral organism. Pallidum is a Latin word which means "pale" and refers to the fact that the bacterium is difficult to stain and appears pale when attempts are made to stain it by Gram's method (see Chap. 1).

The first recorded outbreak of syphilis in Europe occurred in Naples, Italy, in 1494 during a French invasion. Italian doctors described the lesions of the disease they saw on the bodies of French soldiers who had died during the battle of Fornovo, and so it became known as the "French disease". In 1530 the Italian physician and poet Girolamo Fracastoro (Figure) introduced the term syphilis for the disease. He wrote an epic poem entitled "Syphilis or the French Disease" about a character named Syphilus who was a shepherd working for a mythological Greek king called Alcihtous. Syphilus was angry with Apollo because he blamed him for a shortage of water and grass for his sheep and said he would now only worship his king. Apollo then cursed Syphilus and the king's subjects with a disease he named after the shepherd. The people then sacrificed Syphilus to Apollo to try to get the curse removed. Fracastoro used the name syphilis for the disease in his book "On Contagion and Contagious Diseases" which he published in 1546.

Portrait of Girolamo Fracastoro by Titian (1528). National Gallery, London. (Attributed to Titian/Public domain, via Wikimedia Commons. This is a faithful photographic reproduction of a two-dimensional, public domain work of art. The work of art itself is in the public domain for the following reason: This work is in the public domain in its country of origin and other countries and areas where the copyright term is the author's life plus 100 years or fewer. This work is in the public domain in the USA because it was published (or registered with the US Copyright Office) before January 1, 1924).

T. pallidum was first identified by Fritz Schaudinn (a zoologist) and Erich Hoffmann (a dermatologist) in Berlin in 1905. They observed and described the organism in a sample taken from a woman with secondary syphilis.

What Are the Symptoms of the Disease?

Entry of the bacterium through either the mucosal surface or broken skin eventually results in a small lump which then ulcerates. Patients with primary syphilis usually have a single ulcer (known as a "chancre") or multiple lesions which may appear on the genitals or other body sites involved in sexual contact such as the mouth (Fig. 31.1) or anus. In males the usual site affected is the glans penis, while in females it's the vulva or cervix. The patient often also has swollen local lymph glands. The ulcers can appear 10–90 days (usually 2–3 weeks) following infection with the spirochaete and are usually painless, but can be painful, and then heal within a few weeks.

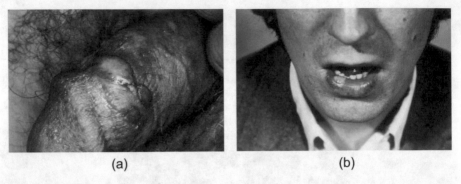

(a)   (b)

Fig. 31.1 Lesions of primary syphilis. (a) Lesions on the penis of a patient with primary syphilis. (Dr. N.J. Fiumara, Centers for Disease Control and Prevention, USA). (b) Oral lesion (chancre) of primary syphilis. (Susan Lindsley, Centers for Disease Control and Prevention, USA)

Approximately 4–10 weeks later, the symptoms of secondary syphilis appear, and these can include fever, headache and a rash (Fig. 31.2a, b) at several body sites including the flank, shoulders, arm, chest, back, palms of the hands and soles of the feet. In warm, moist body sites such as the genitals and perineum, grey-white, painless lesions develop, and these are known as "condylomata lata" – this occurs in about one third of patients (Fig. 31.2c).

Patchy hair loss on the scalp and face may also occur. Secondary syphilis can also affect the liver causing hepatitis, the nervous system causing meningitis and cranial nerve lesions, the kidney causing glomerulonephritis and the eye causing various problems. These symptoms arise as a result of spread of the bacterium throughout the body via the bloodstream.

The symptoms gradually subside, and the patient then enters the latency stage which can last for many years. The first 2 years following the disappearance of the symptoms of primary and secondary stages is known as the "early latent" period, and during this time the patient is still infectious. Up to 25% of people in the early latent phase can suffer a recurrence of secondary syphilis. During the rest of the latency stage, known as the "late latent" period, the patient is no longer infectious although women can still pass on the disease to their unborn child.

15–40% of untreated patients develop tertiary syphilis around 20–40 years after primary syphilis – this is a slowly progressive condition and can affect any organ. It takes three main forms – neurosyphilis, gummatous syphilis and cardiovascular syphilis. However, this is now seen only rarely in developed countries because of effective treatment of the disease in its earlier stages. Neurosyphilis results from invasion of the central nervous system by *T. pallidum*. This occurs in 5–10% of patients with tertiary syphilis and can result in dementia and paralysis. Invasion of the cardiovascular system by the bacterium

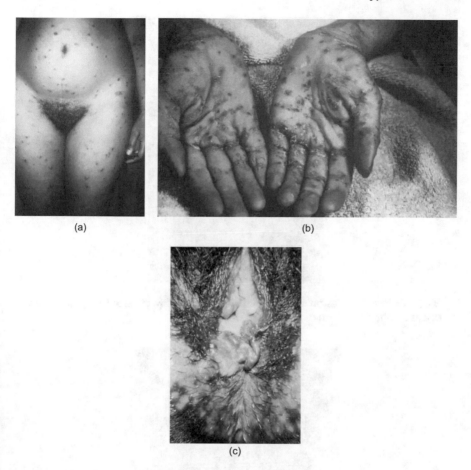

(a)   (b)

(c)

Fig. 31.2 Symptoms of secondary syphilis. (**a**) The torso and upper thighs of a female patient with secondary syphilis showing a typical rash. (Centers for Disease Control and Prevention, USA). (**b**) Lesions on the palms of the hands of a patient with secondary syphilis. (Centers for Disease Control and Prevention, USA). (**c**) A close view of a patient with secondary syphilis showing the presence of condylomata lata lesions of the vagina. Note the knobby, almost pustular appearance of the vaginal lesions. (Joyce Ayers, Centers for Disease Control and Prevention, USA)

results in cardiovascular syphilis which can cause angina, heart failure or bulging of the aorta (aortic aneurysm). Cardiovascular syphilis affects around 10% of patients with tertiary syphilis. The appearance of soft, rubbery growths (known as gummas) on the skin, liver, bones and testes is characteristic of gummatous syphilis (Fig. 31.3). These lesions usually develop 3–10 years following the initial infection. Patients with tertiary syphilis are not infectious.

A woman can pass on syphilis to her unborn child for up to 4 years following initial infection. However, the risks are greatest during the first 1–2 years

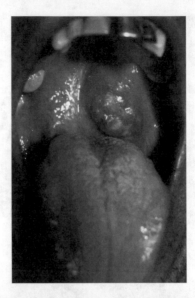

Fig. 31.3 Gummatous lesion on the palate of a patient with tertiary syphilis. (Centers for Disease Control and Prevention, USA)

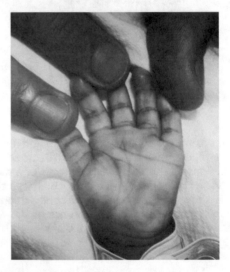

Fig. 31.4 Copper-coloured rash on the palm of a neonate with congenital syphilis. (Centers for Disease Control and Prevention, USA)

after infection, i.e. during the primary, secondary and early latent stages. About 30% of these pregnancies result in foetal death, stillbirth or death shortly after delivery. Congenital syphilis (Fig. 31.4) is classified as "early" (symptoms occur from birth to 2 years of age) or "late" (symptoms occur after the age of 2 years). Symptoms include rhinitis, deformed bones, facial and dental abnormalities, anaemia, enlarged liver and spleen, jaundice, blindness, deafness, meningitis and skin rashes (Fig. 31.4). Screening tests for syphilis should be performed on all pregnant women.

How Common Is It?

Syphilis has a worldwide distribution and is a major public health problem in developing countries. Since the introduction of penicillin for the treatment of the disease, the number of cases in developed countries fell year on year until the middle of the last century. After this time the number of cases started to rise with a significant spike in the 1980s in heterosexuals associated with illicit drug use in inner cities. It's thought that drug use results in more risky behaviour and that budget cuts for STI treatment and prevention programmes may be a contributing factor.

According to the World Health Organization, in 2012 about 18 million people aged 15–49 years of age had syphilis, and the number of new cases each year is about six million. In the USA in 2017, a total of 30,644 new cases of primary and secondary syphilis were reported – an incidence rate of 9.5 cases per 100,000 population. This represented a 10% increase over the previous year. Men are seven times more likely to have the disease than women, and 58% of all reported cases occurred in men who have sex with men. In men, the incidence rate was highest in the age group 25–29 years, while in women it was highest in the age group 20–24 years. The incidence of congenital syphilis has also increased considerably; in the USA in 2017, there were 23.3 cases in 100,000 live births.

In the 1980s in the UK, syphilis was almost eradicated, but, although it's still relatively uncommon, rates are increasing steadily, especially in men in general and in men who have sex with men. In 2018 in the UK, there were 7541 new cases of syphilis (13.1 per 100,000 population) which represented a 5% increase since 2017. Men are almost ten times more likely to have the disease than women. With regard to age group, in women the prevalence of the disease was highest in those aged 20–24 years, while in men the highest prevalence was in those aged 25–34 years. In 2014 in the UK, the incidence of congenital syphilis was 0.1 cases per 100,000 live births.

What Is *T. pallidum*?

T. pallidum is a spiral bacterium 6–15 micrometres long and 0.25 micrometres in diameter. Because it's so thin, it can't be seen in an ordinary light microscope, and a special technique (darkfield microscopy) is necessary to visualise it (Fig. 31.5). It can propel itself along by means of an undulating, wave-like motion. It's a fragile organism that can't withstand dry conditions or disinfectants and can't survive outside the body for long. It prefers an environment with a low concentration of oxygen and is rapidly killed in the presence of normal atmospheric concentrations of oxygen.

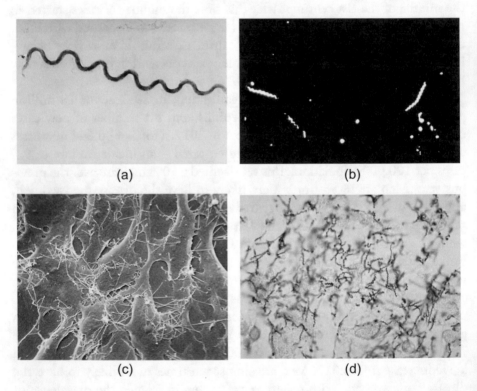

(a)   (b)

(c)   (d)

Fig. 31.5 Microscopic images of *T. pallidum*. (a) Electron micrograph of *T. pallidum*. (Bill Schwartz, Centers for Disease Control and Prevention, USA). (b) *T. pallidum* viewed by darkfield microscopy. (Renelle Woodall, Centers for Disease Control and Prevention, USA). (c). Electron micrograph showing numerous cells of *T. pallidum* on the surface of epithelial cells. (Dr. David Cox, Centers for Disease Control and Prevention, USA). (d) Stained cells of *T. pallidum*. (Dr. Edwin P. Ewing, Jr. Centers for Disease Control and Prevention, USA)

What Happens During an Infection?

Direct contact with an infectious lesion or ulcer is almost always necessary for transmission of the disease, and it can't be acquired from toilet seats or clothes, but it can from shared sex toys.

Unfortunately, all attempts to grow the organism in the laboratory have failed, and this has hampered studies of how it causes disease and how it damages human tissues. Growth in laboratory animals such as rabbits is slow, and reproduction occurs only every 30–33 hours which is approximately 90 times slower than a typical bacterium such as *Escherichia coli*.

Entry of the bacterium into tissues probably occurs via minute abrasions in the genital, anal or oral mucosa that have been caused by sexual activity. Fewer than 10 bacteria have been shown to be able to initiate an infection. It then sticks to proteins in the underlying tissues such as fibronectin and laminin and penetrates into blood vessels by forcing its way between the cells that form the walls of these vessels. This is probably helped by an enzyme that it produces which is able to breakdown the proteins that hold these cells together.

T. pallidum can evade phagocytic cells and the antibodies that are produced against it and so can spread throughout the body via the bloodstream. The tissue damage associated with the disease is caused by the inflammation that the bacterium induces, but little is known about the mechanisms involved.

How Is Syphilis Diagnosed?

Diagnosis of syphilis on the basis of the clinical symptoms is difficult for a number of reasons. The lesions of primary syphilis, which are usually painless, may be missed, especially when they occur in sites such as the cervix or rectum. The rash and other symptoms of secondary syphilis can be faint or mistaken for other diseases.

Because the organism can't be grown in the laboratory or seen using a standard microscope, a suggestive history needs to be confirmed by blood tests or the examination of samples by darkfield microscopy. A number of tests are available, and these detect the presence of antibodies to *T. pallidum* in the serum of the patient – anyone who has ever been infected with the organism should have antibodies against it in their blood.

Serum is the fluid portion of blood left after the cells have been separated from it by spinning (centrifugation) and the fibrinogen (which is responsible for making blood clot) has been removed. The tests are, therefore, known as serological tests and include the fluorescent treponemal antibody absorption (FTA-ABS), microhaemagglutination assay for *T. pallidum* (MHA-TP), *T. pallidum* haemagglutination (TPHA), *T. pallidum* particle agglutination (TPPA) and the immunocapture assay (ICA) tests.

A patient with a positive result should remain positive for life, and therefore the test can't be used to distinguish between an active infection (i.e. one that is currently untreated or has been incompletely treated) and a past infection. These tests become positive around 4–6 weeks after the patient becomes infected.

In order to confirm the diagnosis, and to provide evidence of active disease or a re-infection, a "non-treponemal test" needs to be carried out. Such tests detect the presence, and quantity, of antibodies to cardiolipin which is a compound produced during an active infection with *T. pallidum*. Because they provide a measure of the quantity of antibodies present (known as the "antibody titre"), they can be used to measure disease activity and to monitor the effectiveness of treatment. Two such tests are widely used – the Venereal Disease Research Laboratory (VDRL) test and the "rapid plasma reagin" (RPR) test. The VDRL and RPR tests become positive about 4 weeks after a patient becomes infected with *T. pallidum* and the titre gradually decreases after treatment.

Although serological tests are currently the main approach to diagnosing syphilis, nucleic acid amplification tests (NAATs) such as PCR (see Chap. 1) are being developed and are likely to be employed in the near future. The advantage of NAATs are that they are more specific. Serological tests have a lower specificity because those that test for antibodies against *T. pallidum* can't distinguish between an infection with *T. pallidum* and ones due to other *Treponema* species such as yaws, pinta and bejel. The non-treponemal tests are also not specific for syphilis because cardiolipin is produced during a number of other infections and conditions.

How Is Syphilis Treated?

Prior to the antibiotic era, a number of weird and wonderful treatments for syphilis were tried out (Box 31.2).

Box 31.2 Historical Treatments for Syphilis

In the early sixteenth century, one of the main treatments involved extracts of guaiacum (also known as "holy wood") which was a tree from the Caribbean. Syphilis was thought to have been imported into Europe by Columbus's sailors, so it was widely believed that God would have provided a cure for the disease in the New World. A typical treatment involved "...drink the first potion by the beaker twice a day: in the morning at sunrise and by the light of the evening star. The treatment lasts until the moon completes its orbit and after the space of a month conjoins again with the sun. The patient must remain in a room protected from wind and cold, so that frost and smoke do not diminish the effect of the remedy".

Mercury was also widely used and administered in drinks and applied to the skin in ointments. However, its use gave rise to dreadful side effects (neuropathies, kidney failure, mouth ulcers and tooth loss) and often resulted in the death of the patient. Treatment would last for years and gave rise to the saying, "A night with Venus and a lifetime with mercury". Sweat baths were also widely used because it was thought that the resulting increased salivation and sweating would eliminate the syphilitic poisons.

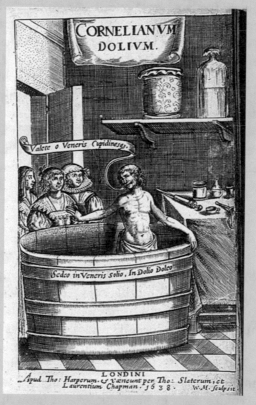

A seventeenth century illustration of a sweating tub used in the treatment of syphilis. (Randolph's Cornelianum dolium. Credit: Wellcome Collection. Attribution 4.0 International (CC BY 4.0))

Care is best managed by specialist sexual health services as full STI screening is required and sexual contact tracing and partner notification need to be offered. In addition, the illness requires several different tests to diagnose, the success of treatment needs to be monitored and affected people require follow up. Prompt diagnosis and treatment of the disease is important in order to minimise progression and thereby avoid the complications that arise from tertiary syphilis. Treatment usually involves the administration of penicillin or doxycycline in those who are allergic to penicillin.

Penicillin is currently the drug of choice for all stages of the disease as well as for congenital syphilis and for pregnant women with the disease. Intramuscular injection is preferred to oral medication as treatment can be monitored, and it achieves effective drug levels more reliably.

Interestingly, despite its long and widespread use for treating the disease, there's no indication that *T. pallidum* is becoming resistant to penicillin. Within a few hours of penicillin administration, constituents of the dying bacteria can stimulate the large-scale release of cytokines which results in an inflammatory response known as the Jarisch-Herxheimer reaction. The symptoms of this include muscle pain, fever, headache and an increased heart rate – these usually disappear within 24 hours.

For patients with either primary, secondary or latent syphilis, the prognosis is good following appropriate treatment, and cure is highly likely. However, for patients with tertiary syphilis, the prognosis is less positive because, although antibiotics are effective at preventing further deterioration, they can't reverse any existing damage. About 20% of untreated patients with tertiary syphilis die of the disease – syphilis is one of the few STIs capable of killing its host.

How Can I Avoid Getting Syphilis?

The use of condoms reduces the genital-to-genital transmission of syphilis although it may have little effect on other routes of transmission such as the orogenital route. It's not yet clear whether male circumcision affects syphilis transmission although it may reduce the risk of other STIs such as HIV.

As people can be infected but not show any symptoms, it's important to practise safe sex, have regular screening and avoid having overlapping relationships. Self-sampling postal kits have been introduced in some areas in the UK looking for syphilis cases in high-risk groups (as well as HIV), so that people who were unaware that they had the infection can be treated thereby reducing onward transmission in the community.

Partners and sexual contacts of known cases should either have treatment or testing as appropriate. Recent contacts, within the last 90 days, are usually treated even if their results are negative. Contact tracing may have to be extended back for many years depending on what phase of infection the affected person is in.

In the past, rare cases have occurred following blood product transfusion and organ transplantation, but screening and refrigeration of the blood supply have since been introduced to prevent this.

Screening for the disease during pregnancy is important to identify and treat asymptomatic women with syphilis so as to prevent transmission to their foetus or baby.

Currently, no vaccine is available against the disease, but research is ongoing to try to develop one.

Want to Know More?

American Academy of Family Physicians. https://familydoctor.org/condition/syphilis/

Centers for Disease Control and Prevention, USA. https://www.cdc.gov/std/syphilis/stdfact-syphilis-detailed.htm

DermNet, New Zealand, https://www.dermnetnz.org/topics/syphilis/

Family Planning Association, UK. https://www.sexwise.fpa.org.uk/stis/syphilis

Mayo Clinic, USA. https://www.mayoclinic.org/diseases-conditions/syphilis/symptoms-causes/syc-20351756

National Health Service, UK. https://www.nhs.uk/conditions/syphilis/

National Institute for Clinical Care and Excellence (NICE), UK. Syphilis, 2019. https://cks.nice.org.uk/syphilis#!topicSummary

Patient Info, UK. https://patient.info/sexual-health/sexually-transmitted-infections-leaflet/syphilis

Terrence Higgins Trust, UK. https://www.tht.org.uk/hiv-and-sexual-health/sexual-health/stis/syphilis?gclid=EAIaIQobChMI5c3E3NPd5QIVmpntCh1qJwbPEAAYASAAEgIGoPD_BwE

US Department of Health and Human Services, https://www.womenshealth.gov/a-z-topics/syphilis

Chandrasekar PH. Syphilis. Medscape from WebMD, 2017. https://emedicine.medscape.com/article/229461-overview

Clement ME, Okeke NL, Hicks CB. Treatment of syphilis: a systematic review. *Journal of the American Medical Association.* 2014 Nov 12;312(18):1905-17. https://doi.org/10.1001/jama.2014.13259. PMID: 25387188

Peeling RW, Mabey D, Kamb ML, Chen XS, Radolf JD, Benzaken AS. Syphilis. *Nature Reviews Disease Primers.* 2017 Oct 12;3:17073. https://doi.org/10.1038/nrdp.2017.73. PMID: 29022569

Salazar JC, Bennett N, Cruz AR. Syphilis infection. BMJ Best Practice, BMJ Publishing Group, 2019. https://bestpractice.bmj.com/topics/en-gb/50

Stamm LV. Syphilis: re-emergence of an old foe. *Microbial Cell.* 2016 Jun 27;3(9):363-370. https://doi.org/10.15698/mic2016.09.523. PMID: 28357375

Tudor ME, Al Aboud AM, Gossman WG. Syphilis. Treasure Island (FL): StatPearls Publishing LLC; 2020. https://www.ncbi.nlm.nih.gov/books/NBK534780/

Part VII

Infections of the Gastrointestinal System

32

Gastroenteritis Due to *Campylobacter*

Abstract A variety of microbes, including viruses, bacteria and protozoa, are able to cause infections of the gastrointestinal tract, i.e. gastroenteritis. The most frequent type of bacterial gastroenteritis globally is campylobacteriosis – a type of food poisoning caused by the bacterium *Campylobacter jejuni* or, occasionally, *Campylobacter coli*.

Those who are infected may have no symptoms at all, but if they do the main symptoms are diarrhoea, abdominal pain and fever – these usually last for 5–7 days. About one third of those infected also experience headache, muscle aches, chills, fever and vomiting about 24 hours before the onset of diarrhoea. Complications are rare but include reactive arthritis which usually affects the knee and ankles and can last for weeks or months. A neurological disorder due to damage to the sensory and motor nerves (known as Guillain-Barre Syndrome, GBS) is an infrequent complication and usually lasts for a few weeks although it may persist for several years.

Campylobacteriosis mainly affects children younger than 5 years and those over 65 years, and, in temperate climates, it occurs more often in the summer. It's a zoonosis, i.e. an infection that is transmitted from animals (or their products) to humans. Most people get the disease by eating contaminated food (such as undercooked poultry) or drinks (such as unpasteurised milk). 50–70% of human infections are contracted from contaminated chickens. It can also be transmitted in water and by contact with human or animal faeces. Often the source of the infection isn't identified.

The disease is diagnosed by finding the organism in your faeces. Most people recover from the infection without specific treatment, but those with other health conditions may require medical attention, and in people with weak

© Springer Nature Switzerland AG 2021
M. Wilson, P. J. K. Wilson, *Close Encounters of the Microbial Kind*,
https://doi.org/10.1007/978-3-030-56978-5_32

immunity, it can be life-threatening. The most common and serious consequence of the disease in most patients is dehydration, and therefore maintaining good hydration is essential. Antibiotics aren't usually needed but may be useful in severe cases and for those who are immunocompromised. The disease can be avoided by maintaining high standards of personal hygiene and safe food-handling practices.

I've only ever had a bad attack of gastroenteritis once, but oh boy I won't ever forget it. It was back in my student days when money was scarce and home hygiene was somewhat lacking in the flat I shared with four other poor and messy individuals. One morning I came down from my room, and I was absolutely starving, ravenous and faint from hunger. But the fridge was empty, and there was nothing in the cupboards. No tin of beans or slice of old bread – nothing. But then I saw it – salvation! On the living room table was somebody's take-away left-overs from the night before. Half of the chicken hadn't been eaten, oh what joy! I happily tucked into what I thought was likely to be the only source of nourishment available to me that day. A few hours later I was gripped with waves of nausea followed by vomiting and impossible-to-describe fluid gushing from the other end. I couldn't face chicken again for a long, long time.

What Is Diarrhoea and What Are the Main Causes?

Diarrhoea means the frequent passing of loose or watery stools (more than three per day) and is a common symptom of an infection of the lower intestinal tract (Fig. 32.1). It's a huge problem globally, affecting more than 550 million people each year. In 2016, diarrhoea was the eighth leading cause of mortality, responsible for more than 1.6 million deaths, more than a quarter of which were in children younger than 5 years.

Not all cases of diarrhoea, however, are due to an infection, and other causes include anxiety, excessive alcohol consumption, food allergy and as a side effect of drugs such as antibiotics, laxatives, non-steroidal anti-inflammatory drugs, antacids and statins.

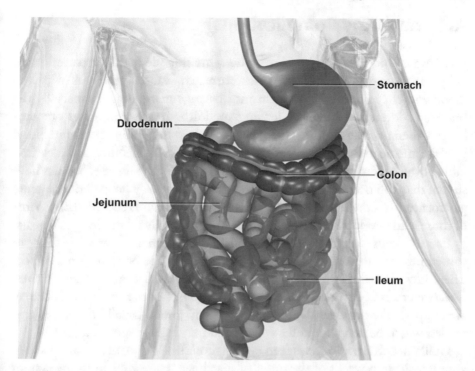

Fig. 32.1   Diagram showing the lower regions of the gastrointestinal system. (Blausen. com staff (2014). "Medical gallery of Blausen Medical 2014" *WikiJournal of Medicine* 1 (2). DOI: https://doi.org/10.15347/wjm/2014.010. ISSN 2002-4436. [CC BY 3.0 (https://creativecommons.org/licenses/by/3.0)])

What Is Campylobacteriosis?

In this chapter we'll discuss diarrhoea due to *Campylobacter* species which is known as campylobacteriosis and is one of the most frequent causes of gastroenteritis in developed countries. It's also a frequent cause of traveller's diarrhoea, i.e. diarrhoea that occurs during, or within 10 days of, a trip abroad. The main symptoms of campylobacteriosis are diarrhoea, abdominal pain and fever. One third of patients experience headache, myalgia, chills, fever and vomiting approximately 24 hours before the onset of diarrhoea. Vomiting occurs in 15% of patients and is most common in infants. The diarrhoea may be bloody in 20–25% of cases, and, at the peak of symptoms, there may be as many as 8–10 bowel movements a day. More than 90% of patients experience fever which can last for up to 1 week. Symptoms usually appear within 2–5 days of exposure to the organism, and the illness typically lasts for 5–7 days, although it may persist for a few weeks.

Are There Any Complications?

If the water and salts lost in the diarrhoea are not adequately replaced, then dehydration can result which, if not corrected, causes problems with the blood and circulation which can in turn damage many organs especially the brain and kidneys, ultimately leading to death. If the symptoms are prolonged, then poor absorption of nutrients can lead to malnutrition, and blood levels of medication can drop. A temporary lactose intolerance and also irritable bowel syndrome (IBS) can follow an episode of the disease. Studies have shown that IBS occurs in as many as one third of those who experience acute gastroenteritis – the condition is known as post-infectious IBS (PI-IBS). Most of those who develop PI-IBS recover, although this may take as long as 6 years.

Bacterial causes of gastroenteritis trigger an inflammatory response that can sometimes lead to various symptoms according to the body region affected. These include arthritis, conjunctivitis, urethritis, carditis and certain rashes. Reactive arthritis occurs 1–4 weeks after the onset of the infection and may affect up to 5% of patients. It can affect many joints, usually the knees and ankles, which become hot and swollen, and may last for several weeks.

Guillain-Barre syndrome is seen rarely (about 0.1% of patients) and usually appears within 6 weeks of the onset of diarrhoea. Here your immune system is abnormally triggered to damage the sensory and motor nerves, and amongst other symptoms, this results in muscle weakness that usually lasts for a few weeks but may persist for several years. In around 15% of cases, the breathing muscles are affected initially which is life-threatening. Most people recover fully, but some have permanent nerve damage, and, in severe cases, the person can become paralysed.

Hepatitis, pancreatitis and peritonitis can result from the direct spread of the organism to nearby organs or structures. In immunocompromised patients, the bacterium may gain access the blood stream and cause a bacteraemia which can lead to sepsis and death. Rarely the bacterium can spread to other sites where it can cause meningitis, bone infections and endocarditis.

How Common Is Campylobacteriosis?

C. jejuni is one of the four main causes of diarrhoeal diseases globally and is considered to be the most common bacterial cause of gastroenteritis worldwide. It affects all age groups but is more common in males, children younger than 5 years and people 65 years and older. Although the infection is generally mild, it can be fatal among very young children, the elderly and immunosuppressed individuals. In temperate climates the disease occurs more frequently in the summer. In 2011, it was estimated that there were around 1.3 million cases of campylobacteriosis in the USA – approximately 14 cases per 100,000 people each year. In the UK there are up to 17 million cases of gastroenteritis each year, and these result in one million medical consultations. Of these, campylobacteriosis accounts for 500,000 cases and 80,000 consultations.

> **Box 32.1 What's in a Name?**
>
> The genus *Campylobacter* gets its name from the Greek words "Kampylos" and "baktron" which mean bent or curved and rod, respectively. *C. jejuni* is so-called because the jejunum (part of the small intestine) is one of the main regions of the intestinal tract that it damages during an infection. The organism was first seen by the Austrian/German physician Theodor Escherich in the stool samples of children with diarrhoea, but it wasn't until 1972 that it was isolated in the laboratory.

Which Microbes Cause Campylobacteriosis?

Campylobacteriosis is so-called because it's caused by bacteria belonging to the genus *Campylobacter*. Members of this genus consist of small Gram-negative curved or spiral bacilli (4 μm long and 0.2–0.5 μm wide) that can move by means of flagella (Fig. 32.2).

The organisms prefer an atmosphere that's low in oxygen, but they can survive normal atmospheric conditions. They aren't found in healthy humans but are members of the intestinal microbiota of birds as well as wild and domesticated mammals including pigs, cattle, dogs and cats. Eleven species of *Campylobacter* cause infections in humans, but 90% of cases of campylobacteriosis are due to *C. jejuni* (Fig. 32.3) – the remaining cases are due mainly to *C. coli*.

Campylobacteriosis is a type of infection known as a zoonosis, i.e. one that is transmitted from other animals (or their products) to humans. Most people get the disease by eating contaminated food (such as undercooked poultry) or drinks (such as unpasteurised milk) – the disease is therefore a type of "food poisoning". It's been estimated that 50–70% of human cases are contracted from contaminated chickens. You can get infected when a chopping board that's been used to prepare raw chicken hasn't been washed before being used to prepare foods that are served raw or lightly cooked, such as salad or fruit. It can also be transmitted in water (contaminated by animal faeces) and by contact with the faeces of an animal (particularly cats and dogs) or an infected human. Flies are known to transmit gastrointestinal infections from

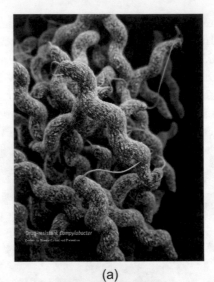

(a)

(b)

Fig. 32.2 Images of a *Campylobacter* species as seen through an electron microscope. Note the curved and spiral shapes **(a)** and the presence of a long, whip-like flagellum **(b)**. ((a) Alissa Eckert and James Archer, Centers for Disease Control and Prevention, USA. **(b)** Credit: David Gregory & Debbie Marshall. CC BY 4.0. Wellcome Images)

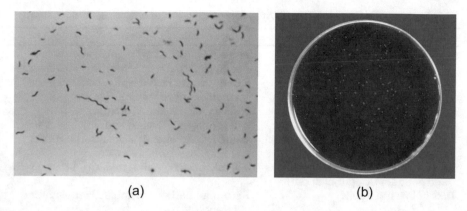

(a)   (b)

Fig. 32.3 Images of *C. jejuni*. (a) Gram stain of *C. jejuni* showing curved and spiral Gram-negative bacilli. (Centers for Disease Control and Prevention, USA). (b). Colonies of *C. jejuni* growing on an agar plate. (Sheila Mitchel, Centers for Disease Control and Prevention, USA)

contaminated faeces and may be responsible for spread between animals and also possibly to humans.

What Happens During an Infection?

In order to cause an infection, a microbe must be able to overcome the extensive range of antimicrobial defence systems present in the gastrointestinal tract, and these are summarised in Table 32.1.

However, a variety of microbes (bacteria, viruses and protozoa) can overcome these defences and cause gastroenteritis, and these are listed in Table 32.2.

The number of live *C. jejuni* required to cause an infection (known as the "infectious dose") can be as low as 500–800. It's sensitive to the acidic contents of the stomach which means that if you are taking antacids, you'll have a less acidic stomach than normal and will therefore be more vulnerable to the infection. Following ingestion, the bacterium penetrates the mucus layer covering the epithelium of the small bowel (jejunum or ileum) and, using its flagella, moves through this and attaches itself to an epithelial cell. This induces an inflammatory response which attracts large numbers of immune cells to the intestinal epithelium. Following attachment, the bacterium then invades the epithelial cell which is eventually killed. *C. jejuni* also produces a toxin that can kill epithelial and other human cells (Table 32.3). The death of large numbers of epithelial cells means that the small intestine can't carry out one of its main functions which is the absorption of water and nutrients.

Table 32.1 Antimicrobial defence systems operating in the gastrointestinal system

Mechanism	Effect
Production of mucus	Prevents microbes adhering to epithelial cells; traps microbes; improves movement of gut contents out of the body thereby expelling microbes
Rapid transit of contents in upper regions	Hinders microbial colonisation in the oesophagus, stomach and small intestine
Shedding of outer layer of epithelial cells (desquamation)	Removes microbes attached to these cells
Production of antimicrobial proteins and peptides by epithelial cells	Kill microbes or inhibit their growth
TLRs of intestinal cells	Recognise pathogens and activate immune defences
Release of cytokines by epithelial cells	Attracts and activates phagocytes
Low pH of stomach and duodenum	Kills or inhibits a wide range of microbes
Bile acids produced by gall bladder	Kills many microbes – particularly Gram-positive species
Stimulation of excessive fluid secretion	Flushes out microbes
Production of antibodies	Blocks adhesion of microbes to epithelial cells

Table 32.2 Microbes responsible for most cases of gastroenteritis in developed countries

Type of microbe	Main examples
Bacterium	*Campylobacter* species, *Salmonella* species, *Clostridium* species, *Shigella* species, *Escherichia coli*
Virus	Norovirus, rotavirus, adenovirus, astrovirus, sapovirus
Protozoan	*Giardia* species, *Cryptosporidium* species

Table 32.3 Virulence factors of *C. jejuni* and their function

Virulence factor	Function
Flagella	Enable the bacterium to move through the mucus layer and reach the underlying epithelium; involved in invasion of epithelial cells
Proteins in the cell wall	Function as chemical anchors (adhesins) which enable the bacterium to adhere to epithelial cells; also involved in invasion of epithelial cells
Toxin	Kills epithelial and other human cells
Capsule	Protects the bacterium from phagocytes; causes the release of inflammatory cytokines which results in the inflammation response
Cell wall components such as lipopolysaccharide	Causes the release of inflammatory cytokines

Approximately 9.3 litres of fluid enter the intestinal tract each day, and the small intestine is responsible for absorbing approximately 90% of this. Failure of the damaged small intestine to absorb fluid results in the profuse diarrhoea that is characteristic of campylobacteriosis and other gastrointestinal infections.

How Is the Disease Diagnosed?

A diagnosis of campylobacteriosis can't be made on the basis of the clinical symptoms alone because they are similar to those resulting from infection by a variety of microbes. A stool sample may be examined for the presence of red blood cells and neutrophils as these are present in 75% of patients with campylobacteriosis. However, a definite diagnosis can only be made by culturing *C. jejuni* from a stool sample, and this usually takes between 2 and 3 days. The isolated microbe can then also be tested to determine which antibiotics it's sensitive to and therefore which would be suitable for treating the disease, should antibiotics be necessary.

How Is Campylobacteriosis Treated?

Most people with campylobacteriosis recover without specific treatment. The most serious consequence of the disease in most patients is dehydration, and therefore good salt and fluid intake is essential. If you're able to drink, then you should drink plenty of water as well as dilute soups and fruit juices. In at-risk groups, such as the frail or elderly, special oral rehydration solutions which contain particular concentrations of sugars and salts should be used in addition, as this effectively replaces the lost fluid and minerals (Box 32.2). You should drink in small, frequent sips.

Box 32.2 Oral Rehydration Therapy

The diarrhoea and vomiting that accompanies gastrointestinal infections can result in appreciable loss not only of water but also of the electrolytes such as sodium, potassium, chloride and bicarbonate ions that are essential for the proper functioning of many body systems. Severe loss of water and electrolytes (particularly sodium ions) results in dehydration which is especially dangerous for young children and older adults. The symptoms of dehydration differ in infants and adults and are summarised in the table below.

Table. Main symptoms of dehydration in infants and adults

Infant	Adult
Dry mouth and tongue	Extreme thirst
No tears when crying	Less frequent urination
Reduced frequency of wet nappies	Dark-coloured urine
Sunken eyes, cheeks	Fatigue
Sunken soft spot on top of the skull	Dizziness
Listlessness or irritability	Confusion

A crucial aspect of the treatment of gastroenteritis is rehydration of individuals who have become dehydrated due to the illness and also replacing electrolytes, particularly sodium ions. Water and sodium ions can't be replaced by simply giving the patient a solution of water and salt (sodium chloride). This is because the system responsible for absorbing sodium ions into the body from the lumen of the intestinal tract (the gut is, effectively, a tube and the central hollow is known as the lumen) is damaged by the illness. Consequently, the sodium ions remain inside the lumen, and this draws more water out of the body and makes the diarrhoea worse. However, if glucose is added to the solution, then glucose molecules are able to pass into the body across the damaged intestinal wall and drag sodium ions with them. The increased concentration of sodium ions inside the body then pulls water molecules from the lumen into the body. In 1978, the prestigious medical journal *The Lancet* stated: "The discovery that sodium transport and glucose transport are coupled in the small intestine so that glucose accelerates absorption of solute and water is potentially the most important medical advance this century". A simple oral rehydration solution, therefore, should consist of water, glucose and sodium chloride. This is usually supplemented with potassium ions which are also lost due to diarrhoea. ORT has been shown to reverse dehydration in 90% of patients with acute diarrhoea. Since the World Health Organization began using ORT in 1978, the annual death rate among children under 5 due to diarrhoea has fallen from five million to fewer than one million. It's estimated that ORT has saved around 70 million lives since its introduction in the late 1970s, and its use has decreased the risk of death from diarrhoea by up to 93%.

If dehydration is particularly severe, then rehydration may have to be administered intravenously – in this case the solution usually contains just water and salt.

If you're unable to take oral fluids, then you may have to be given fluids through a tube passed into the stomach through the nose or intravenously.

There's some evidence suggesting that, when used alongside rehydration therapy, probiotics can shorten the duration of diarrhoea and reduce stool frequency.

Antibiotics aren't needed for most patients. However, they should be considered when patients are very unwell or have severe or prolonged symptoms. Antibiotics are also recommended for immunocompromised patients. Macrolide antibiotics such as erythromycin or azithromycin are usually the most effective. There are reports of increasing resistance to quinolone antibiotics such as ciprofloxacin, so these shouldn't be used unless laboratory tests show that the strain of *C. jejuni* isolated from the patient is sensitive to this.

How Can I Avoid Getting Gastroenteritis?

Appropriate food handling and hygiene practices are essential for preventing the spread of the disease, and these include (i) adequate heating of meat to a minimum internal temperature of 74 °C (165 °F), (ii) washing hands after handling raw meat, (iii) regular washing of kitchen utensils, (iv) keeping raw meat away from other foods, (v) using separate cutting boards for raw meat (including poultry, seafood and beef) and for fresh fruits and vegetables and (v) cleaning all cutting boards, countertops and utensils with soap and hot water after preparing any type of raw meat.

Unpasteurised milk and unpurified water should always be avoided.

You can prevent the spread of infectious diarrhoea by maintaining high standards of personal hygiene including (i) washing your hands thoroughly with soap and warm water after going to the toilet and before eating or preparing food, (ii) cleaning the toilet, including the handle and the seat, with disinfectant after each bout of diarrhoea, (iii) avoiding the sharing of towels, flannels, cutlery or utensils with others and (iv) washing soiled clothing and bed linen separately from other clothes and at the highest temperature possible. Patients shouldn't return to work or school until at least 48 hours after the last episode of diarrhoea.

Want to Know More?

Centers for Disease Control and Prevention, USA https://www.cdc.gov/campylobacter/index.html

European Centre for Disease Prevention and Control https://www.ecdc.europa.eu/en/campylobacteriosis

Food Standards Agency, UK https://www.food.gov.uk/safety-hygiene/campylobacter

Institute of Food Science and Technology, UK https://www.ifst.org/resources/information-statements/foodborne-campylobacteriosis

KidsHealth, USA https://kidshealth.org/en/parents/campylobacter.html

National Health Service, UK https://www.nhsinform.scot/illnesses-and-conditions/infections-and-poisoning/food-poisoning

National Institute for Clinical Care and Excellence (NICE), UK. Gastroenteritis, 2019 https://cks.nice.org.uk/gastroenteritis#!topicSummary

Patient Info, UK https://patient.info/digestive-health/diarrhoea/campylobacter

World Health Organisation https://www.who.int/news-room/fact-sheets/detail/campylobacter

Bonheur JL. Bacterial Gastroenteritis. Medscape from WebMD, 2018. https://emedicine.medscape.com/article/176400-overview

Ehrenpreis ED. Campylobacter infection. BMJ Best Practice, BMJ Publishing Group, 2018. https://bestpractice.bmj.com/topics/en-gb/1175

Facciolà A, Riso R, Avventuroso E, Visalli G, Delia SA, Laganà P. Campylobacter: from microbiology to prevention. *Journal of Preventive Medicine and Hygiene.* 2017 Jun;58(2):E79-E92. PMID: 28900347

Fischer GH, Paterek E. Campylobacter. Treasure Island (FL): StatPearls Publishing LLC; 2020 https://www.ncbi.nlm.nih.gov/books/NBK537033/

Javid MH. Campylobacter Infections. Medscape from WebMD, 2019 https://emedicine.medscape.com/article/213720-overview

Kaakoush NO, Castaño-Rodríguez N, Mitchell HM, Man SM. Global epidemiology of Campylobacter infection. *Clinical Microbiology Reviews.* 2015 Jul;28(3):687-720. https://doi.org/10.1128/CMR.00006-15.

Same RG, Tamma PD. Campylobacter infections in children. *Pediatrics Review.* 2018 Nov;39(11):533-541. https://doi.org/10.1542/pir.2017-0285. PMID: 30385582

Sattar SBA, Singh S. Bacterial Gastroenteritis. Treasure Island (FL): StatPearls Publishing LLC; 2019 https://www.ncbi.nlm.nih.gov/books/NBK513295/

Skarp CPA, Hänninen ML, Rautelin HIK. Campylobacteriosis: the role of poultry meat. *Clinical Microbiology and Infection.* 2016 Feb;22(2):103-109. https://doi.org/10.1016/j.cmi.2015.11.019. Epub 2015 Dec 11. PMID: 26686808

33

Gastroenteritis Due to *Salmonella*

Abstract Salmonellosis is a type of gastroenteritis caused by *Salmonella enterica*. The main symptoms are diarrhoea (which can sometimes be bloody), fever and stomach cramps. Other possible symptoms include nausea, vomiting, headaches and muscle pains. The disease is generally mild and self-limiting and usually resolves within 4–7 days. Complications aren't very common but may include reactive arthritis which most frequently affects the ankles and knees and usually lasts for no longer than a few months. In the very young or old, or those with a weakened immune system, the organism may spread and cause further infection in the heart, bones, joints, brain and gallbladder. Salmonellosis is a zoonosis and can be caught by direct contact with animals carrying *Salmonella*. However, it's mainly a form of food poisoning, and 95% of cases are due to the consumption of contaminated food from infected animals or their products. The foods most frequently affected are poultry, eggs, pork, ground beef, dairy products and peanut products. Diagnosis is usually made by culturing *Sal. enterica* from, or detecting the organism's DNA in, a stool sample. The infection can result in dehydration, and therefore oral or intravenous fluids may be necessary. Patients at high risk of developing a more severe illness may be offered a short course of oral antibiotics. Prevention of salmonellosis involves proper sanitation and hygiene practices, as well as the avoidance of insufficiently cooked or mishandled food.

© Springer Nature Switzerland AG 2021
M. Wilson, P. J. K. Wilson, *Close Encounters of the Microbial Kind*,
https://doi.org/10.1007/978-3-030-56978-5_33

What Is Salmonellosis?

Salmonellosis is an infection caused by *Salmonella enterica* bacteria. The main symptoms are diarrhoea (which may be bloody although it usually isn't), fever and stomach cramps. Some patients also suffer from nausea, vomiting, headaches and muscle pains. Most cases are mild and resolve on their own without specific treatment. However, diarrhoea can result in severe dehydration in vulnerable people, especially in infants and children under 2 years old and in adults over 65 years old. Symptoms usually start within 6 hours–3 days (most commonly 12–36 hours) after infection with the bacterium and generally last 4–7 days. The diarrhoea, without other symptoms, may continue for up to 10 days, and it may take longer (several months) before the bowels return completely to normal.

Box 33.1 What's in a Name?

A *Salmonella* species (*Salmonella typhi*) was first seen in 1880 by the German bacteriologist Karl Eberth in samples of the intestinal tract of patients suffering from typhoid fever. The organism was then grown in the laboratory about 4 years later.

    Salmonella enterica, the bacterium responsible for gastroenteritis, was first grown in the laboratory in 1885 by the American bacteriologist Theobald Smith who was working under Daniel Elmer Salmon in the Veterinary Division of the US Department of Agriculture in Washington, D.C.

Figure (a). Photograph of Professor Theobald Smith. (Credit: Wellcome Collection. CC BY 4.0)

Figure (b). Portrait of Daniel Elmer Salmon in 1905. (C. M. Bell Studio. This work is from the C. M. Bell Studio Collection collection at the Library of Congress. According to the library, there are no known copyright restrictions on the use of this work. Via Wikimedia Commons)

In 1900 the genus was named in honour of Salmon by the French bacteriologist Joseph Leon Marcel Lignières – it has no connection with the delicious fish that is enjoyed by many of us.

The species name, *enterica*, is derived from the Greek word for gut (enteron) and the Latin suffix "-icus" meaning "belonging to". The name *enterica* therefore means "belonging to the gut".

Are There Any Complications of the Infection?

In most people the illness is mild, but in the elderly, young children, those with other health conditions and the immunocompromised, it can be serious and sometimes life-threatening. Complications such as dehydration, malnutrition, Guillain-Barre syndrome, irritable bowel syndrome and reactive arthritis may occur, in common with other types of infectious gastroenteritis, as discussed in more detail in Chap. 32.

Bacteraemia occurs in around 5% of patients with salmonellosis in developed countries, but in some parts of Africa, it's much more common – this is a serious complication and can be fatal. The blood-borne spread can affect many different parts of the body which can result in infection at a variety of sites including the endovascular lining, cardiac valves, bones, joints, meninges, lungs, spleen and the gallbladder. The spread of *Sal. enterica* outside of the intestinal tract most commonly occurs in patients who are very young or old or have a weakened immune system. HIV infection is also a significant risk factor.

Infection of the lining of blood vessels (endarteritis) has a very high mortality, especially when affecting large arteries such as the aorta. Approximately 50% of central nervous system infections are fatal. Bone and joint infections can be difficult to treat. In people with inflammatory bowel disease, the bacteria can invade the bowel wall and cause toxic megacolon and other serious complications. People with sickle cell disease are more susceptible to *Salmonella* infections and complications.

A rare and serious complication, seen mostly in children, is haemolytic uraemic syndrome although this is more usually triggered by *Escherichia coli* rather than *Salmonella*. In this condition a bacterial toxin causes damage to small blood vessels in the kidney, nervous system and gut. This damages the circulating blood cells which then break down. Ultimately this leads to kidney injury, nerve damage and gut problems. The majority of people recover but some can have longer-lasting kidney complications.

How Common Is It?

Sal. enterica is one of the four key global causes of diarrhoeal diseases and is one of the most commonly identified causes of foodborne illnesses in Europe and the USA. It's been estimated that 93.8 million cases of salmonellosis and 155,000 deaths occur globally each year. In the USA salmonellosis affects more than 1.2 million people each year, and this results in 23,000 hospitalisations and 450 deaths. The disease is responsible for 30% of all deaths associated with foodborne disease in the USA. In the UK during the years 2007–2016, the number of cases of salmonellosis averaged 8800 per year – approximately 16 cases per 100,000 people. In the European Union, over 91,000 cases are reported each year, and it's been estimated that the overall economic burden could be as high as EUR 3 billion a year. Infants, young children and the elderly are at highest risk for the disease, and the immuno-compromised, as well as those at the extremes of age, are also at risk of more severe, complicated infections. In temperate climates, most infections occur during the summer months.

Like campylobacteriosis, salmonellosis is a zoonosis, and 95% of cases are due to the consumption of food (meat or other products) derived from infected animals. Another source of the bacterium is the manure from infected animals if this is used to grow vegetables. *Sal. enterica* is the most common foodborne bacterial pathogen in most countries. It's a ubiquitous and resilient microbe and can survive for months in the environment. In many animals the bacterium is part of their normal gut microbiota and so is difficult to eliminate.

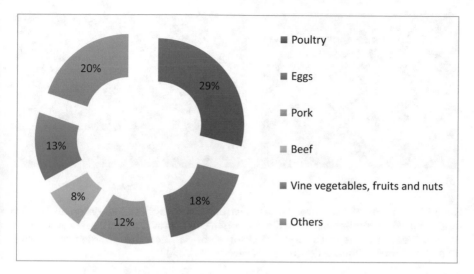

Fig. 33.1 Foods associated with outbreaks of salmonellosis in the USA during the period 2004–2008. "Others" includes sprouts, leafy greens, roots, fish, grains and dairy

The most common sources of the pathogen are poultry, eggs, pork, ground beef, dairy products and peanut products (Fig. 33.1).

The bacterium is spread mainly via the faecal-oral route, but, in addition, infection can result from direct contact with animals carrying *Salmonella*, e.g. amphibians and reptiles (such as iguanas, frogs, turtles, snakes and bearded dragons), chicks, ducklings, kittens and hedgehogs. In the USA, approximately 74,000 cases of salmonellosis each year are attributable to exposure to amphibians and reptiles. Occasionally, contaminated water is the source of the infection, and person-to-person transmission involving food handlers is also possible.

What Happens During an Infection?

Sal. enterica is a Gram-negative bacillus between 2 and 5 µm long and 0.7–1.5 µm in diameter (Fig. 33.2a). It can grow in the presence or absence of oxygen and can move around using flagella. The most common strain responsible for salmonellosis in the UK is *Sal.* Enteritidis, while in the USA a larger range of strains are encountered including *Sal.* Typhimurium, *Sal.* Enteritidis and *Sal.* Newport (note that the naming system used for *Salmonella* species differs from that of most bacteria and is explained in Box 33.2).

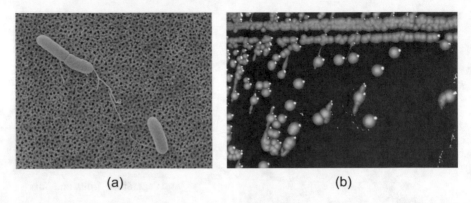

(a)   (b)

Fig. 33.2 Images of *Sal. enterica*. (a) Electron micrograph of *Sal. enterica* showing long, whip-like flagella which enable the organism to move. (Credit: Mark Jepson. CC BY 4.0. Wellcome Image Library). (b) Colonies of *Sal. enterica* growing on an agar plate. (Centers for Disease Control and Prevention, USA)

In healthy individuals it takes about 10^6 bacteria to cause an infection. Many of the bacteria in our food are destroyed by the acidic environment of the stomach, so if this is made less acidic (e.g. by antacid medications), then the number needed may be as low as 10^3. Once it reaches the ileum in the small bowel, the bacterium swims through the mucus layer to reach the underlying epithelial cells. It then attaches itself to an epithelial cell and invades it. Invasion is achieved after a bacterium injects proteins into an epithelial cell using a needle-like device (known as a "type III secretion system"; Fig. 33.3).

These proteins make the cell engulf the bacterium and take it inside (Fig. 33.4). Bacterial adhesion and invasion cause an inflammatory and immune response (described in Chap. 1), but this mobilisation of the antimicrobial defence system is too late to stop an infection from progressing as many of the bacteria will be safely inside the epithelial cells.

While inside the epithelial cell, the bacterium is protected from the body's defence system, and it then grows and reproduces. Eventually the cell ruptures and releases large numbers of new bacteria. The bacteria can then be mopped up by our defence system, particularly by phagocytic cells such as macrophages. However, the killing of large numbers of epithelial cells disrupts the normal functions of the small bowel, including fluid absorption, which results in diarrhoea. *Sal. enterica* can, however, survive for some time inside macrophages, and these cells can re-enter the bloodstream and carry the bacteria to other parts of the body where they may be released and give rise to the various complications described previously. The main virulence factors used by *Sal. enterica* are listed in Table 33.1.

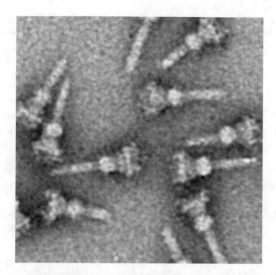

Fig. 33.3 The type III secretion system of *Sal. enterica*. This consists of needle-like structures that can inject proteins into human cells and alter their behaviour. (PLoS Pathog. 2010 Apr 1;6(4):e1000824. doi: https://doi.org/10.1371/journal.ppat.1000824. Topology and organization of the Salmonella typhimurium type III secretion needle complex components. Schraidt O, Lefebre MD, Brunner MJ, Schmied WH, Schmidt A, Radics J, Mechtler K, Galán JE, Marlovits TC

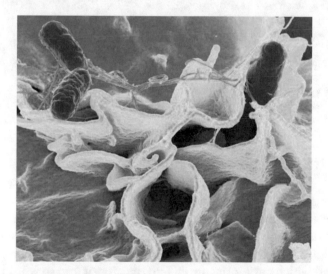

Fig. 33.4 Electron micrograph showing a number of cells of a *Salmonella* species (coloured red) on the surface of a human epithelial cell (coloured yellow). Two of the bacteria are invading the cell – one of these has almost completely disappeared inside the cell. (National Institute of Allergy and Infectious Diseases (NIAID))

Table 33.1 Major virulence factors of *Sal. enterica* and their function

Virulence factor	Function
Flagella	Enable the bacterium to move through the mucus layer and reach the underlying epithelium; induce the release of inflammatory cytokines
Fimbriae	Short filaments on the surface of the bacterium that enable it to adhere to epithelial cells
Type III secretion system	Injects proteins into an epithelial cell which make the cell engulf the bacterium
Capsule	Protects the bacterium from phagocytes; induces the release of inflammatory cytokines
Cell wall components such as lipopolysaccharide	Induce the release of inflammatory cytokines
Multiprotein complex	A set of proteins that enable the bacterium to survive inside epithelial cells and macrophages

Box 33.2 Nomenclature of *Salmonella* Species

The naming of species within the genus *Salmonella* is confusing and has undergone many revisions over the years. Currently, two species are recognised – *Sal. enterica* and *Sal. bongori*. *Sal. enterica* is responsible for most diseases in humans, while *Sal. bongori* causes diseases mainly in reptiles. *Sal. enterica* is divided into 2500 different types on the basis of the antigens they have on their surface – these different types are known as "serovars". Confusingly, these serovars are grouped into six subspecies. In order to specify a particular serovar, it is necessary to stipulate the species, subspecies and serovar so that the correct way to write down the name of one of the most frequent causes of salmonellosis is as follows "*Sal. enterica* subspecies *enterica*, var Enteritidis". To avoid this cumbersome name, bacteriologists have agreed to abbreviate it to *Sal.* Enteritidis. Note that, unlike the system adopted for all other microbes, the second part of the name is capitalised and is not italicised to make this difference clear.

How Is the Disease Diagnosed?

Salmonellosis can't be diagnosed on the basis of the clinical symptoms alone because these are similar to those caused by other pathogens responsible for gastroenteritis. Diagnosis is usually made by culturing *Sal. enterica* from a stool sample, and this takes between 2 and 3 days. The microbe can then be tested to find which antibiotics are effective against it and so would be suitable for treating the infection, if needed. Alternatively, the presence of the organism's DNA can be detected using PCR carried out on a stool sample; this approach is increasingly being employed in microbiology laboratories because it provides results much more rapidly.

How Is Salmonellosis Treated?

In healthy people, uncomplicated salmonellosis is self-limiting, and antibiotics aren't usually necessary – they probably don't shorten the duration of the illness or reduce the severity of symptoms. You may suffer from dehydration and electrolyte imbalance as a result of the diarrhoea, and therefore you should drink plenty of fluids. Those at greater risk ideally should drink oral rehydration solution which also replaces lost electrolytes (sodium and potassium ions). You should drink in small, frequent sips. Intravenous administration of rehydration fluids may be necessary if you can't drink sufficiently because of nausea or vomiting.

Those patients who are more vulnerable or are at high risk of developing more severe disease, bacteraemia or complications should be considered for oral antibiotics. These high-risk groups include infants <1 year of age, immunosuppressed patients, people with vascular abnormalities (such as prosthetic heart valves), people with prosthetic joints, those over 50 years of age, elderly residents in care homes and those with severe colitis and sickle cell disease.

Patients with HIV can suffer a much more serious illness with relapses and recurrences; long-term antibiotic treatment to prevent recurrences may be necessary.

Appropriate antibiotics include fluoroquinolones (such as ciprofloxacin), azithromycin, trimethoprim/sulfamethoxazole and amoxicillin. The choice of antibiotic may have to be modified once the results of antibiotic sensitivity testing on the strain isolated from the patient are available. Unfortunately there has been a worldwide rise in antibiotic resistance in *Salmonella* species, largely attributable to the use of these drugs in animal husbandry – this is a major concern for public health.

Surgery is sometimes required for the various complications including draining of abscesses and the repair or bypass of damaged large blood vessels.

In approximately 0.5% of cases, the patient may become a long-term carrier of *Sal. enterica* without exhibiting any symptoms. Usually this resolves on its own, but some people such as food handlers and care workers may need to be treated with longer courses of antibiotics until the organism has been eliminated. Interestingly, antibiotics may actually increase the duration of the carrier state.

Anti-motility drugs such as loperamide shouldn't be used routinely but may be considered occasionally for adults who have a mild illness and need to return to work or attend a special event, have difficulty reaching the toilet quickly or need to travel. They should not be used in people who have bloody diarrhoea, significant abdominal pain or fever as they can lead to dangerous and even fatal side effects and probably prolong the illness by slowing down the expulsion of the pathogen from the gut.

How Can I Avoid Getting Salmonellosis?

Risk factors for the disease include using antacid medication, sub-total gastrectomy, haemolytic conditions such as sickle cell disease, having had a splenectomy, leukaemia and HIV infection.

Prevention of salmonellosis involves proper sanitation and hygiene, as well as avoiding insufficiently cooked or mishandled food. Preventative measures at all stages in the food chain are carried out including in agriculture, food processing and preparation. National surveillance strategies in livestock and humans allow the monitoring of outbreaks to try to limit the onward spread. In the USA and UK, laboratories and clinicians are required by law to report cases to the Public Health Authorities.

Measures to prevent the spread of the disease include:

- Ensuring food is properly cooked and still hot when served
- Drinking only pasteurised or boiled milk
- Avoiding uncooked or lightly cooked eggs unless these are certified to have come from hens vaccinated against salmonella
- Ensuring eggs are adequately cooked, i.e. until the yolk is set
- Washing hands thoroughly and frequently using soap, in particular after contact with pets or farm animals or after having been to the toilet
- Keeping uncooked meats separate from cooked and ready-to-eat food to avoid cross-contamination
- Washing hands and kitchen utensils thoroughly in hot soapy water immediately after raw meat and poultry have been handled
- Not handling or preparing food while suffering from salmonellosis
- Washing fruits and vegetables carefully, particularly if they're eaten raw
- Not returning to work or school until at least 48 hours after the last episode of diarrhoea
- Avoiding the sharing of towels and flannels
- Washing soiled bed linen and clothes at 60 °C or higher
- Cleaning and disinfecting bathrooms regularly
- Peeling fruit and vegetables during travel in high-risk areas
- Avoiding ice in drinks in high-risk areas unless it's made from safe water
- Boiling or disinfecting drinking water in high-risk areas

Vaccines are available for use in animals, and these may reduce infections in the food chain and, consequently, in humans.

Want to Know More?

American Academy of Pediatrics https://www.healthychildren.org/English/health-issues/conditions/infections/Pages/Salmonella-Infections.aspx

Centers for Disease Control and Prevention, USA https://www.cdc.gov/salmonella/index.html

European Centre for Disease Prevention and Control https://www.ecdc.europa.eu/en/salmonellosis

Food and Drug Administration, USA https://www.fda.gov/food/foodborne-pathogens/salmonella-salmonellosis

Food Standards Agency, UK https://www.food.gov.uk/safety-hygiene/salmonella

KidsHealth, USA https://kidshealth.org/en/parents/salmonellosis.html

Mayo Clinic, USA https://www.mayoclinic.org/diseases-conditions/salmonella/symptoms-causes/syc-20355329

National Health Service, UK https://www.nhsinform.scot/illnesses-and-conditions/infections-and-poisoning/food-poisoning#about-food-poisoning

National Institute for Clinical Care and Excellence (NICE), UK. Gastroenteritis, 2019 https://cks.nice.org.uk/gastroenteritis#!topicSummary

Patient Info, UK https://patient.info/doctor/salmonella-gastroenteritis

World Health Organisation https://www.who.int/news-room/fact-sheets/detail/salmonella-(non-typhoidal).

Ajmera A, Shabbir N. Salmonella. Treasure Island (FL): StatPearls Publishing LLC; 2020 https://www.ncbi.nlm.nih.gov/books/NBK555892/

Antunes P, Mourão J, Campos J, Peixe L. Salmonellosis: the role of poultry meat. *Clinical Microbiology and Infection.* 2016 Feb;22(2):110–121. https://doi.org/10.1016/j.cmi.2015.12.004. Epub 2015 Dec 17.

Chlebicz A, Śliżewska K. Campylobacteriosis, Salmonellosis, Yersiniosis, and Listeriosis as Zoonotic Foodborne Diseases: A Review. *International Journal of Environmental Research and Public Health.* 2018 Apr 26;15(5). pii: E863. https://doi.org/10.3390/ijerph15050863.

Klochko A. Salmonella Infection (Salmonellosis). Medscape from WebMD, 2019 https://emedicine.medscape.com/article/228174-overview

Preziosi M, Fierer J. Salmonellosis. BMJ Best Practice, BMJ Publishing Group, 2017 https://bestpractice.bmj.com/topics/en-gb/817

Whiley H, Ross K. Salmonella and eggs: from production to plate. *International Journal of Environmental Research and Public Health.* 2015 Feb 26;12(3):2543–56. https://doi.org/10.3390/ijerph120302543.

34

Diarrhoea Due to *Clostridium perfringens*

Abstract Food poisoning due to *Clostridium perfringens* normally results in a watery diarrhoea often accompanied by stomach cramps. The disease is generally mild and self-limiting and usually resolves within 24 hours. Complications aren't very common, but in the very young or old, or those with a weakened immune system or underlying health conditions, the symptoms may be more severe and can result in dehydration which may require oral or intravenous fluid replacement. Rare complications include peritonitis and septicaemia. The main sources of infection are cooked meat and meat products that have been kept at an inappropriate temperature for a long time prior to being eaten. Diagnosis is usually made by finding large numbers of *Cl. perfringens* in a faecal sample or by the detection of one of its toxins. The illness can be prevented by good food hygiene practices such as ensuring that food is cooked at a sufficiently high temperature and is subsequently kept at an appropriate temperature prior to being eaten.

What Are the Main Symptoms?

Diarrhoea due to *Cl. perfringens* is the result of consuming food contaminated with the organism and is, therefore, a type of food poisoning. The illness usually starts suddenly, and the main symptoms are watery diarrhoea and abdominal cramps which follow within 6–24 hours (usually around 15 hours) of being infected. It's unlikely that you'll suffer from fever or vomiting, but you may feel tired and experience muscle aches. The symptoms usually last no

© Springer Nature Switzerland AG 2021
M. Wilson, P. J. K. Wilson, *Close Encounters of the Microbial Kind*,
https://doi.org/10.1007/978-3-030-56978-5_34

longer than 24 hours but may occasionally persist, in a less severe form, in some individuals for up to 2 weeks.

Are There Any Complications?

Complications in those who are otherwise healthy are unusual but dehydration can be a problem in the very young or the elderly. In very young or old people, or those who are immunocompromised (such as HIV and cancer patients), the symptoms can be more severe and last for as long as 2 weeks. Possible complications include peritonitis and septicaemia.

Enteritis necroticans (or Pigbel) is an extremely rare and life-threatening complication caused by certain strains of the organism (Type C) that is seen mostly in developing countries in malnourished children. This is characterised by peritonitis and necrosis of the small bowel. It can be treated with antibiotics and may require surgery.

Box 34.1 What's in a Name?

Clostridium perfringens was discovered in 1891 by William Welch while carrying out an autopsy. The bacterium was originally known as *Bacillus welchii*, then *Cl. welchii* and was finally named *Cl. perfringens* in 1898. Early research showed that it was responsible for a number of diseases including gas gangrene, puerperal fever and enteritis.

The genus name *Clostridium* is derived from the Greek word "kloster", which means a spindle. This is because of the spindle-shaped appearance of cells of members of the genus when viewed through a microscope. The species name comes from two Latin words "per" and "frango" which mean "through" and "burst", respectively. This is because one of the diseases caused by the organism, gas gangrene, is accompanied by bursting of the infected tissue due to the copious amounts of gas it produces.

How Common Is the Disease?

In the USA, *Cl. perfringens* is a major cause of food poisoning, responsible for approximately 1 million cases each year, while in the countries of the European Union, there are approximately 5 million cases each year. In the UK in 2018 there were approximately 85,000 cases resulting in almost 13,500 visits to GPs. The number of hospitalisations was 376.

What Type of Microbe Is *Cl. perfringens*?

Cl. perfringens, which used to be called Cl. welchii, is a Gram-positive bacterium. It's rod-shaped (Fig. 34.1), produces spores (Box 34.2) and is an anaerobe, i.e. it only grows in the absence of oxygen. It's widely distributed in nature and is found in soil, water and the gastrointestinal tract of many

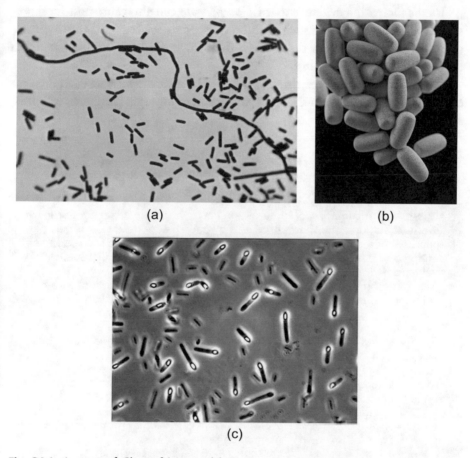

(a)   (b)

(c)

Fig. 34.1 Images of *Cl. perfringens*. (**a**) Gram stain showing vegetative cells of the organism (×1000). (Image courtesy of Dr. Holdeman, Centers for Disease Control and Prevention, USA). (**b**) Scanning electron micrograph of vegetative cells. (Jennifer Oosthuizen and James Archer, Centers for Disease Control and Prevention, USA). (**c**) Unstained cells of *Cl. perfringens* – the bright oval shapes inside the bacteria are spores. (Genome-Wide Transcriptional Profiling of *Clostridium perfringens* SM101 during Sporulation Extends the Core of Putative Sporulation Genes and Genes Determining Spore Properties and Germination Characteristics. PLoS One. 2015; 10(5): e0127036. Xiao Y, van Hijum SA, Abee T, Wells-Bennik MH

animals, including humans. It's often present on raw meat and poultry. It produces more than 20 different toxins, and 7 types of *Cl. perfringens* (Types A–G) are recognised on the basis of which toxins they produce. Strains that don't produce toxins can inhabit the gut without producing any symptoms and contribute to the normal gut microbiota of animals and humans. Most cases associated with gastrointestinal illnesses in humans are caused by Type A strains.

There's increasing evidence that *Cl. perfringens* can also cause non-food poisoning diarrhoeal illnesses such as those that are associated with antibiotic use or sporadic outbreaks. In these cases, the symptoms may be different from those seen in food poisoning.

Some strains of the organism also cause other diseases. Type A strains, for example, can also be responsible for wound infections, gas gangrene, cellulitis, sepsis and cholecystitis.

Box 34.2 Bacterial Spores

A spore is a structure produced by certain bacteria that enable them to survive adverse conditions. The spores produced by bacteria are described as being an endospore because they are formed inside the bacterial cell. The shape, size and position of the spore within the cell are characteristic of a particular species and can be useful for identification purposes.

When a spore-forming bacterium is exposed to adverse conditions such as a high temperature or a lack of water or nutrients, it undergoes a process known as sporulation which results in the formation of a spore. The spore is a protective, resilient structure that contains the bacterium's DNA and other essential cell components. It's a multi-layered structure that provides protection against heat, radiation, dehydration and harmful chemicals.

The spore is a dormant form of the bacterium and is capable of surviving for exceptional lengths of time, sometimes many thousands of years. It's very resistant to heat and can survive boiling but is killed by heating to 121 °C for 15 minutes. Once conditions are again favourable for the growth of the bacterium, the spore is converted to the metabolically active (vegetative) form of the bacterium – this process is known as germination.

Many bacterial species can form spores – those spore-formers that are able to cause disease in humans belong mainly to the genera *Clostridium*, *Bacillus* and *Actinomyces*.

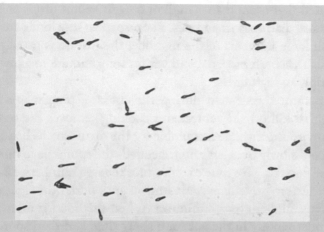

(a) Gram stain of *Clostridium botulinum*. The oval, red-staining structure present in most of these cells is a spore. In most cases, the spore is near one end of the cell. (Dr. George Lombard, Centers for Disease Control and Prevention, USA)

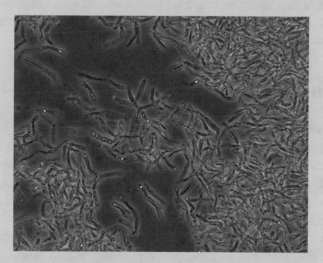

(b) Unstained cells of Bacillus anthracis. Spores are visible as bright, oval structures in some of the cells. (Courtesy of Larry Stauffer, Oregon State Public Health Laboratory)

What Happens During an Infection?

The vast majority of cases of diarrhoea due to *Cl. perfringens* result from the consumption of food contaminated with the organism or, more specifically, its spores. The spores are ubiquitous and frequently present in raw meat. It's been estimated that more than 1 million cells or spores of *Cl. perfringens* are needed to cause diarrhoea in humans. Because most raw foods are unlikely to contain such large numbers of the microbe, the disease is usually associated with food that has been prepared and left for long enough to allow the organism to multiply to high levels.

The temperatures generated during the cooking process are usually not high enough to kill off the spores, and cooking removes the oxygen which makes the environment more favourable for the organism. When food is then allowed to cool slowly or is left unrefrigerated, the spores have time to germinate, and the bacteria have time to reproduce to large numbers. *Cl. perfringens* is one of the fastest-growing bacteria known and, under the right conditions, can double in number every 7 minutes. When this food is eaten, the bacterium multiplies rapidly in the gut for a short time and then sporulates (produces spores) – this is accompanied by the release of a toxin known as *Clostridium perfringens* enterotoxin (CPE). CPE disrupts the outer layers of epithelial cells that line the intestinal tract and can also kill these cells. The resulting damage prevents the gut from absorbing water and nutrients, and this results in diarrhoea.

How Is the Disease Diagnosed?

The characteristic symptoms of the illness help in its diagnosis, but a definitive diagnosis can only be made by identifying the bacterium, or its toxin, in faecal samples from patients. Samples of faeces are cultured under favourable (i.e. anaerobic) conditions in the laboratory, and the number of *Cl. perfringens* found there is counted. Only large numbers (more than 1 million per gram of faeces) are regarded as being significant as the organism is present in the faeces of most healthy individuals. Another option for diagnosis involves detecting CPE in the faecal sample, and this provides results within hours.

Box 34.3 Another Dangerous Clostridium: *Clostridium difficile* (*Clostridioides difficile*)

Cl. difficile often referred to in the media as "*C. diff*" is one of the most frequent causes of healthcare-related infections and of the diarrhoea that sometimes follows antibiotic use ("antibiotic-associated diarrhoea"). Recently it's been re-classified as *Clostridioides difficile*. Major risk factors for colonisation by the organism are antibiotic administration (particularly ampicillin, amoxicillin, cephalosporins, clindamycin and fluoroquinolones) and hospitalisation, and this can result in either symptom-free carriage of the organism or clinical disease. The microbiota of the intestinal tract usually prevents *Cl. difficile* from proliferating and reaching high numbers.

However, if the microbiota is disrupted by antibiotics, or by some medical or surgical procedure, then the bacterium can grow rapidly and cause problems. Illnesses can range in severity from mild diarrhoea to life-threatening pseudomembranous colitis, the symptoms of which include diarrhoea, abdominal pain, fever, nausea and vomiting. Despite successful treatment of an episode of the disease, 15–40% of patients suffer from a recurrence of the infection.

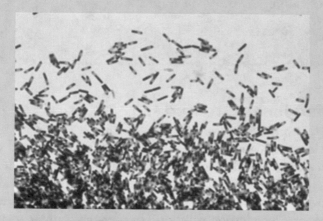

(a) Gram stain of *Cl. difficile* **showing typical rod-shaped cells (darkly stained) and lightly staining/non-staining spores.** (Image courtesy of Dr. Gilda Jones, Centers for Disease Control and Prevention, USA)

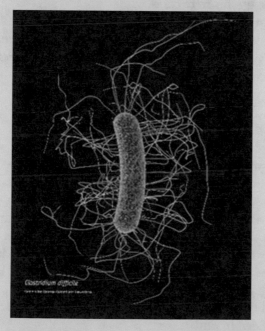

(b) Illustration of *Cl. difficile* **based on photomicrographs showing the typical rod-shaped cell and flagella.** (Image courtesy of James Archer, Centers for Disease Control and Prevention, USA)

(continued)

Box 34.3 (continued)

Its two most important virulence factors are toxin A and toxin B which kill epithelial cells. This disrupts the intestinal epithelium and its ability to absorb water, and this results in diarrhoea. The toxins also induce the production of a variety of cytokines which results in a profound inflammatory response and consequent tissue damage. If required, the disease can be treated with either metronidazole or vancomycin for 10–14 days.

In the USA about a half million people each year suffer from a *Cl. difficile* infection, and in 2011 this resulted in 29,000 deaths. In the UK the number of cases is about 13,000 per year and results in about 100 deaths.

How Is the Illness Treated?

The disease is generally mild and self-limiting in otherwise healthy people, so medication, particularly antibiotics, are not normally necessary in uncomplicated gastrointestinal illnesses. You should drink plenty of fluids in small, frequent sips to avoid dehydration. Those at greater risk of dehydration ideally should drink oral rehydration solution (an over-the-counter remedy, readily available from pharmacies and supermarkets) which also replaces electrolytes (sodium and potassium ions). Intravenous administration of rehydration fluids may be necessary in severe cases.

Antibiotics can be used for complications and suitable choices include penicillins, macrolides (such as erythromycin), chloramphenicol and third-generation cephalosporins (such as cefotaxime), although resistance to some of these is common.

How Is the Disease Prevented?

The disease is foodborne and can't be directly transmitted from one person to another. Institutional settings such as school cafeterias, hospitals, nursing homes, prisons, etc. are the most common circumstances in which the illness occurs. Large events where catered food is served are also at risk. This is because in these situations large quantities of food are often prepared several hours before distribution and are often kept warm for a long time before being served. The most common types of food involved are beef, poultry and meat-containing products (such as gravies and stews). However, vegetable products, including spices and herbs, as well as raw and processed foods have also been implicated in outbreaks.

The illness can be prevented by cooking and keeping food at the correct temperature. Food, particularly beef or poultry, should be cooked to a safe internal temperature and then kept either at (or above) 60 °C or at (or below) 4 °C to prevent the growth of any *Cl. perfringens* that might have survived cooking. Examples of safe internal temperatures include 63 °C for roast beef, veal and lamb and 74 °C for poultry – these temperatures ensure killing of the vegetative cells of *Cl. perfringens*.

Meat dishes should be served hot, within 2 hours after cooking. Leftover foods should be refrigerated at 4 °C or below as soon as possible and certainly within 2 hours of preparation. Small quantities of hot foods can be put directly into the refrigerator, but large cuts of meat, as well as large volumes of soups and stews, should be divided into small quantities prior to refrigeration to ensure that they quickly reach the low temperature in the refrigerator. Leftovers should be reheated to at least 74 °C before serving.

Cl. perfringens is a notifiable organism when it causes food poisoning in the UK and USA, so cases identified by the laboratory or suspected by clinicians are routinely reported to public health authorities in order to detect, monitor, manage and prevent potential outbreaks.

Although direct person-to-person spread of this particular disease doesn't happen, it's important to follow stringent hand and food hygiene when suffering from any gastroenteritis symptoms, especially as the responsible organism is often not known early on. You should also avoid school or work for 48 hours after the diarrhoea symptoms have settled.

No vaccine is currently available for the disease and immunity doesn't follow an infection, so you can get it more than once.

Want to Know More?

Australian Government Department of Health. *Clostridium perfringens* enteritis. https://www1.health.gov.au/internet/main/publishing.nsf/Content/cda-phlncd-clostridium-perfringens-enteritis.htm

British Columbia, Canada. Healthlink, Foodborne Illness: *Clostridium perfringens*. https://www.healthlinkbc.ca/health-topics/te6324

Centers for Disease Control and Prevention, USA https://www.cdc.gov/foodsafety/diseases/clostridium-perfringens.html

Department of Health and Community Services, Canada. *Clostridium perfringens* food intoxication, https://www.health.gov.nl.ca/health/publichealth/envhealth/clostridium_perfringens_food_safety_2011.pdf

French agency for food, environmental and occupational health and safety. *Clostridium perfringens*. https://www.anses.fr/en/system/files/MIC2010sa0235FiEN.pdf

Government of Canada. *Clostridium perfringens* infection, https://www.canada.ca/en/public-health/services/diseases/clostridium-perfringens.html

National Institute for Clinical Care and Excellence (NICE), UK. Gastroenteritis, 2019 https://cks.nice.org.uk/gastroenteritis#!topicSummary

National Institute for Public Health, The Netherlands. *Clostridium perfringens* associated food borne disease. https://www.rivm.nl/bibliotheek/rapporten/330371005.pdf

New Zealand Food Safety Agency https://www.mpi.govt.nz/dmsdocument/11021/direct

Akhondi H, Simonsen KA. Bacterial Diarrhea. Treasure Island (FL): StatPearls Publishing LLC; 2020 https://www.ncbi.nlm.nih.gov/books/NBK551643/

Allan P, Keshav S. Food poisoning. BMJ Best Practice, BMJ Publishing Group, 2018 https://bestpractice.bmj.com/topics/en-gb/203

Gamarra RM. Food Poisoning. Medscape from WebMD, 2018 https://emedicine.medscape.com/article/175569-overview

Raymond Kiu, Lindsay J Hall. An update on the human and animal enteric pathogen *Clostridium perfringens. Emerging Microbes and Infections*, 7 (1), 141, 2018 Aug 6.

35

Gastroenteritis Due to Norovirus

Abstract Norovirus is the most common cause of acute gastroenteritis world-wide. The main symptoms of the infection are diarrhoea and vomiting – these are sometimes accompanied by stomach pain, fever, headache and muscle aches. The diarrhoea is usually watery and not bloody. The vomiting is usually profuse and may be projectile. The illness generally lasts for 2–3 days. It's more common in the winter months and so is also known as "winter vomiting disease". The frequent vomiting and diarrhoea can result in severe dehydration and an electrolyte imbalance. Temporary food intolerances (particularly lactose intolerance) may follow. It's highly contagious and humans are the only known reservoir of the virus although it can survive for a significant length of time in the environment. Person-to-person transmission is the most frequent cause of outbreaks, and this may occur by the faecal-oral route, by ingestion of aerosolised vomit and by indirect exposure via contaminated objects or surfaces. Foodborne and waterborne transmission are also possible. Diagnosis of the infection is usually based on clinical symptoms, but detection of norovirus RNA in stool samples can be used when trying to establish which microbe is responsible for an outbreak of gastroenteritis. Treatment involves replacing fluid and electrolyte loss. Analgesics can be taken to relieve fever and pain. Good hygiene practices, especially hand hygiene, are the most important means of preventing the infection and controlling its transmission.

© Springer Nature Switzerland AG 2021
M. Wilson, P. J. K. Wilson, *Close Encounters of the Microbial Kind*,
https://doi.org/10.1007/978-3-030-56978-5_35

What Are the Main Symptoms of Norovirus Infections?

The main symptoms of infection with norovirus are diarrhoea (in around 88% of cases), vomiting (around 26% of cases) and nausea and stomach pain (around 4%). The diarrhoea is usually watery and generally doesn't contain blood. The vomiting is usually profuse and may be projectile. Other symptoms you might experience include fever (2% of cases), headache and general muscle aches. Typically, the symptoms appear 12–48 hours after acquiring the virus, and the illness generally lasts for 2–3 days. In temperate climates, the infection is more common in the winter months and so is often known as "winter vomiting disease" (Fig. 35.1). It's also called "gastric flu".

Are There Any Complications?

In the majority of cases, the illness is mild, short-lived and self-limiting. Those at risk of serious illness include the elderly, very young, pregnant women, those with impaired immunity and people with other long-term health conditions.

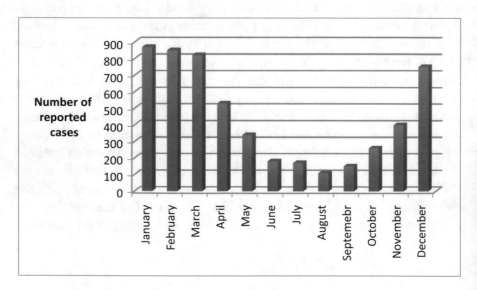

Fig. 35.1 Number of cases of norovirus infections reported each month during 2000–2012 in England and Wales

The frequent vomiting and diarrhoea can result in severe dehydration, the symptoms of which include a decrease in urine production, thirst and dry mouth and feeling lightheaded when standing up. Temporary food intolerances (particularly lactose intolerance) can occur in around 3% of cases, and people with inflammatory bowel disease can experience flare-ups. Malnutrition can result from loss of appetite, vomiting and poor absorption of nutrients. Norovirus infections increase the risk of developing irritable bowel syndrome, and some people take longer to clear the diarrhoea. Unusual and rare complications include colitis and benign infantile convulsions.

Frequent vomiting and diarrhoea can also result in an acidity and electrolyte imbalance in the blood resulting in abnormal levels of sodium, potassium, calcium, chloride or bicarbonate. Blood glucose and medication levels can also fall during the illness. The reduced blood volume also results in impaired blood supply to the vital organs which can start to fail.

If the dehydration and chemical changes in the blood are not corrected, then serious complications result, such as kidney failure, confusion, drowsiness, convulsions, muscle weakness, heart rhythm problems, disseminated intravascular coagulation, respiratory failure, sepsis and ultimately death. Hyperthermia and gastrointestinal bleeding are also sometimes seen.

Box 35.1 What's in a Name?

The first outbreak of gastroenteritis due to norovirus occurred in a school in the American town of Norwalk, Ohio, in 1968. Stool samples from the outbreak were examined by Albert Kapikian and his team at the National Institutes of Health, Bethesda, and in 1972 they found, using electron microscopy, that a virus was the cause of the disease. This was the first time a virus had been shown to be responsible for gastroenteritis, and it was named "the Norwalk virus". Later, the Norwalk virus was found to be only one member of a group of similar viruses, and these were named Norwalk-like viruses or "small round structured viruses". In 1972 an international committee decided on the name norovirus which is derived from the "Nor" of Norwalk and "virus". The illness it causes is known by a variety of names including "winter vomiting disease" and "gastric flu".

How Common Is It?

Norovirus is the most common cause of acute gastroenteritis worldwide and is responsible for about 685 million cases (1 in every 5) of the disease and 200,000 deaths each year. About 200 million cases are among children under 5 years of age, leading to an estimated 50,000 child deaths every year, mostly in developing countries. Each year norovirus is estimated to cost $60 billion worldwide due to healthcare costs and lost productivity. In the USA, norovirus is responsible for 21 million cases (i.e. 60%) of acute gastroenteritis cases each year. It's responsible for 400,000 emergency department visits, 71,000 hospitalisations and 570–800 deaths (mostly in young children and the elderly) each year. In the UK it affects between 1 and 3 million people each year. In developed countries death due to norovirus infection is extremely rare. All age groups are susceptible to the disease but most of the morbidity and mortality occurs at the extremes of age.

How Can I Catch It?

Transmission of the virus occurs by three general routes: person-to-person, via food and via water.

Person-to-person transmission is the most frequent cause of outbreaks and may occur by the faecal-oral route, by ingestion of aerosolised vomit and by indirect exposure via contaminated objects or environmental surfaces (Box 35.2). Virus shedding happens from the onset of symptoms for up to as long as 2 weeks after recovery.

Box 35.2 Vomiting Larry

Infection with the norovirus can induce vomiting, which is often of a projectile nature. This aids the spread of the virus to others either directly, in the form of aerosols, or by contaminating surfaces which can act as reservoirs of the virus. However, there's little information regarding just how much the surrounding environment is contaminated when an individual vomits. In order to establish the extent to which projected fluid can contaminate the environment, a simulated vomiting system was developed at the Health and Safety Laboratory in the UK. This is known as "Vomiting Larry" (Figure). Simulated vomiting was carried out using water containing a fluorescent marker which enabled even small splashes to be detected under ultraviolet light. These experiments showed that splashes and droplets produced during projectile vomiting can travel great distances (>3 m forward spread and 2.6 m lateral spread). The results of the study suggested that areas of at least 7.8 m² should be decontaminated after an episode of projectile vomiting.

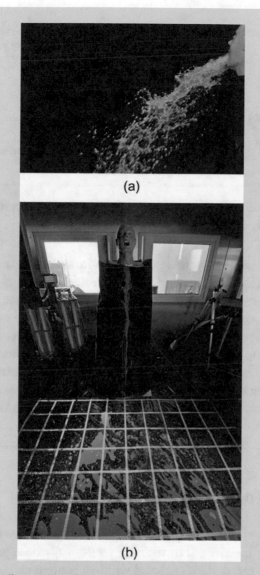

Figure (a) Projectile vomiting from Larry. (b) A fluorescent dye showed how far Larry's vomit had spread. (Images courtesy of Dr Catherine Makison Booth, Health and Safety Executive, UK)

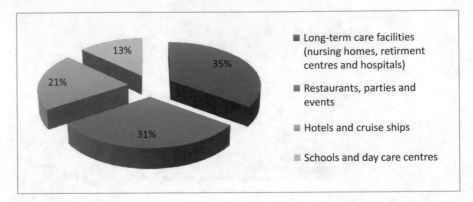

Fig. 35.2 Outbreaks of norovirus infection in various settings in the USA during the period 1994–2006

The virion is extremely stable in the environment, can survive for long periods on different surfaces and resists freezing as well as temperatures as high as 60 °C. It can withstand chlorine-based disinfectants, acids (including vinegar), alcohol, antiseptic hand solutions and high sugar concentrations.

Foodborne transmission usually occurs by contamination of food by infected food handlers. However, it might also occur further upstream in the food distribution system due to contamination with human waste. Shellfish and salads are the most commonly implicated food sources. Contamination of recreational and drinking water can result in large community outbreaks. In a study of outbreaks in the USA, most were found to occur in long-term care facilities (Fig. 35.2). Several outbreaks have been reported on cruise ships.

What Is a Norovirus?

The norovirus is an unenveloped virus (i.e. it doesn't have an outer coating derived from the human cell in which it replicates) whose genetic material is RNA. The RNA molecule is enclosed within a protein capsid that is icosahedral in shape with a diameter of 23–40 nm (Fig. 35.3).

What Happens During an Infection?

The virus is highly contagious with an infectious dose of less than 10 virions. On reaching the small intestine (Fig. 35.4), it binds to the lining of epithelial cells (known as enterocytes) by attaching to the blood group antigens (A, B

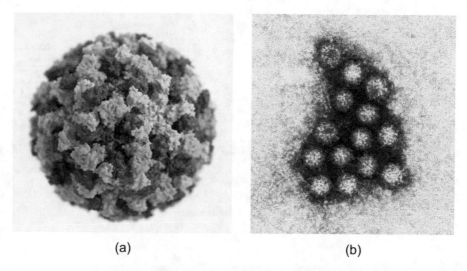

(a)   (b)

Fig. 35.3   Images of the norovirus. (a) Three-dimensional model of the norovirus virion based on images obtained by electron microscopy. The various colours represent the different proteins that form the capsid of the virion. (Alissa Eckert, MS, Centers for Disease Control and Prevention, USA). (b) A large number of norovirus virions as seen by electron microscopy – the image has been digitally colourised. (Credit: David Gregory & Debbie Marshall. CC BY 4.0. Wellcome Image Library)

and O) on their surface. This results in uptake of the virion by the enterocyte and replication within the cell soon follows. Newly synthesised virions then exit the infected cell and can invade other enterocytes.

Attachment to, and invasion of, enterocytes induces an inflammatory response. Invasion of the epithelium and the accompanying inflammatory response result in damage to the villi (which become shorter) and to the microvilli (Fig. 35.4a). These changes reduce the ability of the small intestine to absorb water and nutrients, and this results in diarrhoea. The virus also slows down the normal stomach emptying process, and this is responsible for the vomiting that accompanies the Infection.

Humans are the only known reservoir for norovirus infections. The virus is shed primarily in faeces but is also present in the vomit of infected individuals. It can be detected in faeces for as long as 4 weeks following infection, and it may be present there at a concentration of approximately 100 billion virions per gram.

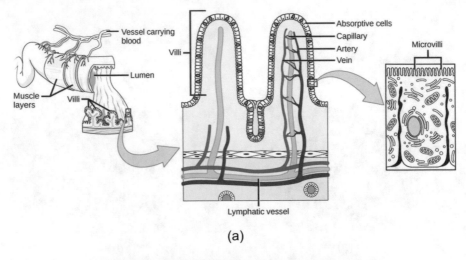

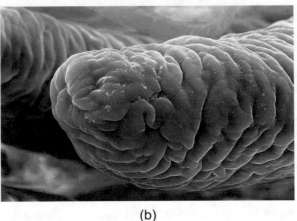

Fig. 35.4 Structure of the mucosa of the human small intestine. (a) The mucosa of the small intestine has numerous finger-like projections (0.5–1.0 mm in length) known as villi which protrude into its hollow interior (lumen). These villi increase its surface area and so enhance absorption of nutrients and water. The gaps between the villi are known as crypts. The epithelial cells covering the villi are absorptive cells known as enterocytes, and their main function is to absorb water and nutrients. Each of these cells has hundreds of hair-like projections known as microvilli which increase the surface area of the cell, thereby increasing its absorptive efficiency. Each villus has a network of capillaries that supplies the blood needed to transport the absorbed nutrients and water to other parts of the body. It also contains a lymphatic vessel for the transportation of absorbed fats. (CNX OpenStax/CC BY (https://creativecommons.org/licenses/by/4.0)). (b) A villus of the small intestine as seen through an electron microscope. (Credit: Liz Hirst, Medical Research Council. CC BY 4.0)

How Is the Infection Diagnosed?

Diagnosis of the infection is usually based on clinical symptoms. It's not usually necessary or practical to carry out laboratory tests on stool or blood samples from all patients with diarrhoea. Blood tests should be considered to assess electrolyte status and kidney function in those patients who have lost considerable volumes of fluid and in high-risk groups such as infants, older people and patients with other major diseases. During outbreaks, stool samples from some individuals are analysed to identify the pathogen, as knowing which microbe is responsible enables suitable measures to be put in place to curtail the outbreak. Detection of norovirus RNA in a stool sample by PCR is the usual diagnostic test used.

How Is the Infection Treated?

Norovirus infections are self-limiting and symptoms usually resolve within 2–3 days. Anti-viral agents aren't usually needed, and treatment is focussed on replacing the fluids and electrolytes you will have lost due to diarrhoea and vomiting. Your fluid intake should be increased and, if required, this can be supplemented by oral rehydration solutions. You may need to take analgesics such as paracetamol and ibuprofen to relieve fever and pain. Severe cases may require anti-emetic medication, admission to hospital for intravenous rehydration and specific electrolyte replacement or dialysis.

How Can I Avoid a Norovirus Infection?

Thorough hand hygiene is the most important means of preventing norovirus infection and controlling its transmission (see Box 1.11 in Chap. 1). Handwashing for at least 20 seconds with running water and plain or antiseptic soap is an effective way of reducing the number of virions on your hands – alcohol-containing hand gels are less effective.

Handwashing is particularly important after going to the toilet, before preparing or touching food or drinks and before eating. Considering the highly infectious nature of norovirus, exclusion and isolation of infected individuals are important means of interrupting transmission of the virus and limiting contamination of the environment. This is particularly important in settings where people reside or congregate such as long-term care facilities, hospitals, cruise ships and college dormitories. In such situations, patients should be isolated during their illness and for a further 48 hours after their symptoms have disappeared.

Infected staff members in healthcare facilities and food handlers should be excluded from work during their illness and for 48–72 hours following resolution of their symptoms. Working adults and schoolchildren who acquire the disease shouldn't return to work or school, respectively, for 48 hours after their symptoms have stopped.

You should use chemical disinfectants to prevent the spread of norovirus from contaminated environmental surfaces, and the most effective of these is sodium hypochlorite solution. You should pay particular attention to the likely areas of greatest contamination such as bathrooms and high-touch surfaces, e.g. door knobs and hand rails.

Immunity to the virus follows infection, but this seems to be short-lived and is specific to the strain. As there is significant genetic variation in the virus, immunity to one type doesn't protect you against the other types. Interestingly, there is also a genetic factor at play in acquiring norovirus infections, with up to 20% of Caucasian populations probably being strongly protected against the common strains due to mutations of certain genes associated with gastrointestinal cell proteins. Your blood group type may also play a role in susceptibility.

A vaccine isn't yet available but is under development. Safe and effective anti-virals aren't yet in widespread use for this microbe but they are being researched.

Although norovirus is not a notifiable disease in the UK and USA, there are systems in place to monitor it and potential outbreaks. These include surveillance of laboratory results and case reporting in certain environments such as healthcare settings and prisons.

Want to Know More?

Centers for Disease Control and Prevention, USA https://www.cdc.gov/norovirus/index.html

Food Standards Agency, UK https://www.food.gov.uk/safety-hygiene/norovirus

Mayo Clinic, USA https://www.mayoclinic.org/diseases-conditions/norovirus/symptoms-causes/syc-20355296

National Health Service, UK https://www.nhs.uk/conditions/norovirus/, https://www.nhsinform.scot/illnesses-and-conditions/infections-and-poisoning/norovirus, https://www.fitfortravel.nhs.uk/advice/disease-prevention-advice/norovirus

National Institute for Clinical Care and Excellence (NICE), UK. Gastroenteritis, 2019 https://cks.nice.org.uk/gastroenteritis#!topicSummary

Patient Info, UK https://patient.info/digestive-health/diarrhoea/norovirus

Royal Society for Public Health, UK https://www.rsph.org.uk/uploads/assets/uploaded/00fa01f0-126c-41de-b0857abc5c4d626b.pdf

Capece G, Gignac E. Norovirus. Treasure Island (FL): StatPearls Publishing LLC; 2019 https://www.ncbi.nlm.nih.gov/books/NBK513265/

Cardemil CV, Parashar UD, Hall AJ. Norovirus infection in older adults: epidemiology, risk factors, and opportunities for prevention and control. *Infectious Disease Clinics of North America.* 2017 Dec;31(4):839–870. doi: https://doi.org/10.1016/j.idc.2017.07.012. Epub 2017 Sep 12.

Currigan J, Grundlingh J. Viral gastroenteritis in adults. BMJ Best Practice, BMJ Publishing Group, 2020 https://bestpractice.bmj.com/topics/en-gb/3000126

Guix S, Pintó RM, Bosch A. Final consumer options to control and prevent foodborne norovirus infections. *Viruses.* 2019 Apr 9;11(4). pii: E333. https://doi.org/10.3390/v11040333.

Khan ZZ. Norovirus. Medscape from WebMD, 2018 https://emedicine.medscape.com/article/224225-overview

Lopman BA, Steele D, Kirkwood CD, Parashar UD. The vast and varied global burden of norovirus: prospects for prevention and control. *PLoS Medicine.* 2016 Apr 26;13(4):e1001999. https://doi.org/10.1371/journal.pmed.1001999. eCollection 2016 Apr.

Robilotti E, Deresinski S, Pinsky BA. Norovirus. *Clinical Microbiology Reviews.* 2015 Jan;28(1):134–64. doi: https://doi.org/10.1128/CMR.00075-14.

Shah MP, Hall AJ. Norovirus illnesses in children and adolescents. *Infectious Disease Clinics of North America.* 2018 Mar;32(1):103–118. https://doi.org/10.1016/j.idc.2017.11.004.

Simons MP, Pike BL, Hulseberg CE, Prouty MG, Swierczewski BE. Norovirus: new developments and implications for travelers' diarrhea. *Tropical Diseases, Travel Medicine and Vaccines.* 2016 Jan 12;2:1. https://doi.org/10.1186/s40794-016-0017-x. eCollection 2016.

36

Gastroenteritis Due to Rotavirus

Abstract Rotaviruses are the most common cause of diarrhoea in children throughout the world although adults can also suffer from the disease. The main symptoms of the disease are severe watery diarrhoea (usually not bloody), vomiting, fever and abdominal pain. Other symptoms may include loss of appetite and dehydration. Most children recover fully after about 1 week. Adults tend to have milder symptoms. The diarrhoea and vomiting can result in dehydration and electrolyte imbalance. Lactose intolerance may occur for a short time after the symptoms have resolved. Treatment involves the administration of oral or intravenous rehydration fluids. Susceptible individuals can become infected following direct contact with infected individuals, by touching contaminated surfaces or by handling contaminated objects. Spread of the infection can be reduced by good hygiene and sanitation practices. However, several vaccines are currently available and have been shown to be highly effective at preventing the disease.

What Are the Main Features of Rotavirus Infections?

Rotaviruses are the most common cause of diarrhoea in children throughout the world although adults can also suffer from the disease. Symptoms of the disease are usually apparent within 2 days of infection. Children who are infected may have severe watery diarrhoea (usually not bloody), vomiting, fever and abdominal pain. The vomiting and watery diarrhoea usually stop after about 3 days but may persist for as long as 9 days. Other symptoms may include loss of appetite and dehydration which can result in (i) decreased urine production, (ii) dry mouth, (iii) feeling dizzy when standing up, (iv) crying with few or no tears, (v) sunken eyes, (vi) weakness and (vii) unusual sleepiness or irritability. Most children in developed countries recover fully after about 1 week. Adults with a rotavirus infection tend to have milder symptoms.

Are There Any Complications?

Adults are less likely to suffer complications than children, and those who have an immunodeficiency are more likely to have severe or prolonged infection and complications.

The most common complications are dehydration and electrolyte imbalance, with rotaviruses generally causing a more severe illness than other viral causes of diarrhoea. Dehydration has been discussed in more detail in Chap. 33. Other gut complications include colitis and intussusception, i.e. where one part of the intestine slides inside an adjoining part. Temporary lactose intolerance may occur after the symptoms have resolved.

The virus can affect other areas of the body beyond the gut, and children can develop seizures, both fever and non-fever related. Other outcomes such as encephalopathy and encephalitis have been seen, similar to other viruses.

> ### Box 36.1 What's in a Name?
>
> The rotavirus gets its name from the Latin word "rota" which means a wheel, and this refers to the wheel-like shape of the virus. The virus was first detected in 1973 by Ruth Bishop and colleagues at the Royal Children's Hospital, Melbourne, Australia, who observed virions in duodenal tissue of children with gastroenteritis. That same year Thomas Flewett and colleagues at the East Birmingham Hospital, Birmingham, UK, observed similar virions in the faeces of patients with gastroenteritis. In 1974 Richard Wyatt and colleagues at the National Institutes of Health, Bethesda, USA, then grew the virus in human intestinal tissue in the laboratory.

Rotaviruses infect nearly every child in the world by the age of 5 years and are globally the leading cause of severe diarrhoea in children aged <5 years. According to WHO estimates, about 125 million children suffer from rotavirus infection each year, and this results in the death of 215,000 children aged under 5 years each year. In the USA, prior to the introduction of immunisation against the disease in 2006, there were 2.3–3.2 million cases of rotavirus infection each year, and this resulted in more than 400,000 doctor visits, more than 200,000 emergency room visits, 55,000 to 70,000 hospitalizations and 20–60 deaths. The direct and indirect costs of the illness amounted to approximately $1.3 billion each year. Since the introduction of a vaccination programme, the number of cases has decreased dramatically, and hospitalisation due to rotavirus infection has decreased by as much as 87% in children aged 1–4 years. In the UK, prior to the introduction of vaccination for infants in 2013, rotaviruses were responsible for approximately 750,000 episodes of diarrhoea, 80,000 general practice consultations and 14,300 hospital admissions in children aged <5 years annually. The cost to the National Health Service was estimated to be £14.2 million per year. During the first year following the introduction of the vaccination programme, there was a 77% decrease in rotavirus infections.

In temperate climates the disease used to be more common in the winter and spring, but the introduction of vaccination programmes has made this pattern less obvious.

Children are most at risk of getting rotavirus infections – especially those in child care centres or other settings where young children congregate. Those between the ages of 6 months and 2 years are most at risk. Elderly adults have a higher risk of getting rotavirus disease as well as adults who care for children with rotavirus disease or who are immunocompromised.

What Are Rotaviruses?

Rotaviruses are round, non-enveloped viruses whose genetic information is encoded by 11 RNA molecules (Fig. 36.1). These are surrounded by a triple-layered coat which is made from 12 different proteins, and protein spikes protrude from this coat. There are 12 different species of rotavirus, and these are designated by the letters A to J – more than 90% of human infections are due to rotavirus A.

What Happens During an Infection?

The disease is highly contagious with an infectious dose of 10–100 virions. On reaching the duodenum in the small intestine, the virus attaches to enterocytes (Fig. 35.4 of Chap. 35) by means of its protein spikes. It's then taken up

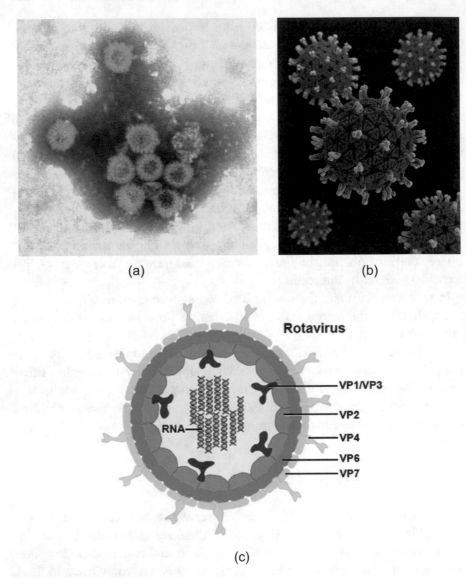

Fig. 36.1 Images of rotavirus. (**a**) Colour-enhanced electron micrograph of rotavirus. (Credit: David Gregory & Debbie Marshall. CC BY 4.0. Wellcome Image Library). (**b**) 3D model of a number of rotavirus virions showing the organism's characteristic wheel-like appearance. (Alissa Eckert, MS, Centers for Disease Control and prevention, USA). (**c**) Diagram showing the structure of the rotavirus virion. Its genetic material consists of 11 RNA molecules. Each virion is surrounded by a coat consisting of three layers made from six different proteins (VP1, VP2, VP3, VP4, VP6 and VP7). A number of protein filaments protrude from the surface of the coat, and these are used to stick to the cells that line the small intestine. (NIAID [CC BY 2.0 (https://creativecommons.org/licenses/by/2.0)])

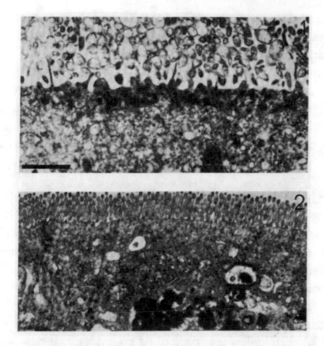

Fig. 36.2 Electron micrograph showing a cross-section through an enterocyte before (bottom image) and after (upper image) infection with rotavirus. In the lower image the surface of the cell can be seen to be covered in a row of tightly packed microvilli, whereas these have been destroyed in the infected cell (bar = 500 nm). (English Wikipedia user Graham Beards [CC BY 3.0 (https://creativecommons.org/licenses/by/3.0)])

by the cell, replicates within it and is released when the cell bursts open. Unlike gastroenteritis due to bacterial pathogens, the inflammatory response to the infection is limited. As with infection by the norovirus, infection of enterocytes results in the shortening of their villi. The microvilli of many of the enterocytes are also damaged (Fig. 36.2). Both of these effects result in a reduced ability of the small intestine to absorb water and nutrients, and, consequently, intestinal fluids are released as diarrhoea. The virus also produces a toxin that causes water loss from intestinal epithelial cells, and this contributes to the diarrhoea. The virus also affects the nervous system – it stimulates the vagus nerve which activates the vomiting reflex.

How Is the Infection Diagnosed?

The cause of gastroenteritis can't be determined on the basis of symptoms alone, although two factors are suggestive of a rotavirus infection: (i) symptoms are often more severe than diarrhoeal diseases due to other microbes, and (ii) in non-equatorial countries, rotavirus infection is more frequent in winter, although this seasonality has been weakened following the introduction of vaccination programmes.

Laboratory tests are not routinely used to confirm rotavirus infection in most cases but may be carried out in severe, complicated or persistent cases and at the beginning of outbreaks to identify the pathogen responsible. Rotavirus RNA can be detected by PCR of stool samples, and antigens of the virus can be detected by an enzyme-linked immunosorbent assay (ELISA).

How Is the Infection Treated?

For people with healthy immune systems who are otherwise well, rotavirus disease is usually self-limiting and lasts only a few days. Treatment involves ensuring adequate fluid and electrolyte intake to prevent dehydration, and oral rehydration solutions may need to be given. In severe cases intravenous rehydration fluids may have to be administered.

Anti-diarrhoeal medication such as loperamide should not be used in children under 12 years unless it is prescribed by a doctor. It may be helpful for symptom control in adults, but you should avoid routine use as there can be serious risks and side effects, which need to be balanced against the advantages. Some studies, but not all, have shown that some probiotics can reduce the duration of diarrhoea accompanying rotavirus infection by 1–2 days.

How Can I Avoid the Disease?

Transmission of the virus occurs mainly by the faecal-oral route, and rotaviruses can also survive for several days in the environment. You may therefore become infected following direct contact with infected individuals, by touching contaminated surfaces or by handling contaminated objects. Spread of the infection can be reduced by good hygiene and sanitation practices including frequent and through hand washing (see Box 1.11 of Chap. 1), disinfection of contaminated surfaces, prompt washing of soiled articles of clothing and avoiding the sharing of towels, flannels, cutlery and utensils. Children should be kept off school or nursery for at least 48 hours after the last episode of diarrhoea and should avoid contact with other children during this time.

Although the above practices are important, another effective preventive strategy is vaccination, and many countries now have a vaccination programme for infants (Fig. 36.3). In the UK since 2013 infants have been given

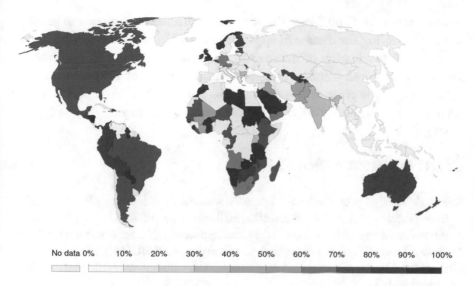

Fig. 36.3 Proportion of 1-year-old children who have received the rotavirus vaccine (2018). Based on data from WHO. (Image courtesy of "Our World in Data" via Wikimedia Commons)

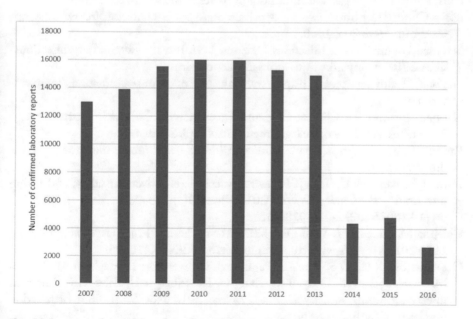

Fig. 36.4 Annual number of confirmed laboratory reports of rotavirus infection in England and Wales. Note the decreased number from 2014 onwards after the introduction of a vaccination programme in 2013

an oral vaccine at 2 and 3 months of age, and this has been shown to be effective at preventing the disease in approximately 75% of recipients (Fig. 36.4). In the USA, vaccination of infants was instituted in 2006, and two vaccines are now widely used – one involves oral administration at 2, 4 and 6 months, while the other is given at 2 and 4 months only. Both of these vaccines have been shown to be highly effective at preventing the disease.

Want to Know More?

American Academy of Pediatrics https://www.healthychildren.org/English/health-issues/vaccine-preventable-diseases/Pages/Rotavirus.aspx

American Academy of Physicians https://familydoctor.org/condition/rotavirus/

American Family Physician https://www.aafp.org/afp/2010/0215/p552.html

Centers for Disease Control and Prevention, USA https://www.cdc.gov/rotavirus/index.html

European Centre for Disease Prevention and Control https://www.ecdc.europa.eu/en/rotavirus-infection/facts

KidsHealth, USA https://kidshealth.org/en/parents/rotavirus-vaccine.html

Mayo Clinic, USA https://www.mayoclinic.org/diseases-conditions/rotavirus/symptoms-causes/syc-20351300

National Foundation for Infectious Diseases, USA https://www.nfid.org/infectious-diseases/frequently-asked-questions-about-rotavirus-2/

National Health Service, UK https://www.nhs.uk/conditions/vaccinations/rotavirus-vaccine/

National Institute for Clinical Care and Excellence (NICE), UK. Gastroenteritis, 2019 https://cks.nice.org.uk/gastroenteritis#!topicSummary

Patient Info, UK https://patient.info/childrens-health/acute-diarrhoea-in-children/rotavirus

Burnett E, Parashar U, Tate J. Rotavirus vaccines: effectiveness, safety, and future directions. *Paediatric Drugs.* 2018 Jun;20(3):223–233. doi: https://doi.org/10.1007/s40272-018-0283-3.

Chiejina M, Samant H. Viral Diarrhea. Treasure Island (FL): StatPearls Publishing LLC; 2019 https://www.ncbi.nlm.nih.gov/books/NBK470525/

Crawford SE, Ramani S, Tate JE, Parashar UD, Svensson L, Hagbom M, Franco MA, Greenberg HB, O'Ryan M, Kang G, Desselberger U, Estes MK. Rotavirus infection. *Nature Reviews Disease Primers.* 2017 Nov 9;3:17083. https://doi.org/10.1038/nrdp.2017.83.

Gonzalez-Ochoa G, Flores-Mendoza LK, Icedo-Garcia R, Gomez-Flores R, Tamez-Guerra P. Modulation of rotavirus severe gastroenteritis by the combination of probiotics and prebiotics. *Archives of Microbiology.* 2017 Sep;199(7):953–961. https://doi.org/10.1007/s00203-017-1400-3. Epub 2017 Jun 20.

Leung AKC. Viral gastroenteritis in children. BMJ Best Practice, BMJ Publishing Group, 2020 https://bestpractice.bmj.com/topics/en-gb/794

Nguyen DD. Rotavirus. Medscape from WebMD, 2018 https://emedicine.medscape.com/article/803885-overview

O'Ryan M. Rotavirus vaccines: a story of success with challenges ahead. *F1000Research.* 2017 Aug 18;6:1517. https://doi.org/10.12688/f1000research.11912.1. eCollection 2017.

Glossary

Acute A disease that appears very suddenly and doesn't usually last for very long.

Adhesin A molecule on the surface of a microbe that binds to a complementary molecule (known as a receptor) on the surface of human cells thereby enabling the microbe to anchor itself and not be easily removed.

Antibiotic A chemical substance, produced by a microbe, which is able to kill, or inhibit the growth of, another microbe.

Antimicrobial peptides A diverse group of peptides with antimicrobial activity that are produced by a variety of mammalian cells. They are an important part of the antimicrobial defence system of humans.

Apoptosis (also known as "programmed cell death") A highly regulated and organised sequence of events that result in the suicide of a human cell. During the process no harmful substances are released that could damage other cells. It is a normal and essential physiological event and is important in human development and antimicrobial defences.

Asymptomatic Not showing signs of any disease.

At-risk groups Those groups of people that are more vulnerable to infectious diseases than the average healthy adult because of some underlying problem which has weakened their immune defences. Examples of such groups include those with diabetes, cancer and HIV infection.

Bacteraemia The presence of bacteria in the bloodstream

Bactericide An agent (chemical, physical or biological) that kills bacteria.

Bacteriocin A compound produced by bacteria that can kill or inhibit other bacteria.

Bacteriostatic A chemical, or set of environmental conditions, that prevents the growth of bacteria without killing them.

Biofilm An aggregate of microbes that is enclosed in a gel-like matrix.

© Springer Nature Switzerland AG 2021
M. Wilson, P. J. K. Wilson, *Close Encounters of the Microbial Kind*,
https://doi.org/10.1007/978-3-030-56978-5

Capsule A jelly-like coating surrounding some microbes. It prevents the microbe from being ingested and destroyed by the body's white blood cells.

Cariogenic Something that promotes the development of tooth decay. This may refer to an organism (such as *Streptococcus mutans*) or a dietary component (such as sucrose).

Cautery The application of high temperatures to a localised area.

Cellulitis An infection of the deeper layers of skin and the underlying tissue.

Chemotherapeutic agent A chemical used to treat infectious diseases but which is not produced by a microbe. Examples include metronidazole, sulphonamides and nalidixic acid. An antibiotic is a particular type of chemotherapeutic agent – one that has been produced by a microbe.

Chronic A disease that lasts a long time – generally longer than 3 months.

Colon (bowel, large intestine) An organ of the human digestive system. It absorbs water and electrolytes from the contents of the digestive tract and also the nutrients produced by the enormous number of microbes that live there.

Cryotherapy The application of low temperatures to a localised area.

Curettage The use of a curette (an instrument with a sharp scoop or hook at its end) to remove tissue.

Cytokine A protein secreted by one type of cell that has a specific effect on another type of cell, i.e. it functions as a signalling molecule. There are dozens of different cytokines, and they are very important in the immune response to infections.

Cytotoxin A compound produced by a microbe that is able to kill one or more types of human cells.

Dental plaque A sticky, colourless coating that forms on the tooth surface. It consists mainly of bacteria and is an example of a *biofilm*.

Deoxyribonucleic acid A complex molecule that, in most organisms, is the hereditary material containing all the instructions an organism needs to develop, live and reproduce. It's made up of chemical building blocks known as nucleotides. Each nucleotide consists of a sugar (deoxyribose), a phosphate group and one of four types of nitrogen bases – adenine, cytosine, thymine and guanine. The order, or sequence, of these bases determines what biological instructions are contained in a strand of DNA.

Dermabrasion A surgical procedure that involves removing the outer layers of the skin by means of a rapidly rotating device. The skin that grows back is usually smoother and younger looking.

Dermatophytes Fungi that can cause infections of the skin, hair and nails because of their ability to utilise keratin.

Diathermy The creation of localised high temperatures using high-frequency electric currents.

Dysbiosis A term used to describe a situation in which the composition of the microbial community inhabiting a body site has changed to such an extent that it is no longer able to exist in a mutualistic (i.e. harmless) association with its host. The associated adjective is dysbiotic.

Dysbiotic See dysbiosis.

Empirical therapy The use of an antibiotic to treat an infection without having direct evidence of which microbe is responsible or to which antibiotic it is susceptible.

Endemic A disease that is regularly found among particular people or in a certain area.

Endocytosis A process by which material is taken into a cell. The material is engulfed by that region of the cell membrane with which it's in contact. The membrane containing the material is internalised and buds off when inside the cell to form a circular vesicle (known as a vacuole) containing the ingested material.

Endotoxin (or lipopolysaccharide) A complex molecule in the cell wall of Gram-negative bacteria that is made from a lipid and carbohydrates. It is toxic to human cells, induces the release of inflammatory cytokines, is antigenic, induces fever in mammals and can cause septic shock.

Enterotoxin A toxin produced by a microbe that has an adverse effect on intestinal function.

Environmental determinant Any factor of an environment that affects the ability of a microbe to colonise, or grow within, that environment.

Epidemic The spread of an infectious disease throughout a population within a short period of time.

Eubiosis A situation in which a mutualistic association exists between a host and its microbial symbionts. A eubiotic microbial community is compatible with a healthy host. The associated adjective is eubiotic.

Exotoxin A molecule secreted by a microbe that can kill or damage human cells.

Expectorant A drug that increases the water content of mucus so making it thinner and easier to cough up.

Fimbriae Hair-like projections on the surface of a microbe that are involved in adhesion of the microbe to surfaces.

Fomite An inanimate object (such as an article of clothing or door handle) that is contaminated with a pathogenic microbe and so can be involved in transmitting the pathogen to other people.

Fungaemia The presence of fungi in the bloodstream.

Glycoprotein A protein that has sugar molecules attached to it.

Host The larger organism in a symbiotic relationship.

Keratinocyte An epidermal cell that produces keratin. It is the predominant cell type found in the epidermis.

Lymphocyte A type of white blood cell that is made in the bone marrow and is found in the blood and in lymph tissue. The two main types of lymphocytes are B lymphocytes and T lymphocytes. B lymphocytes make antibodies, and T lymphocytes help kill tumour cells and infected cells and help control immune responses.

Macrophage A large white blood cell that is involved in the detection, phagocytosis and destruction of bacteria and other microbes.

Microbiota (also sometimes referred to as the *Microbiome*). The microbial community present at a particular site or associated with an organ system or entire organism. It includes all of the bacteria, archaea, fungi, viruses and protozoa present in that community.

Microdermabrasion The physical removal of the outer layers of the skin.

Mid-stream specimen (of urine) This is a sample of urine that has been collected after urine has flowed for a short time, with collection being terminated before the flow has ceased. Such a specimen is considered to be representative of the urine present in the bladder, uncontaminated by urethral organisms which are assumed to have been flushed away by the initial urinary flow.

Mucins Large glycoproteins that are produced by the epithelium and are a major constituent of mucus. Some remain attached to epithelial cells while others are secreted into the mucus layer that covers mucosal surfaces.

Mucolytic A drug that thins mucus by breaking down the large mucin molecules of which it is composed. This makes it less thick and sticky and easier to cough up.

Multicellular An organism that consists of a number of cells. Many fungi and algae are multicellular, whereas bacteria, protozoa and archaea are unicellular, i.e. they consist of a single cell.

Mutualism A symbiosis which is beneficial to both of the organisms involved.

Neutrophil (Neutrophilic granulocyte or polymorphonuclear neutrophil) A type of white blood cell whose main function is to phagocytose and kill microbes. Neutrophils are the most abundant white blood cells in humans.

Nomenclature The system used in the naming of organisms.

Oedema Swelling of the tissues.

Onychomycosis A fungal infection of the nail. Also known as tinea unguium.

Pandemic An epidemic that spreads worldwide or across a large geographical region.

Pathobiont (opportunistic pathogen) A microbe that lives on healthy humans but is also able to cause an infection under certain conditions

Pathogen A microbe that is able to cause disease in a human.

Photomicrograph A photograph taken through a microscope.

Physiology The study of how a living organism functions.

Prebiotic A compound (usually derived from plants) that promotes the growth of microbes considered to be of benefit to human health.

Probiotic A live microbe that is consumed (or applied topically) in order to produce some beneficial effect.

Receptor A component on a host cell or structure that binds to a microbial adhesin thereby enabling adhesion of the microbe.

Rhinorrhoea A runny nose.

Ribonucleic acid Ribonucleic acid is a complex molecule similar to DNA. It's made up of chemical building blocks known as nucleotides. Each nucleotide consists of a sugar (ribose), a phosphate group and one of the four types of nitrogen bases – adenine, cytosine, uracil and guanine. Its main role in most organisms is to convert the genetic information encoded by DNA into proteins. However, in some viruses, it functions as the hereditary material, i.e. it replaces DNA as the molecule containing the microbe's genetic instructions.

Self-limiting A disease that resolves itself without treatment.

Symbiont An organism that is very closely associated with another, usually larger, organism.

Symbiosis The living together of two or more dissimilar organisms.

Toll-like receptors An important group of receptors on human cells that are involved in recognising microbes and their components. They are involved in recognising, and defending us against, pathogenic microbes. They are present on many types of human cells.

Topical medication The application of a medication (cream, ointment) to a body surface (skin, mucosa) rather than by mouth or injection.

Unicellular An organism that consists of just one cell. Bacteria, protozoa and archaea are unicellular. In contrast, some fungi and algae are multicellular.

Uropathogen A microbe that is able to infect the urinary tract. An important example is *Escherichia coli*.

Vaccination The administration of a vaccine consisting of a microbe or a microbial component to an individual which results in that person becoming immune to infections caused by that microbe.

Vasodilation An increase in the diameter of a blood vessel.

Virulence factor Any component of a microbe that enables it to withstand its host's antimicrobial defence systems or that causes damage to its host.

Volatile A substance that readily turns into a gas at room temperature.

Index

© Springer Nature Switzerland AG 2021
M. Wilson, P. J. K. Wilson, *Close Encounters of the Microbial Kind*,
https://doi.org/10.1007/978-3-030-56978-5

Printed in the United States
By Bookmasters